AF746067

Proceedings of the I-ESA Conferences

Volume 8

This series publishes the proceedings of the IESA conferences which began in 2005 as a result of cooperation between two major European research projects of the 6th Framework R&D Programme of the European Commission, the ATHENA IP (Advanced Technologies for Interoperability of Heterogeneous Enterprise Networks and their Applications, Integrated Project) and the INTEROP NoE, (Interoperability Research for Networked Enterprise Applications and Software, Network of Excellence). The I-ESA conferences have been recognized as a tool to lead and generate an extensive research and industrial impact in the field of interoperability for enterprise software and applications.

More information about this series at http://www.springer.com/series/10534

Kai Mertins · Ricardo Jardim-Gonçalves
Keith Popplewell · João P. Mendonça
Editors

Enterprise Interoperability VII

Enterprise Interoperability in the Digitized and Networked Factory of the Future

Universidade do Minho
Escola de Engenharia

InterOP - PtRP
Portuguese Research Pole

InterOP - VLab

ifip

Springer

Editors
Kai Mertins
Knowledge Raven Management GmbH
Berlin
Germany

Ricardo Jardim-Gonçalves
Faculdade de Ciências e Tecnologia
Universidade Nova de Lisboa
Caparica
Portugal

Keith Popplewell
Future Manufacturing Applied Research Centre
Coventry University
Coventry
UK

João P. Mendonça
MEtRICs/Departamento de Engenharia Mecânica
Universidade do Minho
Guimaraes
Portugal

ISSN 2199-2533 ISSN 2199-2541 (electronic)
Proceedings of the I-ESA Conferences
ISBN 978-3-319-30956-9 ISBN 978-3-319-30957-6 (eBook)
DOI 10.1007/978-3-319-30957-6

Library of Congress Control Number: 2016937372

© Springer International Publishing Switzerland 2016
This work is subject to copyright. All rights are reserved by the Publisher, whether the whole or part of the material is concerned, specifically the rights of translation, reprinting, reuse of illustrations, recitation, broadcasting, reproduction on microfilms or in any other physical way, and transmission or information storage and retrieval, electronic adaptation, computer software, or by similar or dissimilar methodology now known or hereafter developed.
The use of general descriptive names, registered names, trademarks, service marks, etc. in this publication does not imply, even in the absence of a specific statement, that such names are exempt from the relevant protective laws and regulations and therefore free for general use.
The publisher, the authors and the editors are safe to assume that the advice and information in this book are believed to be true and accurate at the date of publication. Neither the publisher nor the authors or the editors give a warranty, express or implied, with respect to the material contained herein or for any errors or omissions that may have been made.

Printed on acid-free paper

This Springer imprint is published by Springer Nature
The registered company is Springer International Publishing AG Switzerland

Preface

The digital transformation in business is one of the most challenging trends in the global market and it increases the demand for collaboration among enterprises. Constant changes in inter- and intra-organisational environment will persist in the future. Organisations have to react flexibly and continuously to (imminent) changes in markets and trading partners.

Large companies but also SMEs have to cope with internal changes from both a technical (e.g., new information, communication, software and hardware technologies) and an organisational point of view (e.g., merging, re-organisation, virtual organisations, business disruption, etc).

Nowadays, companies maximise flexibility and speed of response to changing market conditions, by focusing on core innovation and information centred activities, outsourcing capital and labour intensive activities to less advanced economies. They develop knowledge and links with other companies with which they can collaborate to provide the products and services that the market demands—the networked enterprise, or virtual organisation. Successful enterprises in this knowledge economy may be small, but are highly productive, focussed on innovation and capable of flexibly and rapidly sharing information and knowledge to participate in highly competitive networked enterprises. The issue of interoperability within the enterprise is therefore no longer limited to the interoperability between silos of systems within single companies, but has become one of interoperability throughout a value chain.

Enterprises and organizations (of any kind), in today's economic environment recognise the competitive advantage of working in close collaboration with complex networks of suppliers and customers to present product-service combinations to their markets. It has long been recognised that to take part in emerging, potentially opportunistic, collaborative networks, these organisations must ensure that their systems and applications are capable of interoperating across heterogeneous collaborative networks of independent enterprises. The concept of "interoperability" has been defined by INTEROP-VLab as "The ability of an enterprise system or application to interact with others at a low cost in a flexible approach".

Consequently, interoperability of organizations appears as a major issue in successfully building on-the-fly emerging enterprise networks.

Industry's need for Enterprise Interoperability has been one of the significant drivers for research into the Internet of the Future. EI research will embrace and extend contributions from the Internet of Things and the Internet of Services, and will go on to drive the future needs for the Internets of People, Processes, and Knowledge. Industry found at I-ESA'16 an outstanding opportunity to exchange experience and problems on business interoperability in their daily operations.

Initiated in 2005 by two major European research projects of the 6th Framework R&D Programme of the European Commission, the ATHENA IP (Advanced Technologies for Interoperability of Heterogeneous Enterprise Networks and their Applications, Integrated Project) and the INTEROP NoE, (Interoperability Research for Networked Enterprise Applications and Software, Network of Excellence), the I-ESA conferences have been recognized as a tool to lead and generate an extensive research and industrial impact in the field of interoperability for enterprise systems and applications. Organised now by INTEROP-VLab in strong connection with its Poles, I-ESA has spread the basic philosophy developed since the beginning of this research: the solutions for EI are not only technical but are based on a combination of IT, organisation and semantics.

I-ESA brings together the world's leading researchers and practitioners in the area of enterprise interoperability, and it is a unique forum for the exchange of visions, ideas, research results and industrial experiences, dealing with a wealth of interoperability research subjects.

Interoperability for Enterprise Systems and Applications (I-ESA 2016) is the eighth conference after the seven previous successful experiences: Genève (2005), Bordeaux (2006), Madeira (2007), Berlin (2008), Coventry (2010), Valencia (2012) and Albi (2014) and a special edition in Beijing (2009). This time the motto is "Enterprise Interoperability in the digitized and networked Factory of the Future" expecting to offer the forefront of research and the industry expertise/viewpoint in this very important domain of Enterprise Interoperability.

The I-ESA'16 Conference was organized by Escola de Engenharia da Universidade do Minho, on the behalf INTEROP-VLab PrtPl (Portuguese Pole of INTERO-VLab), and the European Virtual Laboratory for Enterprise Interoperability (INTERO-VLab) and sponsored by the International Federation for Information Processing (IFIP).

The program included several keynotes presented by high-level renowned experts from industry, government, and academia:

- Mr. Gerald Santucci, European Commission, EU
- Prof. Klaus-Dieter Thoben, University of Bremen, D
- Pr. Ricardo Jardim Gonçalves, Universidade Nova de Lisboa, PT
- Dr. Ing. Francisco Duarte, Bosch Car Multimedia, PT

World leading researchers and practitioners in the area of Enterprise Integration from government, industry, and academia contributed to this book. As a result,

Enterprise Interoperability VII is a unique anthology presenting visions, ideas, research results, industrial experiences, and problems on business interoperability.

This book is organized into eight parts addressing the major research in the scope of Interoperability for Enterprise Systems and Applications:

I. Introduction
II. Modelling the Enterprise Interoperability
III. Semantics for Enterprise Interoperability
IV. Architectures and Frameworks for Interoperability
V. Services for the Enterprise Interoperability
VI. Ontologies and Concepts for Enterprise Interoperability
VII. Industrial Implementation of Enterprise Interoperability
VIII. Collaborative Supply Networks Interoperability

Berlin
Caparica
Coventry
Guimaraes
February 2016

Kai Mertins
Ricardo Jardim-Gonçalves
Keith Popplewell
João P. Mendonça

Acknowledgements

Escola de Engenharia da Universidade do Minho and INTEROP-VLab would like to thank all the authors, invited speakers, the members of International Programme Committee, International Senior Programme Committee, International Honorary Committee and Steering Committee and participants of the conference that made this book a reality and the I-ESA 2016 a success.

We express our gratitude to all organizations that supported the I-ESA 2016 Conference preparation, especially Escola de Engenharia da Universidade do Minho, INTEROP-VLab, Portuguese pole of INTEROP-VLab PrtPl, the International Federation for Information Processing and all the pole members of INTEROP-VLab.

We are deeply thankful to the organisation support notably Emanuel Miron, Luis Figueiredo, Joao Sousa, Joao Ferrao, Carlos Monteiro, and Sara Cortez for their excellent work for the preparation and the management of the conference and our special thankful to Sophie-Agnès Fensterbank and Emma Barthet for their strong involvement in the support and dissemination activities of I-ESA 2016 Conference.

Contents

Part I
Introduction

An Interoperable Cloud Environment of Manufacturing Control System

Yulia Yadgarova and Victor Taratukhin

Abstract In presented study a new integrated framework for building cloud-based manufacturing environment is described. This approach allows develop future production systems as the class of adaptive distributed systems with virtual cloud model and simulation. Dependency model of equipment interaction is defined and the method of detailed production plan scheduling is introduced. As the result of the investigation architecture and prototype of the system is presented.

Keywords Industrial internet architecture · Learning systems · Cloud manufacturing · Multi-agent systems · Simulation

1 Introduction

Presently the concepts of Industrial Internet, M2M-interaction and Industry 4.0 [1] play a significant role in the process of creation Factories of the Future [2]. Complex industrial production in space and aerospace industries, collaboration of different manufacturers and rapid development of customized products imply changes in the production process. The modern manufacturing lines should be customizable, easy controlled and intelligent to meet complexity demands and to increase quality of the production [3].

Roadmap to the Industry 4.0 concept [1] implies transition of traditional control production process model to the distributed system with collaborative protocols on the different levels. On a business level such systems implement Single information space (with the concept of Virtual Enterprise). Virtual production line is the idea of

Y. Yadgarova (✉)
Bauman Moscow State Technical University, Moscow, Russia
e-mail: y.yadgarova@bmstu.ru

V. Taratukhin
European Research Center for Information Systems (ERCIS), Muenster, Germany

V. Taratukhin
Stanford University, Stanford, CA, USA

© Springer International Publishing Switzerland 2016

K. Mertins et al. (eds.), *Enterprise Interoperability VII*,
Proceedings of the I-ESA Conferences 8, DOI 10.1007/978-3-319-30957-6_1

geographically distributed production units and transparent technological process through these units [4, 5].

Research and development process of virtual enterprise concept was complex and intensive. There are a lot of research groups all over the world investigating this problem. The greatest ones are in USA, for example Advanced Manufacturing partnership [6] that focuses on the Advanced Manufacturing Enterprise concept. Another one is Intelligent machines (IM) which develops smart sensors and controllers also intelligence machines for the FMS (Flexible manufacturing line). Mentioning Europe activities, there is necessary to note global and local projects such as "Factories of the Future" (global), "Industry 4.0" (Germany), "Usine du Futur" (France). According to the "Factories of the Future" [2], classification there are three main concepts:

1. **Digital Factory** concept infers improving quality of the design and production processes as the decreasing expense of modeling and product knowledge management within the production process.
2. **Smart Factory** concept means creation of flexible, customizable production with computer automation systems and robotics.
3. **Virtual Enterprise** concept enforces value added creation by supply chain management and control of distributed manufacturing lines.
4. Also the Digital Factory is the base notion for creation of the Smart and Virtual Enterprises.

On (Fig. 1) detailed description of each type is presented.

The purpose of our investigation is to create framework, methods and prototype platform for the Smart Factory concept.

Industrial internet consortium initiative [7] defines common principles of building reference architectures of the interacted devices. Some principles (reliability, scalability, usability, maintainability, portability and so on) were taken as a

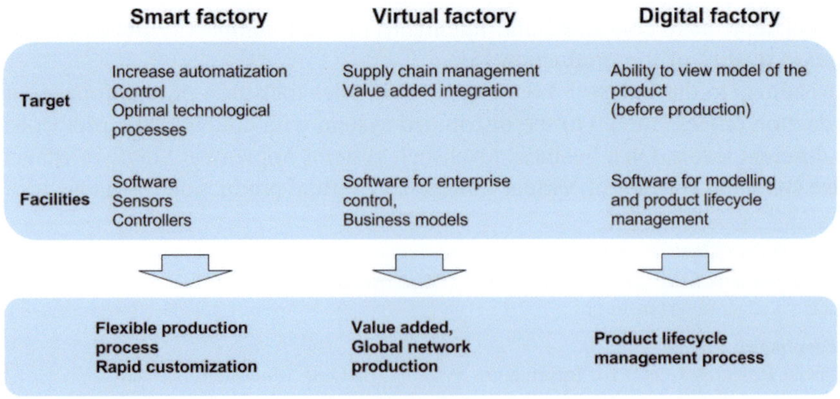

Fig. 1 Concepts of the factory-of-the-future

basis of our architecture. Moreover there are several critical characteristics that system should implement alongside the standard [8].

Now the cloud systems are being integrated in plant landscapes. In the [9] researchers provide system architecture of cloud robotics system based on Machine-to-Machine and Machine-to-Cloud communications. At that work several cloud architectures have been compared (Peer-based, Proxy-based and Clone-based) also robustness, interoperability and mobility of such design solutions have been evaluated. Although the final option of result architecture depends on the application of the system. In the cloud robotics model we have several robots and ad-hock network between them [10]. During the production process there are different states when each unit of equipment can be critically important. Because of that in the current research the clone-based architecture of the system is presented.

Another challenge at constructing the cloud-based environment of production line is the integration of Information technologies domain (IT) and operation technologies (OT) [7]. In Information Technologies domain all system characteristics are represented by digital world. Symbol-grounding problem is the main problem of the information technology domain- representing a meaning of the symbols to machine. On the other case in Operation technology domain (OT) "control" should be applied directly to the physical process without any attempting to the models. In the IoT (Internet of Things) concept (and especially in Industrial Internet domain) it's vital to use the link between IT-domain and OT-domain. One of the main instruments for minimizing the difference between OT-state and IT-model is protocol of interaction between real-time devices and computer world. The first part of the study highlights the formal model of interaction of devices. In the second part the implementation of cloud system based on test equipment and stack of Java technologies is described.

2 Cloud Manufacturing Framework

A set of activities to create the cloud-based environment is complex. The whole investigation process was divided into several stages. Firstly the problem was to create clone of the real-world system with real-time interaction and minimization of the latency. The initial architecture of the physical system is presented on Fig. 5. The Flexible manufacturing line consisted of several devices: miller and lathe tools, conveyor, two robots, storage and robocara. All presented devices had to be exposed into cloud. The research project implied decreasing structure differences between real device and it's model clone. Secondly the thing was to provide interaction and self-organization of clones in the cloud. Achieving this goal would give strong evidence for feasibility of such architectures and in perspective would bring a big impact in building unmanned production environment. In fact the first issue is described in detail at [9] while the second have not properly investigated yet.

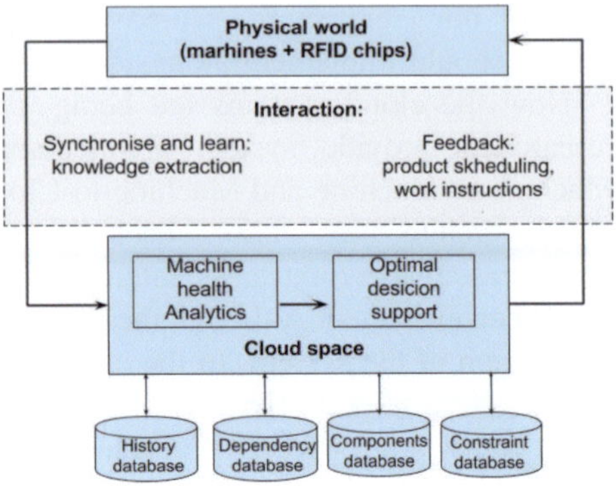

Fig. 2 The cloud manufacturing framework

As the solution for the first issue we tie a RFID chip with work instructions information on the manufacturing part. Also each device and machine tool had a link with cloud environment through driver's software. As a result it can be stated that the stroke model of the production system in the cloud had been obtained. Now each physical device has its own representation model (clone) based on signals and driver's status. Meanwhile the part arrives to the shop floor information about its components is extracted from process instructions in the chip. After these activities the process of runtime configuration starts. The runtime schedule based on dependencies and constraints of the devices begins configuring (Fig. 2). During this process (described below) additional devices (that are needed for the operation) are extracted from the dependency database. Another (Constraint) database stores process attached rules (time constraints and sequences). Moreover, the essential part at self-organization process is history database, where analytics for all devices are stored. Process can be automatically rescheduled according to historical data.

Dependency database consists of rules that link two or more devices. To start processing on the machine tool, part should be transported on the work space by the robot. Meanwhile to grab and move the part robot has to perceive it in the processing zone. So, robot need conveyor to transport part in that zone.

Based on dependency database information, relation model of shop floor constraints of the equipment links, history the concrete work instructions and time schedule are arrived on the virtual clones in the cloud model (Fig. 3).

Constraint database are used after Dependency storage to indicate limitations between devices. Overall schema of the production plan scheduling in flexible manufacturing system is presented at Fig. 4.

Next step of the cloud control process is simulation. The main impacts of the modelling are to ensure that there are no gaps between devices interaction and to predict system failures and troubles. In case of time/structure gap the system asks master of the shop floor and offers a several variants. Below the most essential branches of the research are described in detail.

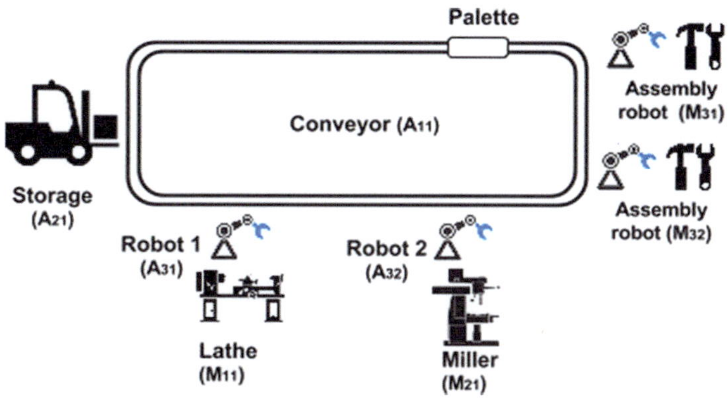

Fig. 3 Schema of the flexible manufacturing line

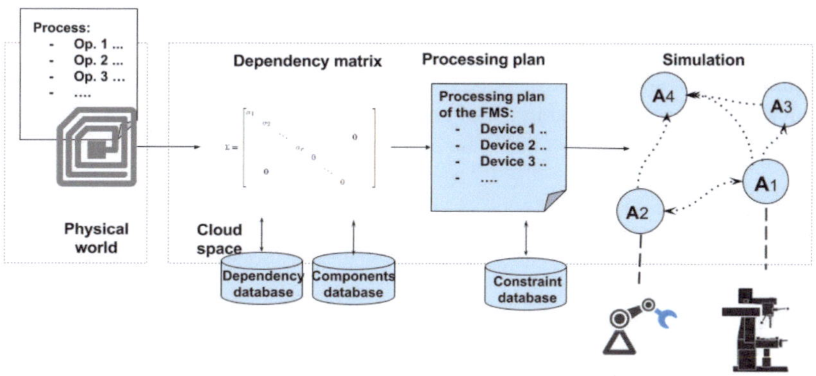

Fig. 4 Sequence of the processing

2.1 Architecture of Clone-Based Model

The target system was developed based on clone-based architecture [9] and implements twin for each equipment unit. At once each clone was represented by the actor [11] that could be able to cooperate with other actors and to solve scheduling and operation problems at the shop.

The final environment consists of static Actor-Dependency model, presented above. Each equipment unit have a virtual clone (Ai, actor), that have a dependency with other units (Fig. 5). These dependences are defined by the user during developing the cloud structure of the manufacturing line.

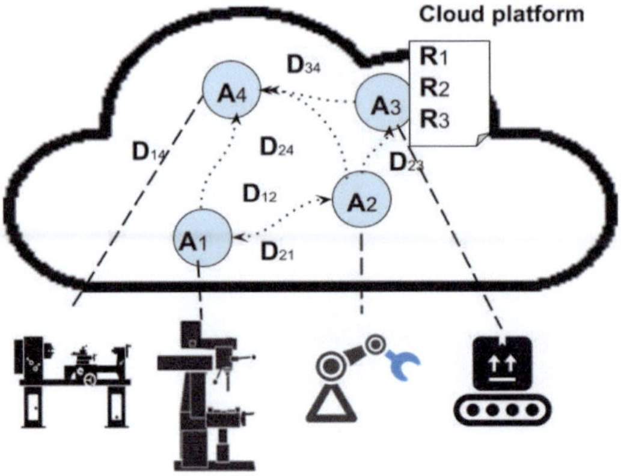

Fig. 5 Dependencies and components

2.2 *Operation Processing Plan Generation*

The cloud model represents different equipment units. In this system each equipment unit is presented as actor with dependencies (Fig. 6).

The production process starts when the task arrives to the workshop. At the time point t states of the system's units are presented by the equation:

$$\begin{aligned} O(t) = \{M_{jk}(t)/j = 1, \ldots, J;\ k = 1, \ldots, K_j, \\ A_{mn}(t)/m = 1, \ldots, M; \quad n = 1, \ldots, N_m \end{aligned} \tag{1}$$

M_{jk}	The main equipment (machine tools, assembly robots, work centers);
A_{mn}	Auxiliary equipment (manipulators, conveyors, robocars, storage);
J	Number of types of the main equipment;
K_j	Number of units of the main equipment (j—type);
M	Number of types of the auxiliary equipment;
N_m	Number of units of the auxiliary equipment (m—type);

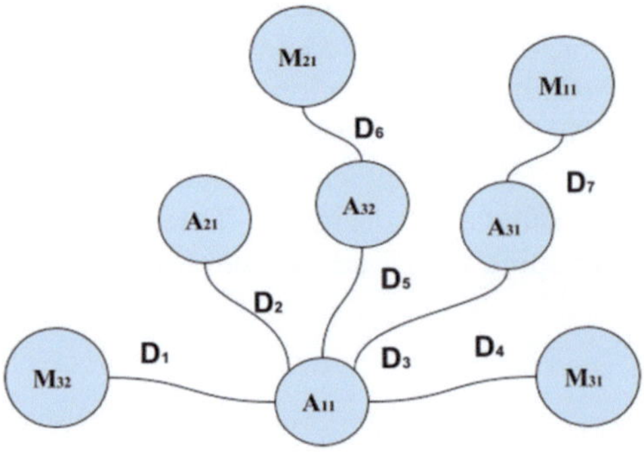

Fig. 6 Dependency graph of the equipment

$$F^D = \begin{array}{c|cccccccc} & M_{11} & M_{21} & M_{31} & M_{32} & A_{11} & A_{21} & A_{31} & A_{32} \\ \hline M_{11} & 0 & 0 & 0 & 0 & 0 & 0 & 1 & 0 \\ M_{21} & 0 & 0 & 0 & 0 & 0 & 0 & 0 & 1 \\ M_{31} & 0 & 0 & 0 & 0 & 1 & 0 & 0 & 0 \\ M_{32} & 0 & 0 & 0 & 0 & 1 & 0 & 0 & 0 \\ A_{11} & 0 & 0 & 1 & 1 & 0 & 1 & 1 & 1 \\ A_{21} & 0 & 0 & 0 & 0 & 1 & 0 & 0 & 0 \\ A_{31} & 1 & 0 & 0 & 0 & 1 & 0 & 0 & 0 \\ A_{32} & 0 & 1 & 0 & 0 & 1 & 0 & 0 & 0 \end{array} \qquad (1.2)$$

Fig. 7 Dependency matrix for the flexible manufacturing line

The k—unit of main equipment of the j—type $M_{jk}(t)$ is characterized by several parameters (properties and restrictions).

The production process is defined by the dependency graph of the equipment (Fig. 6) as well as Dependency matrix F^D (Fig. 7).

Below the matrix for production line presented at (Fig. 3) is defined.

Dependency graph for relevant flexible manufacturing system is presented on the Fig. 6.

There are two types of devices are presented in the matrix: main and auxiliary. Indeed each device of particular type has restrictions that help system to build a correct operation plan. Restrictions resemble rules in this case. For instance, all auxiliary devices' dependencies of type A→M (dependency from Auxiliary to the main type works from one way).

Dependency matrix is static and forming while user of the system defines structure of the production line.

Case:

Order incoming to the shop floor is presented as:

$$Z_i(t) = \langle M_i, \alpha_j, t_i, T_i^t \rangle \qquad (3)$$

where

- M_i Sequence of the main equipment unit types (route map);
- α_j Processing conditions (time of the processing, machine tools' parameters, CNC-program name etc.);
- t_i Time of the order income;
- T_i Planned time of issue;

Auxiliary devices dependencies are calculated by applying sequence of processing (input parameter). For instance, if the sequence of processing consists of M_1 and M_3 then the first step is to define equipment units of this type to meet target function.

As a target functions there are several types that can meet production requirements. The target function can be minimization of processing time or cost, also minimization of equipment downtime, etc. It depends on the final target and can also be multipurpose.

At the further steps of defining operation sequence auxiliary dependencies should be obtained. In this case, dependencies A_{11} and A_{31} was found by using Dependency matrix. According to matrix data and restrictions, the overall operation production sequence presented as:

$$A_{11} \rightarrow A_{31} \rightarrow M_{11} \rightarrow A_{31} \rightarrow A_{11} \rightarrow M_{31} \quad (4)$$

3 Prototype of the System

Prototype of the system was developed using Flexible Manufacturing line in research lab (Fig. 3).

3.1 Architecture of the System

The Cloud-based manufacturing model consists of two parts: local part including device drivers (connectors), communication server, reactive planner (to perform real-time actions) and cloud part based on web technologies (Fig. 8).

Capabilities of local part include control physical device and communication with the cloud server. All functions that driver can manage are transmitted to the cloud, as with monitored statuses. Protocol of interaction between cloud and device is based on Web-Sockets [12]. The presented architecture guarantees fast delivery of the messages indeed if device would be disconnected from control system.

The cloud part of the system consists of several architecture blocks (Fig. 8):

- communication web-socket server
- presentation and settings model
- process planer and modeler
- databases (dependency, components, constraint)
- decision support system
- analytics model

During the device control process the cloud component of the system performs simulation within the server; on the next steps commands will be transmitted to the physical system.

Characteristics of the clone model (functions, properties and dependencies) are defined by the user and stored in the particular database. Interaction between several agents is rested on the properties and dependencies.

At Fig. 9 user interface of the system is presented. There are conveyor dependencies definition process and tracking conveyor state.

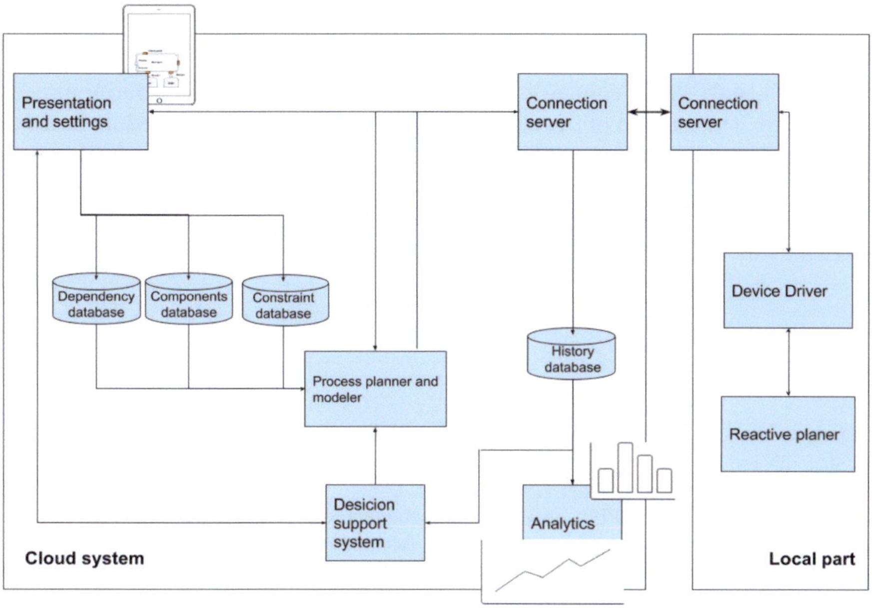

Fig. 8 Architecture of the system

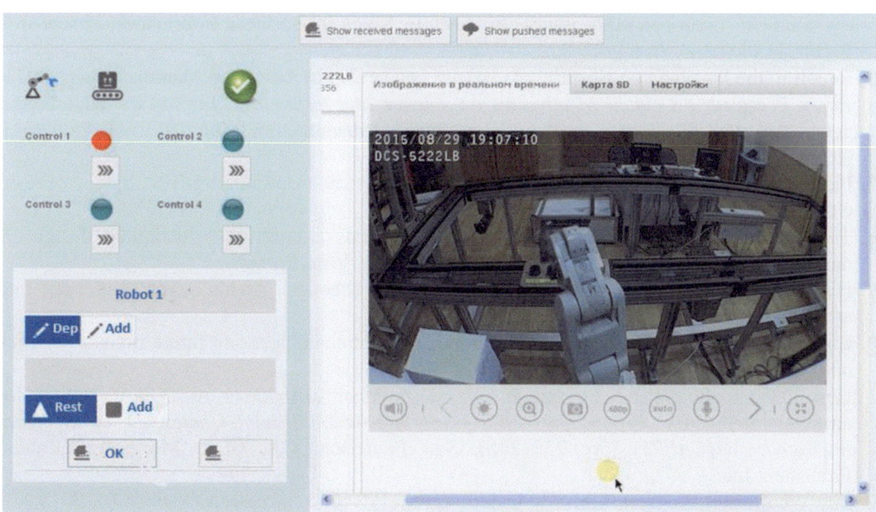

Fig. 9 Interface of the system

4 Conclusions

At research study the integrated framework for cloud-based manufacturing process is described. The method of developing detailed production plan schedule based on the dependency model is described. Presented investigation provides evidence for efficiency of such architectures.

We hope that obtained results can be the basis for design future production systems as the class of adaptive distributed systems.

References

1. Industry 4.0 Smart Manufacturing for the Future. http://www.gtai.de/GTAI/Content/EN/Invest/_SharedDocs/Downloads/GTAI/Brochures/Industries/industrie4.0-smart-manufacturing-for-the-future-en.pdf. Accessed October 3, 2015.
2. Factories of the Future PPP: Towards Competitive EU Manufacturing. http://ec.europa.eu/research/press/2013/pdf/ppp/fof_factsheet.pdf. Accessed October 3, 2015.
3. Brettel, M., et al. (2014). How virtualization, decentralization and network building change the manufacturing landscape: An industry 4.0 perspective. *International Journal of Science, Engineering and Technology* 8(1), 37–44.
4. Lee, J., Kao, H.-A., & Yang, S. (2014). Service innovation and smart analytics for industry 4.0 and big data environment. *Procedia CIRP*, *16*, 3–8, ISSN 2212-8271.
5. Yadgarova, Y., Taratoukhine, V., & Skachko, E. (2015). Multi-agent product life cycle environment. Interoperability issues. In: Lecture Notes in Business Information Processing, Berlin, Heidelberg: Springer.
6. Report to the President on Ensuring American Leadership in Advanced Manufacturing. (2011, June). *President's Council of advisors on science and technology*. http://www.whitehouse.gov/sites/default/files/microsites/ostp/pcast-advanced-manufacturing-june2011.pdf. Accessed October 3, 2015.
7. Industrial Internet Reference Architecture. http://www.iiconsortium.org/IIRA.htm. Accessed October 4, 2015.
8. Taratoukhine, V., & Bechkoum, K. (1999). Towards a consistent distributed design: A multi-agent approach. In: *Proceedings of Information Visualisation.*
9. Hu, G., Tay, W. P., & Wen, Y. (2012). Cloud robotics: Architecture, challenges and applications. *IEEE Network, 26*(3), 21–28.
10. Chibani, A., et al. (2013). Ubiquitous robotics: Recent challenges and future trends. *Robotics and Autonomous Systems, 61*(11), 1162–1172.
11. Hewitt, C., Bishop, P., & Steiger, R. (1973). A universal modular ACTOR formalism for artificial intelligence. *Proceedings of the 3rd International Joint Conference on Artificial Intelligence (IJCAI'73)* (pp. 235–245). San Francisco, CA, USA: Morgan Kaufmann Publishers Inc.
12. The WebSocket Protocol. https://tools.ietf.org/html/rfc6455. Accessed October 10, 2015.

A New Framework for Strategic Risk Analysis in a Global Pump Manufacturing Network

A. Niknejad, D. Petrovic, K. Popplewell, F.W. Jäkel and S. Pajkovska-Goceva

Abstract This paper presents a new risk analysis framework applied to a global production network for pump manufacturing considering strategic decisions regarding alternative suppliers and markets. External and internal risk scenarios are defined and alternative network configurations are evaluated considering the constructed risk scenarios. Inoperability of individual nodes in the global production networks caused by identified risks are determined by taking into account propagation of risks due to the interdependencies between nodes. Fuzzy arithmetic is applied to track the level of uncertainty inherent in the model parameters and the outcomes. It is demonstrated how recommendations can be made with regard to the network configuration and handling of the uncertainty in the results.

Keywords Resilient production networks · Collaborative supply networks modelling · Risk management · Global production networks · Inoperability input output model · Fuzzy arithmetic

1 Introduction

Global Production Networks (GPNs) are networks of globally interconnected actors, such as suppliers, production facilities, logistics providers and customers that facilitate the provision of products and services [1]. Due to the diversity in global conditions and the inherent complexity of these networks, they are subject to different risks, such as political risks, economic risks, insolvencies, accidents, delays and so on. Risks affect certain nodes or regions of the network directly, while other nodes are affected through the interdependencies within the network and as a

A. Niknejad (✉) · D. Petrovic · K. Popplewell
Faculty of Engineering, Environment and Computing, Coventry University, Coventry CV1 5FB, UK
e-mail: ali.niknejad@coventry.ac.uk

F.W. Jäkel · S. Pajkovska-Goceva
Fraunhofer IPK, Pascalstraße 8-9, 10587 Berlin, Germany

© Springer International Publishing Switzerland 2016
K. Mertins et al. (eds.), *Enterprise Interoperability VII*,
Proceedings of the I-ESA Conferences 8, DOI 10.1007/978-3-319-30957-6_2

result of risk propagation. These risks and their propagation need to be taken into account when strategic decisions, such as choosing the key partners, are being made. In this paper, we consider a real world pump manufacturing network and consider three alternative GPN configurations. These GPN configurations are then investigated and analysed with regard to risks.

In Sect. 2, the relevant literature, and the Interoperability Input Output Model used in this research, are briefly introduced. In Sect. 3, the framework for the strategic risk evaluation in GPNs is proposed, while, in Sect. 4 the three alternative GPN configurations for the pump manufacturing network are presented. Furthermore, in Sect. 5 a number of external and internal risk scenarios are introduced by providing the relevant details such as risk likelihood and its impact. Section 6 provides the results of the inoperability model and the expected loss of risk, defined here as the expected reduction of financial revenue arising from the anticipated risk. Finally, Sect. 7 provides concluding remarks and future directions.

2 Literature Review

Supply chain risk has been extensively studied in recent years. We review some of the relevant literature. Reference [2] considers the processes for risk management in complex supply networks with strategic collaborations. Reference [3] provides a classification of supply chain risk by conducting a comprehensive survey of the literature. Furthermore, Ref. [4] conceptually examines the levels and dimensions of risk propagation. Additionally, Ref. [5] uses both network theory and Monte Carlo simulation to investigate bottleneck identification in supply networks. Also, in an interesting study about project management risks, Ref. [6] describes a real case of utilising causal maps to engage with the stakeholders to develop a comprehensive risk profile. We refer interested readers to [7] for a comprehensive review of supply chain risks literature.

Our analysis of impact of risk and its propagation is based on Inoperability Input Output Model (IIM) [8]. IIM examines risks and inoperability, i.e., deviation from the planned productivity level, within the economic sectors. Among the IIM literature, Ref. [9] introduces a dynamic variation of the IIM while others incorporate fuzzy uncertainties [10–13]. To the best of our knowledge, Ref. [14] is the only other application of IIM on supply networks, where the effect of disruptions on an example network is investigated using the IIM model and a multi-criteria method for interdependencies. However, uncertainties in the GPN and dynamism in the disruptions have not been considered.

In this paper, a pump manufacturing network is studied by considering the strategic decisions that need to be made with regard to its suppliers and customers. We apply a fuzzy dynamic variation of the IIM to analyse the impact of risks described in the risk scenarios on the alternative GPN configurations.

3 Strategic Risk Evaluation in Global Production Networks

The purpose of strategic risk evaluation of GPNs is to determine the suitability of alternative GPN configurations with respect to the identified risk criteria. The risk criteria that can potentially affect the company are defined in a set of risk scenarios. The analysis is done using all risk scenarios imposed on each of the GPN configurations and the results obtained for each scenario are aggregated to determine risk indicators including average inoperability and expected loss of risk, of each of the GPN configurations.

The framework of this evaluation is shown in Fig. 1.

The framework requires inputs including the interdependency values of all GPNs, perturbation impact and timings for all scenarios, the intended revenue of each node, resilience of the nodes to risk, regions and locations of each node within regions and also likelihood of each scenario. Furthermore, the framework evaluates each GPN by determining the inoperability value of nodes for each risk scenario as well as the loss of risk due to the inoperability. We will discuss each of these inputs and outputs further in the following sections.

Most inputs and outputs of the framework, such as interdependencies, perturbation impact, intended revenue, resilience, likelihood, inoperability and loss of risk, are assumed to be uncertain. To model these uncertainties, we utilise triangular fuzzy numbers in the form of [a, b, c] where 'a' represents the lowest possible value, 'b' identifies the most likely value and 'c' is the highest possible value. It is possible to use linguistic terms, such as low, medium and high, to describe the inputs which are then translated into triangular fuzzy numbers.

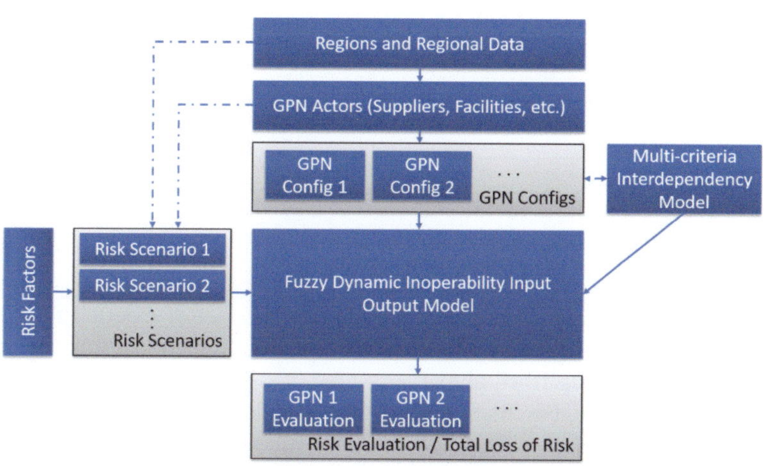

Fig. 1 Framework of the risk evaluation of GPNs

GPN configurations rely on GPN actors data (such as suppliers, production facilities, etc.) that is, in turn, dependent on regional data on various actors' locations. In addition, a multi-criteria interdependency model is used to determine the rate of dependency between GPN actors. The criteria included in the method include trade volume, inventory, substitutability of the product or supplier, distance and collaboration agreement. Risk scenarios are based on risk factors that represent the conceptual categories of risk, and are defined over either regions or actors.

A novel Fuzzy Dynamic Inoperability Input Output Model (FDIIM) is developed that uses fuzzy arithmetic to track uncertainty, from inputs to derived outcomes. It determines the inoperability of individual nodes in a GPN by considering the initial perturbations as well as the propagation of the perturbations to the related nodes. The level of inoperability shows the deviation of the node operation from its intended operation level, and is also used to estimate the financial loss of risk. FDIIM is presented in Appendix.

4 Alternative Configurations for the GPN

The framework for the risk evaluation is described using as an example a pump manufacturing GPN and considering three possible configurations. The first configuration includes four suppliers for base constituents, hydraulic components, control unit and electric motor in Europe and an additional supplier for electric motors in Asia. All the supplies are delivered to the main assembly plant in Europe, either directly or through a delivery company for electric motors. Then, the finished products in the assembly plant can be delivered to customers in either Asia or Europe. See Fig. 2 for a visualisation of the network configuration.

Configuration 2 has a single difference compared to Configuration 1 where the electric motor supplier in Asia is being excluded from the network and only the supplier in Europe is being utilized for sourcing electric motors. The network configuration is presented in Fig. 3.

Configuration 3 considers the case that the products are only sold in the European market. This configuration is shown in Fig. 4.

4.1 Regions

Regions play an important role in determining the impact of external risks on the GPNs. Nodes can be differently affected based on their region and this is identified in the configurations described above.

In the proposed case, two main regions are considered: Europe and Asia. Nodes within each region will be affected by the risks relevant to that particular region.

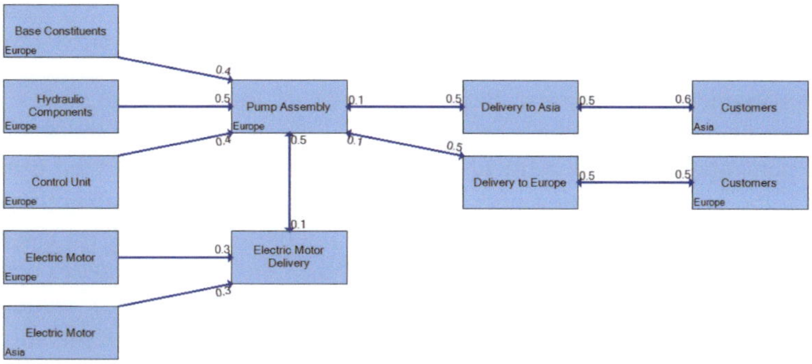

Fig. 2 Configuration 1 for the pump GPN

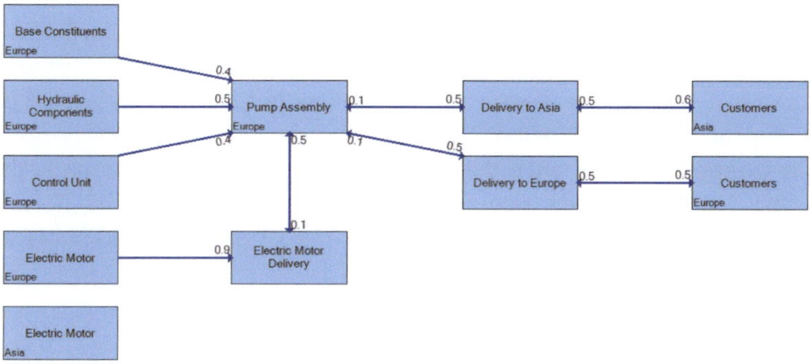

Fig. 3 Configuration 2 for the pump GPN

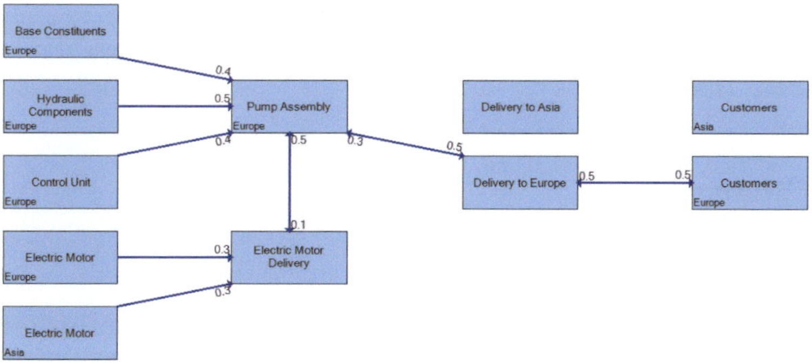

Fig. 4 Configuration 3 for the pump GPN

4.2 Interdependencies

The interdependencies measure the dependency of the dependent nodes on the supporting nodes. The numbers on the actors' links in Figs. 2, 3 and 4 show the relative percentage of the dependent node operation that can be affected per unit of inoperability in the supporting node. Calculating this information directly through statistical methods is extremely difficult, if not impossible. Instead, we use a fuzzy multi criteria method that allows the experts to use their judgements to determine the interdependency values and their confidence in the estimated values. Both estimated values and confidence are described using linguistic terms, including very low, low, fairly low, medium, fairly high, high and very high, and modelled using fuzzy sets. The fuzzy values and the corresponding confidence are aggregated and defuzzified into scalar values which best represent obtained aggregated fuzzy values. These defuzzified results are shown as labels on the arrows in Figs. 2, 3, and 4.

It is interesting to point out that value of interdependency has an inverse relationship with the substitutability of the supplier. In the provided example, in Configuration 1 where both electric motor suppliers in Europe and Asia are present, the interdependency between them and the electric motor delivery is set to 0.3, while, in Configuration 2, when the electric motor supplier in Europe is the only option, the interdependency rate is 0.9. A similar issue can be observed in Configuration 1 and 3, in the interdependency rate of the pump assembly and the two markets.

5 Risk Scenarios

Two types of risks are considered: those directly affecting the individual actors and those that are due to regional issues that can have an equal impact on all actors within the region. For each of these two types of risks, a set of risk scenarios are constructed. These scenarios are introduced in the following sections.

5.1 External Risks

With regard to Europe, two risk scenarios are constructed:

1. Economic Issues: the risk of Europe being hit with economic issues, such as recession or inflation is considered. It is estimated that each company operating in Europe will have a perturbation with an impact of [0.7, 0.8, 0.9] (triangular fuzzy number) for 40 periods with a likelihood of [0.02, 0.03, 0.04]. As mentioned earlier, user estimates the parameters of the model by entering triangular fuzzy numbers which determine the lowest possible value, the most likely value and the highest possible value of the parameter.

2. Compliance Risk: The risk of a new legislation which can significantly affect the operation of partners in Europe is considered. A perturbation impact is estimated as [0.3, 0.4, 0.5] for 10 periods with a likelihood of [0.05, 0.1, 0.15].

Also, two risk scenarios relevant to Asia are defined:

1. Social Unrest: This risk is related to possible social unrest as a result of political conflicts or economic problems. The scenario includes a perturbation with an impact of [0.7, 0.8, 0.9] for 20 periods with a likelihood of [0.05, 0.1, 0.15].
2. Embargo: this is risk of countries within Asia to be put under embargo. It includes a perturbation with an impact of [0.8, 0.9, 1] for 30 periods with a likelihood of [0.01, 0.02, 0.03].

5.2 *Internal Risks*

Internal risks affect individual nodes within the network. These could be related to individual suppliers, production facilities, logistics providers or customer markets. The following internal risk scenarios are considered:

1. Strike in the supplier of Electric Motor in Asia: It includes a perturbation with an impact of [0.9, 1, 1] for 20 periods on the electric motor supplier in Asia with a likelihood of [0.05, 0.1, 0.15].
2. Transport accident in delivery to Asia: It includes a perturbation with an impact of [1] for 10 periods on 'Delivery to Asia' with a likelihood of [0.1, 0.2, 0.3].
3. Temporary unavailability of pump's base constituents: It includes a perturbation with an impact of [0.7, 0.8, 0.9] for 20 periods on 'Base Constituents' with a likelihood of [0.1, 0.2, 0.3].
4. Custom issue in Asia affecting Delivery to Asia: where the deliveries to Asia are delayed or impounded by customs. It includes a perturbation with an impact of [0.1, 0.2, 0.3] for 10 periods on 'Delivery to Asia' with a likelihood of [0.1, 0.15, 0.2].
5. Insolvency of European customers: It includes a perturbation with an impact of [0.8, 0.9, 1] for 10 periods on 'Customers—Europe' with a likelihood of [0.005, 0.01, 0.015].

6 Analysis of Results

As mentioned earlier, the FDIIM is applied to each of the network configurations for all risk scenarios. This results in determining the inoperability of each node within the network for each of the risk scenarios over the time horizon. These results are aggregated in two ways: first, the average inoperability of nodes for all scenarios and over the time horizon is calculated, which gives an indication of the average susceptibility of the node in the configuration under consideration to the risks. In the second, the expected loss of risk in each GPN configuration is determined. Loss of risk is the financial impact of the risks that is calculated in the inoperability model as the product of the intended revenue of the operations of a node and the calculated inoperability of the node (presented in Appendix).

6.1 Nodes Inoperability

In Table 1, the average value of inoperability of all nodes and the configurations aggregated over the time horizon and all the risk scenarios are presented. It can be noted that in all the three configurations, pump assembly, delivery to Europe and customers (Europe) are among the most affected. However, in the first two configurations that include supply of pumps to customers in Asia, customers (Asia) and Delivery to Asia have had a higher inoperability, while, in Configuration 3, customers (Asia) and delivery to Asia have very little inoperability which is due to isolated impact of risk scenarios that directly affect Asia and Asian customers. Additionally, suppliers of base constituents, hydraulic components, control unit and electric motor have the same inoperability values across the GPN configurations. This is due to the fact that these suppliers are assumed to have no dependency on the other nodes in the configurations and they are only affected by direct perturbations, which is the same for all configurations.

6.2 Expected Loss of Risk

For the purpose of identifying financial loss, only nodes that are adding value to the company will be considered. In this example, only electric motor supplier in Europe

Table 1 Average inoperability of nodes for all three GPN configurations

Node	Configuration 1	Configuration 2	Configuration 3
Base constituents (Europe)	[0.09, **0.1**, 0.11]	[0.09, **0.1**, 0.11]	[0.09, **0.1**, 0.11]
Hydraulic components (Europe)	[0.06, **0.07**, 0.08]	[0.06, **0.07**, 0.08]	[0.06, **0.07**, 0.08]
Control unit (Europe)	[0.06, **0.07**, 0.08]	[0.06, **0.07**, 0.08]	[0.06, **0.07**, 0.08]
Electric motor (Europe)	[0.06, **0.07**, 0.08]	[0.06, **0.07**, 0.08]	[0.06, **0.07**, 0.08]
Electric motor delivery	[0.03, **0.07**, 0.13]	[0.05, **0.08**, 0.11]	[0.03, **0.07**, 0.13]
Electric motor (Asia)	[0.11, **0.12**, 0.13]	[0.11, **0.12**, 0.13]	[0.11, **0.12**, 0.13]
Pump assembly (Europe)	[0.11, **0.15**, 0.24]	[0.1, **0.13**, 0.19]	[0.11, **0.15**, 0.22]
Customers (Europe)	[0.11, **0.13**, 0.2]	[0.11, **0.13**, 0.17]	[0.11, **0.13**, 0.19]
Customers (Asia)	[0.13, **0.17**, 0.26]	[0.13, **0.17**, 0.24]	[0.07, **0.08**, 0.09]
Delivery to Asia	[0.11, **0.18**, 0.32]	[0.11, **0.16**, 0.27]	[0.02, **0.02**, 0.02]
Delivery to Europe	[0.09, **0.14**, 0.25]	[0.08, **0.13**, 0.2]	[0.09, **0.14**, 0.22]

The most likely values of inoperability are presented in bold

Table 2 Expected loss of risks for all three configurations

GPN	Expected loss of risk
Configuration 1	[1,723,488 €, **5,110,298 €**, 14,796,154 €]
Configuration 2	[1,623,458 €, **4,428,031 €**, 11,924,483 €]
Configuration 3	[1,836,773 €, **4,947,313 €**, 12,288,945 €]

The most likely values of inoperability are presented in bold

and pump assembly that are subsidiaries of the main company are considered to be generating revenue. Using experts opinion, intended revenues for electric motor supplier in Europe and pump assembly are estimated as [190,000 €, 200,000 €, 210000 €] and [990,000 €, 1,000,000 €, 1,010,000 €] respectively. Table 2 shows the expected loss of risk, aggregated considering all the risk scenarios for each of the configurations.

The expected loss of risk is lower for Configuration 2 in comparison with Configuration 1. So, the use of both suppliers in Europe and Asia for electric motors is not justified as it increases the risk. This is especially unacceptable, as having two suppliers in comparison with just one is usually done at a financial cost, mainly with the goal of significantly reducing risks.

Also, we see a lower loss in Configuration 3 in comparison with Configuration 1, which means focusing on European Customers only reduced the risk. However, it is important to point out that this analysis only considers the risk perspective while other criteria, such as revenue, could compensate for the increased risks.

Furthermore, as it can be observed from both Tables 1 and 2, the uncertainty levels are quite high in the analysis. For example, the highest possible value of inoperability for pump assembly in Configuration 1 is almost 70 % higher than the most likely value. This problem is even more obvious in the loss of risks, as the highest possible loss of risk for the Configuration 1 is near three times as much as the most likely value. This uncertainty is due to the uncertainty in the inputs of the model, including the interdependency values, impact, likelihood and intended revenue. More precise data would generate results with smaller uncertainty associated with them!

7 Conclusions

In this paper, strategic risk analysis for a pump manufacturing network has been investigated. The analysis is based on the FDIIM we developed, which determines the propagation of risks due to interdependencies between nodes, and uses fuzzy arithmetic to track uncertainty levels in the parameter values. To illustrate the framework for strategic risk analysis proposed, three GPN configurations are defined and analysed with respect to a set of external and internal risk scenarios. The configurations differ in using an alternative supplier for electric motors and also in the market they supply. It is demonstrated how the framework can be used to decide by GPN configuration with respect to risk, for example, which suppliers to use, which market to target and so on. In addition to risk analysis, GPN configurations should be analysed considering an economic aspect. This is a direction of our current research.

Acknowledgments This work has been funded by the European Commission through the Project FLEXINET: Intelligent System Configuration Services for Flexible Dynamic Global Production Networks (Grant Agreement No. 608627), which is gratefully acknowledged.

Appendix: Fuzzy Dynamic Inoperability Input Output Model

A vector representation of the fuzzy dynamic inoperability input output model function is as follows:

$$\tilde{q}(t+1) = \widetilde{K}\widetilde{A^*}\tilde{q}(t) + \widetilde{K}\widetilde{c^*}(t) + \left(I - \widetilde{K}\right)\tilde{q}(t) \tag{1}$$

where $\tilde{q}(t+1)$ is the vector of fuzzy inoperability of nodes at time period $t+1$, $\widetilde{K}$ is the fuzzy diagonal matrix of resilience, $\widetilde{A^*}$ is the fuzzy interdependency matrix and $\widetilde{c^*}(t)$ is the fuzzy perturbation of nodes for the risk scenario under consideration at time period t. Resilience represents the speed that the node can recover from disruptions.

The expected loss of risk for all risk scenarios is calculated as follows:

$$\tilde{Q} = \widetilde{x^T}\sum_{s=1}^{S}\widetilde{p_s}\sum_{t=1}^{T}\widetilde{q_s(t)} \tag{2}$$

where $\tilde{Q}$ is the fuzzy loss of risk for the GPN configuration, $\widetilde{x^T}$ is the transposed vector of the fuzzy intended revenues of the nodes, S is the number of risk scenarios, $\widetilde{p_s}$ is the fuzzy likelihood of risk scenario s, T is the number of time periods in the time horizon and $\widetilde{q_s}(t)$ is the fuzzy inoperability vector of nodes in scenario s at time period t.

The FDDIM method have been described in more details in [15].

References

1. Coe, N. M., Dicken, P., & Hess, M. (2008). Global production networks: Realizing the potential. *Journal of Economic Geography, 8*, 271–295. doi:10.1093/jeg/lbn002
2. Hallikas, J., Karvonen, I., Pulkkinen, U., & Virolainen, V. (2004). Risk management processes in supplier networks. *International Journal of Production Economics, 90*, 47–58. doi:10.1016/j.ijpe.2004.02.007

3. Rangel, D. A., de Oliveira, T. K., & Leite, M. S. A. (2014). Supply chain risk classification: Discussion and proposal. *International Journal of Production Research*, 1–20. doi:10.1080/00207543.2014.910620
4. Ghadge, A., Dani, S., & Kalawsky, R. (2011). Systems thinking for modeling risk propagation in supply networks. In *2011 IEEE MTT-S International Microwave Workshop Series on Innovation Wireless Power Transmission Technology System, Applications* (pp. 1685–1689). doi:10.1109/IMWS.2011.6116790
5. Mizgier, K., Jüttner, M., & Wagner, S. (2013). Bottleneck identification in supply chain networks. *International Journal of Production Research, 51*, 1477–1490.
6. Ackermann, F., Howick, S., Quigley, J., Walls, L., & Houghton, T. (2014). Systemic risk elicitation: Using causal maps to engage stakeholders and build a comprehensive view of risks. *European Journal of Operational Research, 238*, 290–299. doi:10.1016/j.ejor.2014.03.035
7. Heckmann, I., Comes, T., & Nickel, S. (2015). A critical review on supply chain risk—Definition, measure and modeling. *Omega, 52*, 119–132. doi:10.1016/j.omega.2014.10.004
8. Santos, J. R., & Haimes, Y. Y. (2004). Modeling the demand reduction input–output (I–O) inoperability due to terrorism of interconnected infrastructures. *Risk Analysis, 24*, 1437–1451. doi:10.1111/j.0272-4332.2004.00540.x
9. Haimes, Y. Y., & Horowitz, B. M. (2005). Inoperability input–output model for interdependent infrastructure sectors. I: Theory and methodology. *Journal of Infrastructure Systems, 11*, 67–79.
10. Panzieri, S., & Setola, R. (2008). Failures propagation in critical interdependent infrastructures. *International Journal of Modelling, Identification and Control, 3*, 69–78.
11. Setola, R., De Porcellinis, S., & Sforna, M. (2009). Critical infrastructure dependency assessment using the input–output inoperability model. *International Journal of Critical Infrastructure Protection, 2*, 170–178. doi:10.1016/j.ijcip.2009.09.002
12. Oliva, G., Panzieri, S., & Setola, R. (2011). Fuzzy dynamic input–output inoperability model. *International Journal of Critical Infrastructure Protection, 4*, 165–175. doi:10.1016/j.ijcip.2011.09.003
13. Oliva, G., Setola, R., & Barker, K. (2014). Fuzzy importance measures for ranking key interdependent sectors under uncertainty. *IEEE Transactions on Reliability, 63*, 42–57. doi:10.1109/TR.2014.2299113
14. Wei, H., Dong, M., & Sun, S. (2010). Inoperability input–output modeling (IIM) of disruptions to supply chain networks. *Syst Eng, 13*, 324–339. doi:10.1002/sys
15. Edelbrock, M., Niknejad, A., Schlosser, S, Petrovic, D., Otto, B., & Popplewell, K. (2015). FLEXINET deliverable [D2.3]: Design specification for business model innovation.

Cloud and Services Testing Applied in Manufacturing

F. Alves, C. Coutinho, R. Peña-Ortiz, X. Xu, A. Nieto, J. Ahola, L. Saari and A. Katasonov

Abstract Software testing is a complex and thorough process, which verifies software correctness from bare simple code units to the intricate complexity of fully integrated systems. Such major development contributor element should be planned from the beginning of the project (perhaps even before the code itself), and should play a role as important as the development itself. The use of the latest, fit, and most complete tools is important to ease the testing activities, but in most cases the selection of the best tool is specific to each particular development environment. This article presents an overview of the state-of-the-art in testing methodologies and

F. Alves · C. Coutinho (✉)
Caixa Mágica Software, Rua Soeiro Pereira Gomes, Lote 1 - 4 B, 1600-196 Lisbon, Portugal
e-mail: carlos.coutinho@caixamagica.pt

F. Alves
e-mail: fernando.alves@caixamagica.pt

R. Peña-Ortiz
Instituto Tecnológico de Informática, Camino de Vera, S/N, Edif.8G—Acc. B—4ª Planta, 46022 València, Spain
e-mail: rpenya@iti.es

X. Xu · A. Nieto
Tampere University of Technology, Korkeakoulunkatu 10, 33720 Tampere, Finland
e-mail: xiangbin.xu@tut.fi

A. Nieto
e-mail: angelica.nieto@tut.fi

J. Ahola · L. Saari
VTT Technical Research Centre of Finland, Vuorimiehentie 3, Espoo, Finland
e-mail: jukka.ahola@vtt.fi

L. Saari
e-mail: leila.saari@vtt.fi

A. Katasonov
VTT Technical Research Centre of Finland, Kaitoväylä 1, Oulu, Finland
e-mail: artem.katasonov@vtt.fi

© Springer International Publishing Switzerland 2016

K. Mertins et al. (eds.), *Enterprise Interoperability VII*,
Proceedings of the I-ESA Conferences 8, DOI 10.1007/978-3-319-30957-6_3

tools, particularising the use of such tools in the context of the H2020 European project C2NET, which handles, among others, Cloud environments and Web Services.

Keywords Software testing · Validation · Verification · Testing methodologies · Best practices

1 Introduction

The discipline of Software Testing is a development necessity since developers are human and make mistakes. Some are irrelevant others are expensive or dangerous. Software needs to be checked in order to generate a correct product. Part of these mistakes comes from bad assumptions and blind spots; thus, the validation of the produced software has also an effect of creating a knowledge base of previous faults, preventing or avoiding them to be committed subsequently in the future.

More than simply finding defects, software testing aims at preventing new flaws, improving the confidence of developers and consumers, and ultimately, making sure the results meet the business and user requirements. Errors and defects may be originated by flaws in the specification or in the implementation of the system, or by errors in the use and operation of the system, or even from environmental conditions.

The International Software Testing Qualifications Board [1], one of the most important associations for the development of testing body of knowledge in Europe and worldwide, numbers seven principles for testing:

1. Testing shows presence of defects: Performing tests may or not prove the existence of defects. However, detecting no defects (or in the extreme case, performing no tests) is no proof that there are no defects;
2. It is impossible to do an exhaustive testing. For any reasonably complex project, the testing universe has simply too many possible configurations. Therefore, risk analysis is critical and provides strong arguments for testing software;
3. As soon as development begins, code should be tested to avoid discovering problems later on the development life cycle, which would increase repair costs;
4. The Pareto (20/80) rule is applicable: the majority of bugs is contained in a small number of modules, or these show the most operational failures;
5. Testes should be renewed frequently to avoid the pesticide paradox [2], where the same set of tests will eventually stop finding defects;
6. Testing is a context-dependent task, i.e., to different types of systems are applied different tests. For example, machinery control software is tested differently from web-services;
7. The system has to fulfil its purposes or fixing it is a waste of resources. If the system is not compliant to its requirements, the system will always be incorrect.

All defects of a software system can be categorised by two main metrics, their severity (which measures how much the defect affects the system—usually going from cosmetic to critical) and their priority (the urgency that the defect needs to be fixed, typically rated in terms of customer requirements, and ranging from low to high). With these qualitative ratings, a matrix may be determined that helps the developers of the system to prioritise the correction of the defects, providing greater transparency to developers, testers and customers, and aligning their expectations.

Finally, an important care should be taken regarding the human factors of testing, starting from the analysis of the mind-set and different points of view of the developers and the testers, to the balance of performing self-tests or carrying an independent test campaign, and the communication that takes place between the tester and developer. Often these factors make the difference between success or failure in the correction of defects, or in the effort needed to achieve a proper system.

This article intends to describe in a state-of-the-art fashion the policies, tools and methods used for developing and testing software components. We present a summary of best practices for testing, targeted for any software development, and complemented with examples for the C2NET project (described in a brief summary below). Furthermore, we list a selection of technologies, tools and any other relevant artefacts that can be useful, hereafter called "assets".

The C2NET project

The Horizon2020 C2NET project is an ambitious initiative towards providing cloud-enabled tools for supporting the supply chain network of manufacturing and logistic assets. The purpose of the project is to allow factories to master the complexity and data security of their supply network, by storing products, processes and logistic data, and optimising the manufacturing assets through collaborative computation of production plans, resulting in efficient delivery plans that are available to decision makers to monitor, visualise, control, share, and collaborate.

For that purpose, the C2NET comprises the validation using multiple pilots, either in the roles of suppliers for automotive parts, or in the field of transformation of metal-mechanics, towards the production and distribution in the field of dermo-cosmetics. In all these fields, C2NET intends to provide a set of cloud-based services that need to be properly tested and validated. The pilots in C2NET are:

1. In the Automotive area, will automatically generate optimised plans and carry out collaborative production planning and synchronised replenishment planning of car components;
2. In the Dermo-cosmetics area, will facilitate the collaboration for an agile production and distribution of dermo-cosmetics;
3. In the Metalworking SME's network, will enhance collaborative purchases and transportation;
4. In Original Equipment Manufacturing (OEM), will focus on the interactions and transactions between the manufacturer of hydraulic and lubrication products and the supply chain interactions necessary to provide ordered products to customers.

2 Testing Methodologies and Components

In this section, we describe the main testing types and practices in use. These are useful for an overview of the various types of tests required for the validity and correct functioning of any software.

The software development life cycle is divided in different phases, such as requirement gathering and analysis. Each phase should be tested accordingly, using different testing levels and methodologies. Subsequently, we present multiple types of testing, and below we provide a more complete description of the most important types:

- **Static Testing**: Is done without running any code; only observation is performed;
- **Unit testing**: Tests conducted by the developers to in order to guarantee that their code is working correctly and meets its requirements;
- **Component testing**: Also called module testing, it is different from unit testing as in unit testing developers test their own code (typically a small method or function), whereas component testing is meant to test the component as a whole;
- **Integration testing**: Integration testing is performed to evaluate the correctness and behaviour of the integration of two or more modules.
- **System integration testing**: This type of testing involves all related systems, and tests if data integrity is maintained, as well as accessing the operation in coordination with other systems;
- **System testing**: In system testing the testers test the compatibility of the application with the system;
- **Acceptance testing**: It is performed to ensure that the specified (user or contractors') requirements are met;
- **Alpha testing**: Once the development process ends, this type of testing verifies the whole system, and is performed at the developers' site;
- **Beta testing**: Same as Alpha testing, but at the customers' site, just before the launch of the product.

Next, we detail some of the most important types of testing, in the scope of the C2NET project needs.

2.1 Unit Testing

Unit tests are intended for finding bugs in a method or a code block. They are also used to perform regression tests to verify that the changes do not introduce new bugs, and code refactoring. In test-driven development, unit tests are created from the use-cases before the application code is written. Additionally, unit tests are often created as needed for the testing harness, when new issues arise. Writing unit tests is a time consuming task, but it helps to limit the number of defects in the code and validates the functionality of the test cases. Many continuous integration testing software support automated unit testing of the whole application. Automated unit testing is no substitute for manual unit testing and typically best results are achieved by an ideal balance between automated and manual testing [3, 4].

Unit tests cover:

- Code with "business logic";
- Core code that is accessed by other modules;
- Code that seem to gather a lot of bugs;
- Code written by multiple developers.

Unit testing is particularly useful to reduce the bugs testers and users experience during integration testing.

2.2 *API Testing*

API testing is often part of the integration testing as the different tiers and layers need to work together. The test cases are combinations of API requests and responses, and in transactional operations a predetermined sequence of calls. Web services and software library components are typical targets for API testing. Interoperability, functionality, security and scalability are important aspects of API testing.

API testing may comprise a large set of different testing and techniques:

- **Random manual testing**: Manually perform calls to the methods exposed by thc API, in order to verify that all the listed resources can be listed, created and deleted appropriately;
- **Usability testing**: This testing validates the functionality and user-friendliness of the API, and checks if the API supports cross platform appropriately;
- **Security testing**: This testing is meant to disclose possible security issues in a system, such as proper encryption of data sent over the network;
- **Automated testing**: To run a script that tests the API automatically. This script should be run frequently;
- **Documentation**: A formal verification that the API's documentation provides adequate information for a correct interaction with it. This is typically performed at development's end.

2.3 *Integration Testing*

System integration is the process that interconnects various components into a single system. In large projects, typically the subsystems are developed by different people, which contributes for increasing the difficulty of merging the different components. Integration (itself) and integration testing are techniques that merge and test the merging of subsystems, respectively.

As specified above, unit tests enable the detection of defects in the unit/component being tested. However, unit tests have a limited reaching, not covering tests across external boundaries or between units [5]. For example, unit tests do not catch bugs in stored procedures [6], databases, other services, or even simple files. Since these interactions are also vital parts of a system, their adequate testing is critical. Integration testing accesses the correctness of the interactions among all dependencies. These dependencies can be mocked to create an isolation environment for the integration testing.

Integration testing covers:

- Verifying software modules work in unity;
- Verifying new changes in module development;
- Verifying Interfaces of the software modules with the database;
- Verifying external hardware interfaces, if any;
- Checking inadequate exception handling.

The following sub-sections are some guidelines in the actual creation of integration tests.

Data Setup and Tear Down

Integration tests need to properly set up and tear down data. A test sets up its relevant data (such as logging in or placing entries on a database), and tears down that data once the test is complete to avoid polluting the system with "dummy" data. Also, never assume the data has already been deployed, since this reduces the safety of the test. There are several methods to set up and tear down data, and several places to execute these actions.

Options for Data Setup and Tear Down:

- **Code**: A first option is to write code that performs the setup and tear-down logic. The code itself does the three steps:
- Placing data,
- (the actual) testing, and
- Removing the exact same data placed during step 1.
- **Use DB Backup/Restore**: It is also possible to use scripts, or a database backup that contains all of the data needed to execute the tests. The script will follow the same three steps mentioned above;
- **Centralising Data Setup**: During a multiple service project it is recommended to centralise the data setup and tear-down logic into a single assembly, since logic from all projects is likely required to be connected.

String Tests

String tests perform a set of actions required to test a complex functionality, such as "Login => Send email => Logout". These tests try to mimic the interaction of a user with the system.

Assembling a test plan document

Integration testing should be thoroughly planned before execution. The testing team will use this plan as a reference, and to set a logical test sequence. Besides, this plan can be used to document the testing process and to maintain coherence among the team's efforts. Finally, managers can consult the document for a complete overview of the process, allowing them to allocate resources accordingly.

Completing unit testing prior to integration testing

Integration testing should not be performed with untested, possibly buggy code. It is not productive to test the integration of modules if these are not guaranteed to function correctly alone.

Automating test processes when possible

As with unit testing, integration testing should be automated. Manually checking for defects is tedious, time-consuming, and in some cases, ineffective. By performing tests automatically, the teams can continue development without the burden of spending large amounts of time performing tests. Besides, automated tests have a higher chance of identifying small defects automatically, streamlining the production of corrections. Test automation tools should be considered for any project.

2.4 Acceptance Testing

To accomplish the acceptance testing, each pilot generated three or four validation scenarios. The set of scenarios include at least one scenario for planning and one for manufacturing (or executing the plan) for each Pilot. These validation scenarios were mapped with the functional requirements, to analyse if the identified validation scenarios cover the functional requirement classes adequately. Each validation scenario was described with a storyboard and step descriptions. In addition, acceptance criteria of each validation scenario were defined. Those will describe what kind of performance or outcome is acceptable in each corresponding scenario after the implementation (see Fig. 1).

Fig. 1 One frame from a storyboard

The Validation scenarios for the C2NET project were defined using storyboard pictures with written step descriptions. The storyboards were drawn using the software "Storyboard That®"[1] tool.

The users of the validation scenarios are the end-users of the Pilots. That is why this testing phase will cover also the Beta testing. In C2NET, also the business impact of implemented collaborative manufacturing network solution is important. The business impact will be evaluated with Key Performance Indicators.

3 Testing Tools

In this section, we present a set of criteria used to select testing tools. It is worth noting that tools may adapt to the specific context, especially if they are open source. Nevertheless, modifying a testing tool implies testing the testing tool, which may add a considerable overhead to a project's execution.

3.1 Quality Criteria

The selected testing tools should meet the quality requirements presented in the ISO/IEC 9126 standard, which proposes an evaluation model composed of six characteristics: functionality, reliability, usability, efficiency, maintainability and portability; and corresponding sub characteristics for software product quality [7]. In addition, the quality of vendor or producer of the testing tool needs to be considered. For instance, general vendor qualification, tools support, licensing and price.

3.2 Functional Criteria

Functional criteria for selecting testing tools depends on the testing needs, namely, what type of testing tasks will be performed and what testing methods are required. Besides, the developed testing application should be able to use the following testing methods:

- **Black box testing**: examine the functionality of the APIs without peering into its internal structures;
- **Automation testing**: compares the outputs predicted and produced by the test tool, and controls the execution of tests;
- **Functional test**: examine the functionality of the APIs;

[1]https://www.storyboardthat.com.

- **Integration test**: examine the functionality of several interdependent APIs;
- **Scenario test**: examine the C2NET component APIs could complete a user scenario;
- **Stress test**: tests APIs beyond the limits defined by the API, providing a robustness appraisal;
- **Load test**: measure the response of APIs when putting demand on them;
- **Compatibility test**: evaluate the C2NET component's compatibility with the computing environment;
- **Regression Test**: examine new bugs after changes such as enhancements.

3.3 Selecting Testing Tools

The first step to select tools for a project is to matching the criteria with the project. Secondly, and besides the quality and functional criteria, the following factors may also be considered:

- Support various programming languages;
- Support cross platform (desktop, mobile), OS (Linux, Windows);
- Support different testing types listed above;
- Quality of results;
- Readability of the results;
- Open source;
- Easy debugging and logging;
- Ease of maintenance of the debug scripts;
- Data Driven Capability—The ability to externally store and load data from its own storage.

These criteria may be especially relevant for some projects, or can be used to clean up ties.

Next, an analysis was performed on presenting a list of candidates for testing various types of interfaces. First, we presented a list of the tools considered for the C2NET, followed by a summary of the tools in Table 1, and then performed an alternatives analysis over the tools selected. These serve as an example of applying criteria to a specific situation.

Web application testing

- WatiR/WatiN: http://watir.com/http://watin.org.
- TestCafe: http://testcafe.devexpress.com.
- Selenium: http://www.seleniumhq.org.
- Wrk: https://github.com/wg/wrk.

Table 1 Tool summary

Tool name	Description
Watir/Watin	WatiR is composed of a set of Ruby libraries intended for automating web browsers. WatiR's tests are easy to read and maintain, since it is a simple and flexible tool It tries to mock the way users interact with browsers: it clicks links, fills in forms, presses buttons. WatiR also checks results, such as whether expected text appears on the page WatiR connects to databases, reads data files and spreadsheets, exports XML, and structures code as reusable libraries WatiN provides the same functionalities as WatiR, but is implemented in .NET
TestCafe	TestCafe enables to runs tests on browsers that support HTML5 on either Windows, Mac or Linux platforms. It uses scripts to communicate directly with the DOM and intercept user's actions. TestCafe also includes a built-in traffic analysis tool that detects missing resources, response codes and JavaScript errors
Selenium	Selenium is composed of various software tools, and each performs a different type of testing. The set aims at covering all the needs of web testing. Besides complex UI interactions, it allows the comparison of a predicted output with an actual one
Wrk	Wrk is a HTTP benchmarking tool, that generates traffic to test the load balancing of HTTP servers. It is possible to use LuaJIT scripts to perform response processing, and custom reporting
Grinder	The Grinder is a load testing framework for distributed testing through the use of injector machines. Test scripts are written in Python, and can call out to Java code, allow to test any network protocol. The Grinder comes with a plug-in for testing HTTP services, and recording these
JMeter	Apache JMeter is a well-known open-source alternative to test functional behavior and measure performance. JMeter has a modular structure, extensible through plugins, thus being augmented by Apache or by the community
JBehave	JBehave is a framework for Behaviour-Driven Development (BDD). Being a derivate from test-driven development (TDD) and acceptance-test driven design, BDD aims at making these practices easier for everyone
Cucumber	Cucumber's main functionality is providing to developers a means to describe software development in plain text, according to a set of rules. The software is tested against those rules, for comparison of predicted and actual behaviour
Jasmine	Jasmine is an open-source testing framework for JavaScript. It does not influence the application nor the IDE, thus not altering the normal application's behaviour
SoapUI	SoapUI is a free and open source cross-platform web Functional Testing application. SoapUI predicts the execution of automated functional, regression, compliance, and load tests. SoapUI provides complete test coverage and supports all the standard protocols and technologies

(continued)

Table 1 (continued)

Tool name	Description
Fressia	Fressia is an open-source framework for testing automation. It is set up to be usable out of the box. Although it is a recent tool, it is a useful, general purpose testing tool
Arquilian	Arquillian is a extensible testing platform for JVM that allows the creation of automated integration, functional and acceptance tests for Java. Arquillian can be used to bundle test cases, dependent classes and resources. It is also capable of deploying archives into containers, execute tests in the containers, and capture results to create reports
JTest	JTest is an automated Java software testing and static analysis software. It includes functionalities for Unit test-case generation and execution, static code analysis, data flow static analysis, and metrics analysis, regression testing, run-time error detection. JTest can detect run-time errors, such as race conditions, exceptions, resource and memory leaks, or security attack vulnerabilities
TestNG	TestNG is a testing framework for Java. It covers unit, functional, end-to-end, integration, and other types of testing. It provides complex features, such as annotations, running tests in big thread pools with various policies available, code testing in a multi thread safe, flexible test configurations, or data-driven testing support for parameters
JUnit	JUnit is a unit testing framework for Java. It is one of the main and most used testing frameworks for Java. JUnit is linked as a JAR at compile-time and can be used to write repeatable tests
JWalk	JWalk is designed as a unit testing toolkit for Java. It has been designed to support a testing paradigm called Lazy Systematic Unit Testing. The JWalkTester tool performs any tests of any compiled Java class, supplied by a programmer. It is capable of testing conformance to a lazy specification, by static and dynamic analysis, and from hints by the programmer behind the code
Mockito	Mockito is an open source testing framework for Java. Mockito allows programmers to create and test mock objects in automated unit tests for TDD or BDD
Powermock	PowerMock is a Java Framework for unit testing. It runs as an extension of other Mocking frameworks like Mockito but provides more powerful capabilities. PowerMock utilizes a custom classloader and bytecode manipulator to enable mocking of static methods, removal of static initializes, constructors, final classes and methods and private methods
Visual Studio Unit Testing Framework	Microsoft Visual Studio is an IDE used to develop computer programs for Microsoft Windows, as well as web sites, web applications and web services. Built-in tools include a forms designer for building GUI applications, web designer, class designer, and database schema designer. It can be enhanced by plugins. It provides debugging support for all phases of development
NUnit	JUnit equivalent, for C#

(continued)

Table 1 (continued)

Tool name	Description
AssertJ	AssertJ is a Java library that provides a fluent interface for writing assertions, in order to improve test code readability and make maintenance of tests easier AssertJ can be integrated with JUnit or TestNG
Hamcrest	Hamcrest is available to various languages. Its main objective is to provide flexible tests through the matching (comparison) of objects

Load testing

- Grinder: http://grinder.sourceforge.net/index.html.
- JMeter: http://jmeter.apache.org.

Behaviour testing

- JBehave: http://jbehave.org/.
- Cucumber: https://cucumber.io/.
- Cucumber-js: https://github.com/cucumber/cucumber-js.
- Jasmine: http://jasmine.github.io/.

Unit and other types of testing

- SoapUI: http://www.soapui.org.
- Fressia Project: http://fressia.sourceforge.net.
- Arquilian: http://arquillian.org.
- JTest: http://www.parasoft.com/product/jtest.
- TestNG: http://testng.org.
- JUnit: https://github.com/junit-team/junit.
- JWalk: http://staffwww.dcs.shef.ac.uk/people/A.Simons/jwalk.
- Mockito: http://mockito.org.
- Powermock: https://github.com/jayway/powermock.
- Visual Studio Unit Testing Framework: https://msdn.microsoft.com/en-us/library/hh694602.aspx.
- NUnit: http://www.nunit.org.
- AssertJ: http://joel-costigliola.github.io/assertj/.
- Hamcrest: http://hamcrest.org/.

To test the web interfaces in the C2NET platform, we proposed the use of the SoapUI tool. SoapUI is an open source web service testing application for both SOA and REST. It performs web service inspection, invoking, development, simulation and mocking, and functional testing, load and compliance testing. Being developed in Java, it is supported by Java IDEs like Eclipse, IntelliJ and Netbeans. It can test connections to databases, messaging functionalities, and rich internet applications among others. It also possesses options for scripting and a command line interface, ideal for test automation. The tool can be augmented with the use of plugins.

SoapUI provides flexible tests, since it can, for example, convert functional tests into load tests. This allows at the same time to test if a functionality is correct, and its efficiency. It includes a mocking functionality, where it simulates systems that may not have been implemented yet. This allows the assessment of the correctness of a functionality's logic. Also, it can generate error messages automatically for a completer testing.

Complete reports are generated automatically and can be exported into various formats, such as PDF or CSV. It is also possible to overview the results as an HTML file.

For the remainder of the C2NET platform, including the cloud components, we suggested the use of the TestNG framework. TestNG is a Java unit testing framework that aims to overcome many limitations of JUnit. It is designed to simplify testing requirements such as functional testing, regression and end-to-end.

It includes simplified annotations, making it easier for testers to understand them. Reports are generated in HTML for ease of handling and compatibility, and offer various levels of complexity. Multiple test cases can be executed at a time by creating a simple XML file. Also, parametric testing is supported.

TestNG can also be used to run failed test-cases, to group test cases, and to declare dependencies (for example, if a method depends on another); these are main advantages over JUnit. It is also possible to define which test cases are ran using a simple annotation, and their execution order.

TestNG also provides static code analysis, that evaluates the possibility of occurrence of an exception. This method will give the details of the exceptions that are expected to be presented by that particular method.

4 Conclusions and Future Work

Testing and evaluating systems and software is a thorough activity that needs to be started very early in the system development life cycle, and accompanies this cycle until its delivery. Testing has its own life cycle, with preparation phases that precede the actual assessment of the target system, which should be customised according to the target system needs and requirements.

In this article, we presented a study of the state-of the-art on testing and its tools, and provide an example of selecting tools on the context of the C2NET project. As future work, there is the need to perform the actual validation of the C2NET platform and tools using the selected methodologies and tools described.

Acknowledgments This work has been partly funded by the European Commission through the Project H2020 C2NET: Cloud Collaborative Manufacturing Networks (Grant Agreement No. 636909). The authors wish to acknowledge the Commission for their support. We also wish to acknowledge our gratitude and appreciation to all the C2NET Project partners for their contribution during the development of various ideas and concepts presented in this paper.

References

1. International Software Qualification Testing Board (Online). Available: www.isqtb.org
2. Beizer, B. (1990). Software Testing Techniques (2nd ed.). Van Nostrand Reinhold, 115 Fifth Avenue, New York, New York 10003
3. Culp, A. (2011, June). General service testing best practices (Online). Available: https://msdn.microsoft.com/en-us/library/vstudio/hh323702(v=vs.100).aspx
4. Narayanan, U. (2015, October). Top Five Best Practices for Writing Unit Test Scripts (Online). Available: http://www.developer.com/mgmt/top-five-best-practices-for-writing-unit-test-scripts.html
5. Culp, A. (2011, June). Integration testing (Online). Available: https://msdn.microsoft.com/en-us/library/vstudio/hh323698(v=vs.100).aspx
6. Gaines, L. (2014, September 03). 5 Integration testing best practices to improve software development (Online). Available: https://www.vectorcast.com/blog/2014/09/5-integration-testing-best-practices-improve-software-development
7. Illes, T., Herrmann, A., Paech, B., & Rückert, J. (2005). Criteria for software testing tool evaluation—A task oriented view. In *iSQI (Hrsg) Proceedings of the 3rd World Congress for Software Quality*, Munich.

A Self Sustainable Approach for IoT Services Provisioning

Raquel Melo-Almeida, Carlos Agostinho and Ricardo Jardim-Goncalves

Abstract Internet of Things (IoT) paradigm is opening a large set of opportunities for things to make part of the processes that monitor our industries, support our decisions and control our lives. For that kind of power things must be reliable, i.e., they have to be able to handle on their own, without human interaction, means to still deliver when faults are detected. This work presents an approach that aims to increase the reliability and trust of constrained IoT devices in the effort of pushing the technology further and increase its usability in the application scenarios that IoT is supposed for. It will also be integrated in European Union's Horizon 2020 C2NET project, as an essential key in the monitoring of pilots production lines.

Keywords IoT · Sustainable systems · Self-organisation · Reliability · Services provisioning

1 Introduction

Internet of Things, a global network of RFIDs, as firstly described by Kevin Ashton in 1999 [1], is nowadays seen as the present and tomorrows future, where billions of things are connected and capable to communicate amongst them and within the Internet.

In this, 'things' refer to any object on face of the Earth, whether it is a smart communicating technological device or a non-communicating, so called constrained, device [2]. From a smartphone, laptop to a leaf of a tree or a bottle of

R. Melo-Almeida (✉) · C. Agostinho · R. Jardim-Goncalves
Departamento de Engenharia Electrotécnica, Faculdade de Ciências e Tecnologia,
FCT, Universidade Nova de Lisboa, 2829-516 Caparica, Portugal
e-mail: ram@uninova.pt

R. Melo-Almeida · C. Agostinho · R. Jardim-Goncalves
Centre of Technology and Systems, CTS, UNINOVA, 2829-516 Caparica, Portugal

© Springer International Publishing Switzerland 2016

K. Mertins et al. (eds.), *Enterprise Interoperability VII*,
Proceedings of the I-ESA Conferences 8, DOI 10.1007/978-3-319-30957-6_4

beverage, anything can be part of World Wide Web. This is one of the reasons why IoT cannot be seen only as a hardware or software paradigm, but also as a societal challenge that will tackle the privacy of information to a completely higher level, since there are a lot of different scenarios refereed as possible applications for IoT.

- Elder and Healthcare [3–8]: IoT biggest challenge, reduce the costs spent by European governments in the heavy bill of public health and elder care by helping patients monitor their diseases at home, promote prevention of diseases by usage of wearable devices, and support emergencies for elders left alone at home.
- Smart homes and cities [9–12]: Most discussed scenario for IoT in which sensors help us understand the state of our Environment, promote energy efficiency and waste management. Even the safety of citizens can be improved with the help of IoT.
- Manufacturing and Supply-Chain [13–15]: All of these are willing to bet on IoT to help them manage their business, for monitoring their productions, reduce waste and non-conformities of products, accurately track and trace packages on the move, and even the trucks that transport them, knowing in real time every step of the process when receiving or sending an order.

The few presented scenarios are expecting maybe a little too much from IoT. How can we rely on this, small, constrained devices for the real time monitoring of our most important things?

In the future it is expected that IoT networks become capable of gaining such type of intelligence and power that will allow them the ability of adapting and organising themselves without having any superior or external entity managing them. IoT devices will be embedded in several stable or unstable environments, in which they will have to perform with exactly the same precision. A Complex Adaptive System has the same type of behaviour. Those are systems composed by a large set of agents, in which each individual acts according to its rules or principles in response to local interactions with other agents, adapting through a mechanism of self-organisation [16].

In this paper a different approach for self-organization and sustainability maintenance in IoT networks that aims to increase its reliability is presented. This approach, presented in the next chapters, basis on the usage of three essential components that will act at the network level, the middleware level and at the client level. The presented work is an approach that can be used in production monitoring processes, and is currently in implementation under the C2NET project [17]. The goal of the C2NET project is to develop collaborative cloud tools that could support the management and production strategies of such companies and reduce their costs related with the excess of waste and non-conform products.

2 IoT Services Architecture

An IoT device never work solo. Instead all the networked devices should act in a way that seen by the outside them seem to be only one. This feature is actually the concretization of IoT. Each device benefits from being networked to share its processing skills and reduce its individual energy costs [2]. For that devices must work closely together and be aware that they have alive neighbour devices with services that they can count on.

2.1 IoT-Based Architecture

2.1.1 IoT-A

IoT-A [18], is a European Project that proposes the creation of an architectural reference model (ARM) and the definition of its building blocks. It is envisioned in this project that the solution presented create the foundations that should be embraced in a future Internet of Things. IoT-A combines architectural principles top-down reasoning and also simulation and prototyping guidelines to explore the designed architecture (Fig. 1).

IoT Architectural Reference Model consists of three parts:

- **The IoT Reference Model**: provides the highest abstraction level for IoT. It promotes a common understanding of the IoT domain.
- **The IoT Reference Architecture**: is the reference for building compliant IoT architectures. It provides different views and perspectives on architectural aspects that concerns IoT stakeholders.

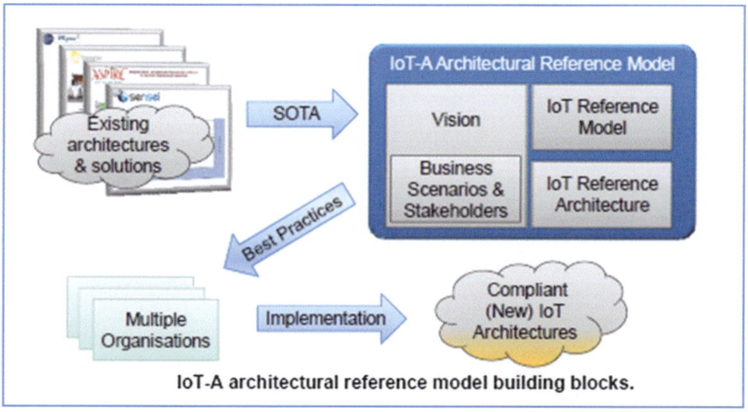

Fig. 1 IoT ARM building blocks [19]

- **The Guidelines**: the first two give the models, views and perspectives for designing the architecture, while this item has the importance to guide the architect during the process.

2.1.2 FIWARE Approach

FIWARE [20] Generic Enablers allow Things to become available, searchable, accessible, and usable context resources fostering FIWARE-based Apps interaction with real-life objects. The innovative idea is that all Things or IoT resources are exposed just as other NGSI Context Entities, simplifying the work as developers will just need to learn and use the same NGSI API used in FIWARE to represent all Context information. Considering the complete scenario, IoT GEs are spread over two different domains:

- **IoT Backend**. It comprises the set of functions, logical resources and services hosted in a Cloud datacenter. Up north, it is connected to the data chapter ContextBroker, so IoT resources are translated into NGSI Context Entities. South-wise the IoT backend is connected to the IoT edge element, that is all the physical IoT infrastructure.
- **IoT Edge**. It is made of all on-field IoT infrastructure elements needed to connect physical devices to FIWARE Apps. Typically, it comprises: IoT end-nodes, IoT gateways and IoT networks (connectivity). The IoT Edge and its related APIs will facilitate the integration of new types of gateways and devices, which are under definition in many innovative research projects, and warranty the openness of FIWARE IoT architecture (Fig. 2).

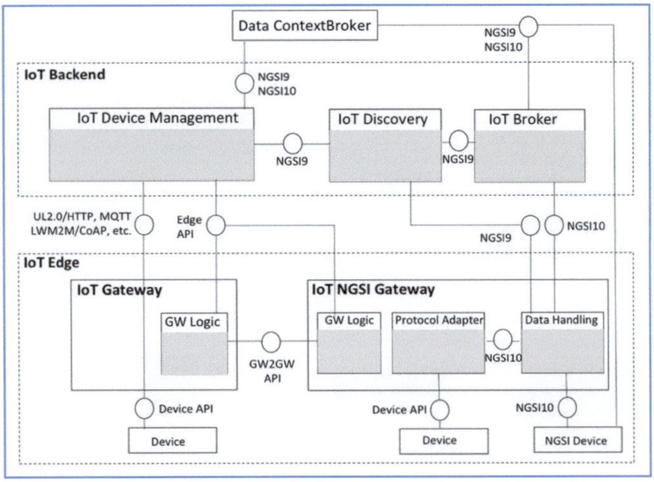

Fig. 2 FIWARE IoT GEs [21]

2.2 IoT Layered Architecture for Service Virtualization

In the management of an IoT network one must be aware that it could be desirable to know in real time all the services available in the network, even if the devices that were available five minutes ago delivering the service are now disconnected or were "moved" inside the network.

IoT services should be available to others to use, if one have a bunch of candies in the top shelf of the kitchen cabinet and do not reach there, then what is the purpose of having them? In this case the network should grant easy and secure access to those through standardized interfaces.

To be understandable, the network should perform a translation between requests from outside and inside technologies, handling the negotiation of how interoperability is going to be achieved.

A middleware architecture that meet the described requirements should be similar to the one displayed in Fig. 3. that is essentially composed by the following layers:

Exposition Layer: Top layer of the architecture that handles network-client interactions. It should abstract the way that the network of services is implemented and be able to expose all the inherent services.

Service Management Layer: Responsible for understanding the networked services. Also it should match client requests and responses grating that interoperability is maintained.

Network Access Layer: In this, the physical link to the IoT network of devices is handled.

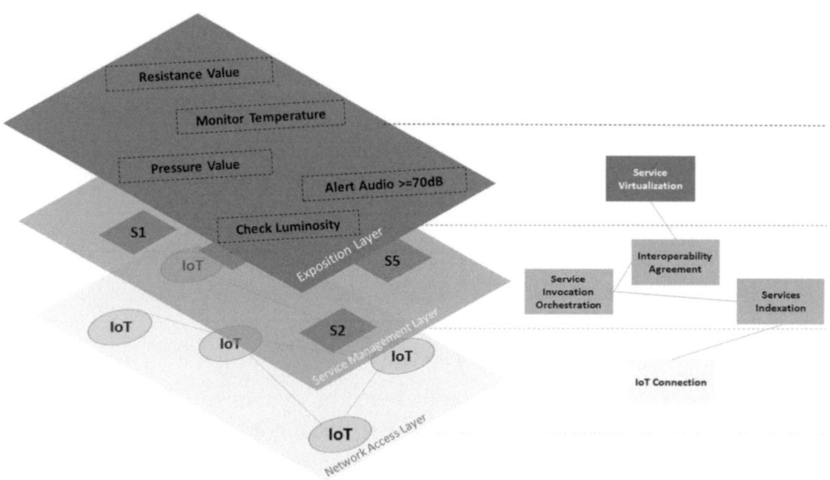

Fig. 3 Layered architecture

Each layer can be decomposed in the following components:

- **Service Virtualization**: The available services (at each moment in the IoT network) are gathered and virtualized as Cloud Services, granting ease of access to clients using the Service Virtualization component. This component should keep is coherency even if changes are performed, so that the next layers of the architecture receive the same information and is able to perform.
- **Service Invocation Orchestration**: This component is responsible for managing the client invocations for a service. A track is kept in this module that is able to match client invocations and their associated responses keeping the accuracy of the system.
- **Interoperability Agreement**: The interoperability agreement is responsible for performing the translation between outside world (client side) and inside world (IoT network). It should be granted with the most common translations, but kept open so that the designer of the architecture can easily implement new translations when needed.
- **Services Indexation**: This component should always be aware of which services are available at each moment and where they are located inside the IoT network. When a new service is present in the network it should be registered in this component.
- **IoT Connection**: The IoT Connection is responsible for performing and handling the physical bounding to the network of services.

3 Self-sustainability in IoT

Looking at the layers perspective presented in the previous section it is clear that the bottom part will be connected to the IoT network, the top layer is the connection to the services consumer and the middle makes the connection between both. So adding some characteristics like organization, aggregation and intelligence knowledge to the different layers can achieve the proposed goal.

3.1 Self-organization in IoT Networks

One possible approach to design an IoT network keeping the focus of enriching it with the ability of self-organization can be based on dynamism. It is not possible for such system to be statically organized. Since they cannot be organized once in time and use the same organization in the living period of the system, since that during time it suffers changes. It have to be able to handle the mobility of the devices, it should rapidly accept new devices in the network and handle the fact that the probability of the device to be disconnected is high. So dynamism implies that the network can be actively self-organized adapting itself to the constant changes in the inside and surrounding environment.

Building monitoring applications on top of dynamic IoT networks imply that one should keep in mind that the devices performing the task should not be known to the application, i.e., the application knows that the network is able of providing a set of services, but it is not aware of which device is responsible for collecting and delivering the associated data.

Using a method similar than the one presented in [22] can enable this type of abstraction. It relies in a system based in subscriptions that combines the spatial location of devices disposed for monitoring specific geographically defined places with the services that they have to offer.

When a requester wants to monitor a determined location, it sends a subscription to the area in which it wants to retrieve responses in opposition of sending it directly to a provider. That way all the devices that are enclosed in the area receive the subscription and can reply with responses, like presented next.

This is possible since the devices are aware of their coordinates and their sensing circular range that is defined according to the Cartesian equation of the circle:

$$\left(x_{provider} - x_{subscription}\right)^2 + \left(y_{provider} - y_{subscription}\right)^2 \leq r_{provider}^2$$

3.2 Service Aggregation for Location

Combining the architecture presented in Fig. 3. and the network design presented in Fig. 4. it becomes clear that the self-organization of the network approach can add some value to other architectural layers.

The Indexation of Services can benefit from such approach since that if each service that the IoT Network has to offer is aggregated with the location of itself inside the network in terms of its geographical coordinates, then it will be possible to understand which services are available in determined areas.

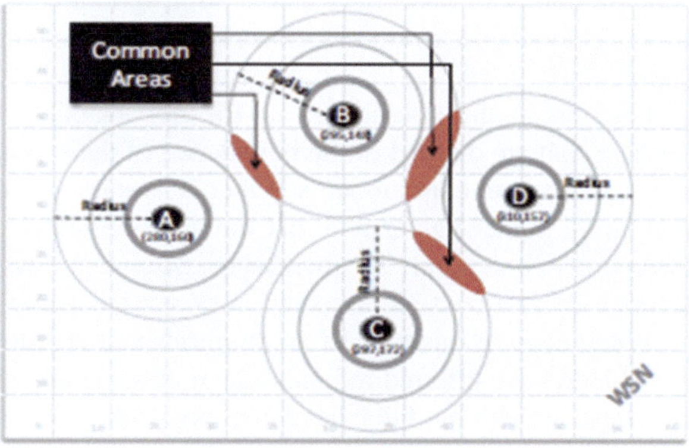

Fig. 4 Subscription approach [22]

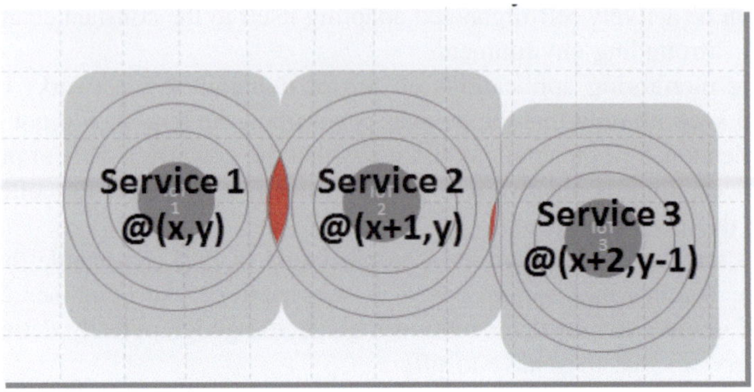

Fig. 5 Services location

Using this information, then it is possible to aggregate services per location. As an example, looking at Fig. 5 one can see that at the red area Service 1 and Service 2 are available. I.E, when sending a subscription to the red area, one can know that the temperature of the room (Service 2) and the luminosity of the same (Service 1) are available. On the other little red area one can also know that the air conditioner inside the room can be on or off (Service 3).

3.3 Intelligent Service Assessment

Using the analogy made with Complex Adaptive Systems, this is just moving up in the complexity scale of the system [23] presented in Fig. 6. Instead of having "simply" devices providing information according to their location that have the lowest complexity level, the same were combined and now are dynamically interacting to compose services that abstract the devices or set of devices that are sourcing the service, level two of complexity.

But if one thinks that the sensor, or set of sensors that is measuring the temperature drains all his battery and is not responding anymore. No other temperature sensor is available inside the room, so the temperature (Service 2) is not available anymore.

Looking at the neighbours of the not available service, there are two other services that are defined in the boundaries of the room, a luminosity service and the status of the air conditioner service. It is easy to infer that if the luminosity is presenting higher values and that the air conditioner is off then the temperature of the room will certainly be increasing. So we can easily associate the temperature (service 2) as a possible association of service 1 and service 3.

When making this type of assumptions and creating a rule system in which it is possible to have inferred redundancy of information with the surrounding objects, it is, when something fails have the possibility to grab the same information at

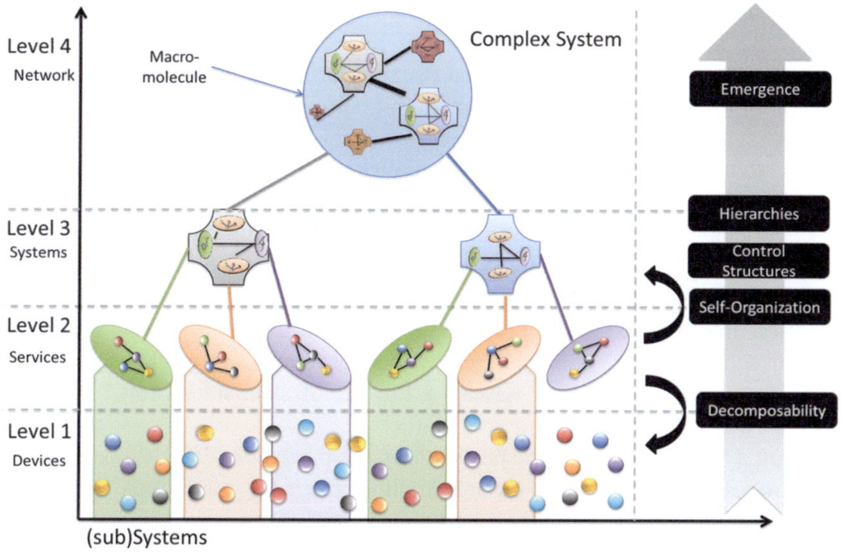

Fig. 6 Complexity levels (based on [23])

different sources the reliability of this systems to monitor really important issues can be superior. Comparing in the complexity scale the level is posed at 3. When the services are granted with self-organization it is then possible to have control structures and hierarchies that makes it easier to create rules.

4 C2NET Use Case

C2NET goal is to add value in the enhancement of SMEs overall competitiveness, innovation and adaptability in networked enterprise partnership scenario. Providing cross-country and inter-enterprise interchanges, building networked enterprises that are supported by stable relationship schemas and modern co-operation and co-ordination business paradigms. This will grant end consumers' advantages, mainly in terms of diminishment of products time-to-market and costs.

The C2NET Metalworking Networked SME's industrial pilot will be used as a practical use-case for the conceptual approach presented in the previous section. The network is composed by two SMEs that work in the transforming industries in which they buy steel at suppliers and transform it to finished products ready to sell to the market. In one the expertise is on producing steel tubes, in the other the expertise is related with the creation of steel shelfs.

Both companies want to reduce their production costs, cutting down the quantity of waste and non-conform products that are generated during production times. Waste is defined as the leftovers of the products that are generally created when the machines are not maximizing the usage of the material, and non-conform products are also the faulty products that do not fit in the standard ordered by the client. Those products generally are generated because machines lost the calibration or the precision or even the material is not at the higher quality and suffer deformation in the process.

Even though the companies have quality of service inspectors, production managers etc., the access to the information suffer a big delay. The quality inspector is generally only one person that have to check every product to see if they are conform, and he works in rounds not being able to spent the entire working hours looking at the same machine. When he detects a problem he has to inform the production manager that a problem have arise and then the production manager takes corrective actions. If the companies do not detect the problems in time real enough they suffer the risk of having a full set of steel transformed into straws instead of tubes.

Having a full set of sensors disposed along the production line will certainly help in the process of having real time information about the production, but still, the companies will not take the risk of producing at night as an example, when nobody will be at the factory because the companies cannot really trust in IoT for doing the full job. Because of the same issue, that if a sensor fails then there is not possible to have access to the information that the specified device was gathering.

5 Conclusions and Future Work

IoT has all the potential to grow up and stand for competition with large, well-established, expensive and not portable systems. Working all together small devices can have big powers. But there is still a journey to be taken in terms of security and trust ability in those devices.

Increasing the reliability in those devices can be possible when the abstraction from devices is made. If one think about it, it is not important to know which device does what inside a room, it is only necessary to understand that inside a room one has a set of services that can used for whatever one needs. But since the services are not independent from the sensors it is important to have inferred replication of information inside a system. That could be achieved if a set of rules is pre settled that will clearly state the alternative path to take for accessing the same information using different means.

It its authors truly believe that introducing the presented concept in C2NET project will enhance the reliability of IoT in the industrial world and therefore it can be replicated at the different application areas. Having systems that can be considered more reliable will be translated in a trust increase by part of the user/consumer. In the future, it is the authors desire to discover means to increase the trust levels on IoT systems directly at the network level and not only at an application level.

Acknowledgments This work has been partly funded by the European Commission through the European Union's Horizon 2020 Programme (H2020/2014-2020) under grant agreement: C2NET nº 636909. The authors wish to acknowledge the Commission for their support.

References

1. Ashton, K. (2009). That 'internet of things' thing. *RFiD Journal, 22*(7), 97–114.
2. Gubbi, J., Buyya, R., Marusic, S., & Palaniswami, M. (2013). Internet of things (IoT): A vision, architectural elements, and future directions. *Future Generation Computer Systems, 29* (7), 1645–1660.
3. Istepanian, R. S. H., Hu, S., Philip, N. Y., & Sungoor, A. (2011) The potential of Internet of m-health Things m-IoT for non-invasive glucose level sensing. In *Proceedings of Annual International Conference on IEEE Engineering in Medicine and Biology Society EMBS* (pp. 5264–5266).
4. Yu, L., Lu, Y., & Zhu, X. (2012). Smart hospital based on internet of things. *Journal of Networks, 7*(10), 1654–1661.
5. McCue, T. (2015). $117 bilion market for internet of things in healthcare by 2020. *FORBES*. Available http://www.forbes.com/sites/tjmccue/2015/04/22/117-billion-market-for-internet-of-things-in-healthcare-by-2020/
6. Miori, V., & Russo, D. (2012). Anticipating health hazards through an ontology-based, IoT domotic environment. In *Proceedings of 6th International Conference on Innovative Mobile and Internet Services in Ubiquitous Computing (IMIS)* (pp. 745–750).
7. McCullagh, P., & Augusto, J. C. (2011). The internet of things: The potential to facilitate health and wellness. *UPGRADE: The European Journal for the Informatics Professional.*
8. Jara, J., Alcolea, F., Zamora, M., Gómez Skarmeta, F., & Alsaedy, M. (2010). Drugs interaction checker based on IoT. In *2010 Internet Things (IoT).*
9. Smarthings. (2015). *SmartThings|Smart Home*. Available http://www.smartthings.com/
10. Chan, M., Campo, E., Estève, D., & Fourniols, J. Y. (2009). Smart homes—Current features and future perspectives. *Maturitas, 64*(2), 90–97.
11. Vlacheas, P., Giaffreda, R., Stavroulaki, V., Kelaidonis, D., Foteinos, V., Poulios, G., et al. (2013). Enabling smart cities through a cognitive management framework for the internet of things. *IEEE Communications Magazine, 51*(6), 102–111.
12. Zanella, A., Bui, N., Castellani, A., Vangelista, L., & Zorzi, M. (2014). Internet of things for smart cities. *IEEE Internet Things Journal, 1*(1), 22–32.
13. Bi, Z., Da Xu, L., & Wang, C. (2014). Internet of things for enterprise systems of modern manufacturing. *IEEE Transactions on Industrial Informatics, 10*(2), 1537–1546.
14. Liu, T., Yuan, R., & Chang, H. (2012). Research on the internet of things in the automotive industry. In *2012 International Conference on Management of e-Commerce and e-Government* (pp. 230–233).
15. Han-jiang, Z., & Fang, G. (2013). The study of a dual-channel automotive supply chain based on internet of things. In *2013 International Conference on Management Science and Engineering (ICMSE)* (pp. 650–658).
16. Agostinho, C., & Jardim-Goncalves, R. (2015). Sustaining interoperability of networked liquid-sensing enterprises: A complex systems perspective. *Annual Review in Control, 39*, 128–143.
17. C2NET Project. (2015). Available http://c2net-project.eu/
18. Internet of Things—Architecture. (2016). Available http://www.iot-a.eu/public

19. Bauer, M., Boussard, M., Bui, N., Carrez, F., Giacomin, P., Haller, S., et al. (2012). Introduction to the architectural reference model for the internet of things. In *Internet-of-Things Architecture—IoT-A Deliverable D1.3—Updated Reference Model IoT v1.5*.
20. F. (Core Platform for the FutureInternet). (2015). *FIWARE*. Available http://catalogue.fiware.org/chapter/internet-things-services-enablement
21. FIWARE. (2015). *Internet of things (IoT) services enablement architecture*. Available https://forge.fiware.org/plugins/mediawiki/wiki/fiware/index.php/Internet_of_Things_(IoT)_Services_Enablement_Architecture
22. Maló, P., Almeida, B., Melo, R., Kalaboukas, K., & Cousin, P. (2013). Self-organised middleware architecture for the internet-of-things. In *Proceedings—2013 IEEE International Conference on Green Computing and Communications and IEEE Internet of Things and IEEE Cyber, Physical and Social Computing, GreenCom-iThings-CPSCom 2013* (pp. 445–451).
23. Agostinho, C. M. M. (2012). *Sustainability of systems interoperability in dynamic business networks*.

Part II
Modelling the Enterprise Interoperability

Requirement Pattern Elicitation Approach of Massive Customers in Domain Oriented Service Requirement Engineering

Zhiying Tu and Xiaofei Xu

Abstract Nowadays, software service engineering has become one of the important development trends of the IT world. More and more software resources are developed and reconstructed as online services. These services appear as XaaS (e.g., IaaS, PaaS, SaaS, BPaaS) architecture with huge service capacity to serve massive customers. Under this servitization revolution, various kinds of services from multi-domains and multi-networks are converged as a huge complicated service network or eco-system, which can be called as Big Service. Even though this big service ecosystem contains abundant services resources, it is still very difficult for the customers to select the most suitable services to satisfy their requirements. Meanwhile, different individual customer has diverse service requirements. Thus, how to develop the most applicable service solution by reusing the existed heterogeneous service resources to meet massive individualized customer requirements is a key issue in the big service ecosystem. With the aim of solving this issue, Xiaofei Xu proposed a new paradigm of software service engineering called RE2SEP (Requirement-Engineering Two-Phase of Service Engineering Paradigm). This paradigm divides the software service development lifecycle into two different phases (service oriented requirement engineering and domain oriented software engineering) that can be executed simultaneously. The research work presented in this paper is under the service oriented requirement engineering phase. Our goal is to propose an approach that can detect the accurate requirements rapidly from massive customers. This approach is based on requirement pattern elicitation, so this paper will firstly present the requirement pattern definition, and then present the methodology and algorithm of requirement pattern elicitation.

Keywords Software service engineering · Big service · Service oriented requirement engineering · Domain oriented service engineering · Software reuse · Requirements pattern

---

Z. Tu (✉) · X. Xu
School of Computer Science and Technology, Harbin Institute of Technology, Harbin, China
e-mail: tzy_hit@hit.edu.cn

© Springer International Publishing Switzerland 2016

K. Mertins et al. (eds.), *Enterprise Interoperability VII*,
Proceedings of the I-ESA Conferences 8, DOI 10.1007/978-3-319-30957-6_5

1 Introduction

With the servitization revolution, software service has become one of the indispensable part of the IT world. More and more software resources have been or are going to be developed and reconstructed as online services. These services appear as XaaS (e.g., IaaS, PaaS, SaaS, BPaaS) architecture with huge service capacity to serve massive customers [1]. Various kinds of services from multi-domains and multi-networks are converged as a huge complicated service network or ecosystem, which can be called as Big Service [2]. Every day, different individual customers or IT systems have to invoke different services to accomplish their mission or satisfy their needs. Even though this big service ecosystem contains abundant services resources, it is still very difficult for the customers to select the most suitable services to satisfy their requirements. Also, different customers may have diverse service requirements. For example, in the weekend, one customer wants to drive his family to Disneyland. Firstly, he needs to weather report service to know whether the weather is fine. Then, he needs the ticket reservation service to book the ticket. Afterward, he needs the traffic navigation service to find the right way with less traffic jam. Finally, he needs to park service to find the place to park the car. For this travel plan, this customer needs various kinds of services from different service platform. Quite a lot service platform provides the same service, such as weather forecast service, and then this customer has to make a decision, and use different websites or apps. Sometimes, it is a complex job, especially if he travels to a city he is not familiar to. Thus, how to develop the most applicable service solution by reusing the existed heterogeneous service resources to give an intelligent service recommendation, which can automatically adjust to meets massive individualized customer requirements, is a key issue in the big service ecosystem. With the aim of solving this issue, Xiaofei Xu proposed a new paradigm of software service engineering called RE2SEP (Requirement-Engineering Two-Phase of Service Engineering Paradigm) [2, 3]. This paradigm divides the software service development lifecycle into two different phases, service oriented requirement engineering (SORE) and domain oriented software engineering (DOSE), which can be executed simultaneously. The research work presented in this paper is under the service oriented requirement engineering phase. Our goal is to propose an approach that can dig out the accurate customer requirement propositions rapidly from massive customers. This approach is based on requirement pattern elicitation, so this paper will firstly present the requirement pattern definition, and then present the methodology and algorithm of requirement pattern elicitation.

This paper is organized as follows: Sect. 2 introduces the related works; Sect. 3 gives the motivation of this research and formulates the research problems; Sect. 4 describes a requirement pattern elicitation approach with six steps. The pseudo code of the elicitation algorithms are introduced in Sect. 4 as well; Sect. 5 illustrate the experiment result to show the achievement and the drawbacks that will be fixed up in the further study; Sect. 6 concludes this work and points out the future work.

2 Related Works

2.1 Service Modeling Works

The traditional software engineering process follows a simple or repetitive top-down approach in accordance with software's lifecycle. For example, the typical waterfall model [4] follows the software lifecycle, respectively, requirements, software design, implementation and testing phases to deal with software code. Most of the other process models for software engineering [5] improve the waterfall model by introducing recursive or evolutionary process based on the basic software lifecycle. For example, the spiral model [6] repeats the waterfall engineering process to improve the engineering results continuously until satisfaction.

The idea of Component-Based Software Engineering (CBSE) firstly became prominent in 1968 [7]. CBSE regards the definition of components as part of the starting platform. It is a reuse-based approach to defining, implementing and composing loosely coupled independent components into systems. A component represents software package that encapsulates a relatively independent set of related functions or data. By composing the components that cohesively implement different modules of a given system, the system can be finally obtained. The components are regarded as significant assets that could be reused and substituted in applications by defining standard interfaces [8].

Model Driven Architecture (MDA) [9] is launched by Object Management Group (OMG) in 2001 as a type of model-driven engineering of software systems which provides guidelines for structuring specifications as layered models. MDA is a kind of domain engineering, which is expressed as models and produces code from human understandable specifications through model transformations from Computation Independent Model (CIM) to Platform Independent Model (PIM), and then to Platform Specific Model (PSM). MDA attempts to confront the force of change by separating and relating plat-form-independent models and platform-specific models using transformation techniques. The model transformation and implementation become the key issues of software development.

Service Oriented Architecture (SOA) brings the servitizational revolution of software engineering, which allows different software applications' functionalities to be accessible to each other as services via a protocol. In 2004, IBM published Service-Oriented Modeling and Architecture (SOMA) [10], which extends traditional Object Oriented Programming (OOP) and Component Based Software Engineering (CBSE) methods to include concerns relevant to and supporting SOA. SOMA is a service-oriented modeling, analysis and design method. It covers the phases of identification, specification and realization of services, service components, and processes that can be used to compose services. In service modeling phase, there are some new techniques such as goal-service modeling, service model creation and a service litmus test to help determine the granularity of a service.

In 2008, Michael Bell proposed a holistic modeling language for Service development, which is called Service-Oriented Modeling Framework (SOMF). It provides tactical and strategic solutions to enterprise problems by employing disciplines and a universal language [11].

To provide a solution for establishing service systems responding to customer requirements in an agile style, Service Model Driven Architecture (SMDA) was proposed by Xu in Harbin Institute of Technology [12]. In the SMDA approach, the MDA approach is applied for service development, and the USML extended from UML is applied for describing service models. Multi-dimensional service model definition and transformation, and service components reuse are the main features of the SMDA.

As Big Service is constructed through service convergence from multi-domains and multi-networks presented by Xu [2, 3]. In the Big Service ecosystem, the local service layer encapsulates the IT infra-structure, physical and digital resources into services through IoT and virtualization, and organizes them together with the local services provided by individuals or organizations as fundamental services for Big Service. The domain oriented service layer aggregates and composite the local services according to their related business domain, demands and relationships, forming a domain oriented services communities or IoS. This layer becomes the backbone of the Big Service ecosystem, which contains huge number of partial service solutions or service patterns in the service domains. The demand oriented service solution layer constructs customized service solutions by means of convergence of the domain oriented services to meet the massive individualized customer requirements and to create value. Moreover, the IT infrastructure layer at the bottom provides the basic condition and support for big services, while the client layer at the top deliver services to the customers based on customers' requirements and value proposition.

Service context modeling and context-awareness for web services received much concern in past few years. Sheng [13] have presented ContextUML that specifies classes for context types, sources, and context services. ContextUML can model context-aware objects within a service (services, operations, messages) and the binding of context-aware mechanism. Traditionally, context information modeling was always separated from the service modeling and development. Instead, ContextUML considers it as an integrated part of the service modeling.

2.2 Frequent Pattern Elicitation Work

In data mining research area, there are many researchers are working on eliciting the frequent pattern and finding out the association rules between data set. Agrawal et al. [14, 15] propose Apriori algorithm and association detection algorithm to dig out the interesting relations among the frequent transaction items in the market basket data to give the decision how to bind different products for sale. Cai et al. [16] propose an algorithm of mining association rules with weighted Items. Silverstein et al. [17] introduce an algorithm of generalizing association rules to dependence rules. Han et al. [18] proposes a frequent pattern algorithm for mining Frequent Patterns (FP) without Candidate Generation.

Some researchers also working on improving the existing algorithm in some specific aspect or using the frequent pattern finding algorithm or association rule detection algorithm in some specific domain. Lucchese et al. [19] introduce an improved algorithm—fast and memory efficient mining of frequent closed item sets algorithm. Shen et al. [20] improve the existing FP algorithm to mining dynamic association rules between the items with comments. Palmer et al. [21] use the existing association detection algorithm in the virtual enterprise knowledge discovery work and customer relation management work. Lim et al. [22] use the existing association detection algorithm in business performance measurements, decision support systems and online analytical processing.

3 Motivation and Problem Formulation

3.1 Motivation

RE2SEP is a new paradigm of software service engineering that adapts to the big data environment. Xu mentioned that in big data era, more and more abundant open and reusable data, software and service resources exist in the Internet and cloud. He defined Big Service concept with several critical new features: Massive, Complex, Cross, Convergence, Customer, Credit and Value (MC5V for short), which are corresponding to the 5Vs (Volume, Variety, Velocity, Veracity, Value) features of big data [2]. He believed that this phenomenon is changing not only the way of software and services applications, but also the approaches of software and services development. The software service engineering will gradually enter into the era of the maximal reusing of Internet-accessible abundant open services sources, such as web services, semantic web, virtualized services, cloud services, etc. [3]. Thus, the primary purposes of RE2SEP are how to rapidly gather the massive customer requirements and accurately predict individual customer requirement propositions from those; and how to find the suitable existing services to composite a corresponding service solution with the highest QoS.

RE2SEP is a bidirectional approach with combination of two phases of service oriented requirement engineering and software engineering. Its core technique lies in not only the service oriented software engineering reusing the abundant open source in multi-domains/communities on the Internet, but also the matching between service requirements and service solutions using service context as a mediating facility [3]. It consists of two different engineering processes with opposite directions, which can be executed simultaneously. On the one hand, the DOSE process starts from available physical and virtualized software services in a certain domain and domain knowledge, and focuses on how to provide and leverage these services more sufficiently and wisely. On the other hand, the SORE process begins with the individualized customer requirements and the typical virtualized

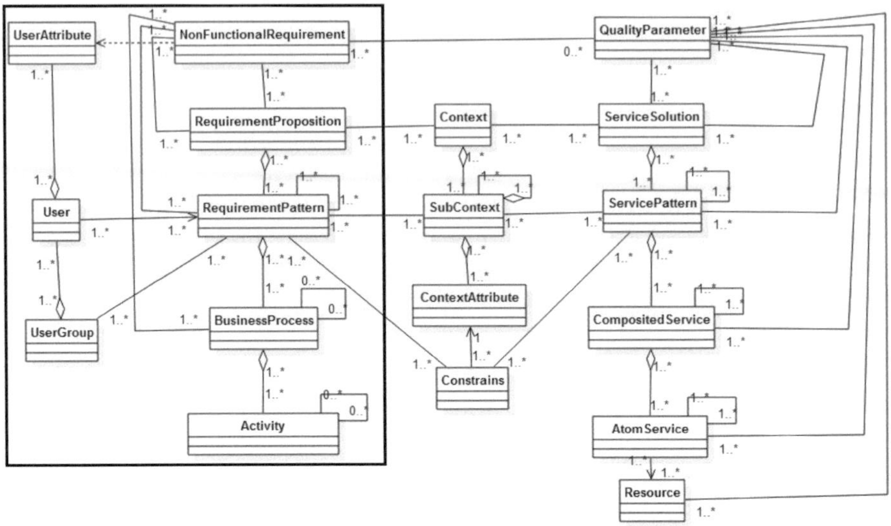

Fig. 1 Meta-model of RE2SEP

partial user requirement description, and focuses on deriving application-specific service requirement propositions to ask the adaptive service solutions.

The research work presented in this paper aims at contributing on the SORE process of RE2SEP. In this process, RE2SEP proposes a new term called "requirement pattern". Requirement pattern is defined as a modularized piece of description of customers' service requirements, which represents a common routine or relatively stable service process or sub-solution in domain business services. Requirement pattern is a set of common requirements of most customers, by which a complete customer requirement proposition is composited. Each requirement pattern must contain one or more business process, which is regularly composed of a series of activities. We have defined the Meta-model for requirement pattern as models in the black box shown in Fig. 1. The main task of the approach presented in this paper is to detect users' requirement pattern.

In a specific domain, each individual user has various requirements for the services. However, not all these requirements need to be satisfied simultaneously. Under certain preconditions, part of these requirements must be fulfilled at the same time. These preconditions includes constrains from physical world, such as time, location, weather, and etc., and user related non-functional requirement, expected service quality, affordable price, and etc. In our research, we define the set of frequently simultaneous required items as requirement pattern (as the class "RequirementPattern" shown in Fig. 1), which is suite for the specific scenario with certain number of constrains from physical world (as the class "constrains" shown in Fig. 1) and user related non-functional requirements (as the class "NonFunctionalRequirement" shown in Fig. 1). Each RequirementPattern must contain several business processes that can achieve one/some specific requirements in the requirement pattern. Each business process has a series of activities that are sequentially ordered to accomplish required tasks. An original customer perspective might trigger a series of eventful activities to achieve it. The activity choice criteria is the scenario (time, location, weather, and etc.), and user related non-functional requirement, expected service quality, affordable price, and etc. Thus, "business process" is part of the decision network (graph) for customer perspective accomplishment. One thing has to be mentioned is that one activity is not only belongs to one business process. Once, it combines with other different activities, and then they can form another business process. It is the same for the relationship between "BusinessProcess" and "RequirementPattern". One individual customer requirement proposition could be a set of several "RequirementPattern" or a set of several "RequirementPattern" and "BusinessProcess".

Up to now, the "RequirementPattern" mentioned here is only for individual user. In order to make this "RequirementPattern" as a priori knowledge of related domain, we prefer this "RequirementPattern" is not only for one individual user but also for others who have common interests. Thus, we cluster users into various kinds of user group by analysis their attributes, such as age, sex, and etc., or their behaviors. If one "RequirementPattern" is a common pattern for most of the users in one user group, then this "RequirementPattern" is the requirement pattern of this user group.

3.2 Problem Formulation

The description of the terminologies of SORE model of RE2SEP is shown in the Table 1. We define the users, user attributes, activities, business processes, requirement patterns, and the relationships among them within the service domains.

1. Service Domain Collection (*D*)
   $D = \{D_1, \ldots, D_K\}$, D indicates all service domain collection, and the total number of domains is *K*.
2. User Collection (*U*)
   $U = \{U_1, \ldots, U_N\}$, U indicates all user collection, and the total number of users is *N*.
3. User property/User Attribute Collection (*UA*)
   $UA = \langle UA_1, \ldots, UA_M \rangle$, *UA* indicates all user attribute collection, and the total number of user attributes is *M*.
4. User Attribute (UA^{D_k}) considered in $D_k(1 \le k \le K)$
   $UA^{D_k} = \langle UA_m | UA_m \in UA, 1 \le m \le M \rangle, UA^{D_k} \subseteq UA$, UA^{D_k} includes the user attributes considered in service domain D_k.
5. Activity Collection (*A*)

Table 1 Terminologies in SORE of RE2SEP

Terms	Description of terms
K	The total number of service domains
D_k	Service domain k
N	The total number of users
U_n	User n
UA_m	User Attribute m
UA^{D_k}	User Attributes considered in D_k
X	The total number of activities
A^{D_k}	Activities considered in D_k
BP_y	Business process y
RP_z	Requirement pattern z
$ReProp^{D_k}$	Requirement Proposition considered in D_k

$A^{D_k} = \{A_1, \ldots, A_X\}$, A^{D_k} indicates the activity collection, and the total number of activities is X.

$A_x^{D_k} = \langle AName, ADescription, AInput, AOutput \rangle$, $A_x^{D_k}$ includes the name, description, input and output of activity.

6. Business Process (BP)

   $BP^{D_k} = \{BP_1, \ldots, BP_Y\}$, BP^{D_k} indicates the business process collection considered in service domain D_k and the total number of business processes is Y.

   $BP_y^{D_k} = \langle AL, E^{AL}, BPDescription \rangle$, AL is the list of requirements, E^{AL} is the relationship between activities in AL, *and* $BPDescription$ *is* the requirement description. $E^{AL} = \{(A_i, A_j) | A_i, A_j \in AL\} AL \subseteq A^{D_k}$.

7. Requirement Pattern (RP)

   $RP^{D_k} = \{RP_1, \ldots, RP_Z\}$, RP^{D_k} indicates the requirement pattern collection that is considered in service domain D_k, and the total number of requirement patterns is Z. $RP_z, 1 \le z \le Z$ is requirement pattern z.

   $RP_z^{D_k} = \langle BPL, E^{BPL}, RPDescription \rangle$, RP is constituted by a set of business processes. BPL is the list of business processes used in $RP_z^{D_k}$, E^{BPL} is the relationship between BP in BPL. $RPDescription$ indicates description of $RP_z^{D_k}$.

$$E^{BPL} = \{(BP_i, BP_j) | BP_i, BP_j \in BPL\}, BPL \subseteq BP^{D_k},$$

8. Requirement Proposition ($ReProp$)

$$ReProp^{D_k} = \langle RPL, BPL, E, RePropDescription \rangle$$

$ReProp$ is constituted by a set of RPs and BPs. $BPL \subseteq BP^{D_k}$, BPL is the list of business processes used in $ReProp^{D_k}$, $RPL \subseteq RP^{D_k}$, RPL is the list of requirement patterns used in $ReProp^{D_k}$.

$E = \{(BP_i, BP_j) \cup (RP_i, RP_j) \cup (BP_i, RP_j) | BP_i, BP_j \in BPL \wedge RP_i, RP_j \in RPL\}$, E is the relationship between BP and RP. *RePropDescription* includes description of $ReProp^{D_k}$.

4 Requirement Pattern Elicitation Approach

The requirement pattern elicitation approach contains six steps, as shown in Fig. 2. This section will describe the main task of each step.

4.1 Frequency Analysis of Individual Customer Activities

In a specific domain, different users may use different ways to accomplish their tasks, even it is the same task. This means that their requirement propositions may have different compositions of functional (activity) requirements. Meanwhile, some of this activity compositions or part of them are always the customer's routines, which means some of the activity requirements are requested repeatedly everyday by one user. Thus, we firstly analyze the frequency of individual customer activities.

We installed an app on the volunteer's mobile phone, which can capture different users' service requests every day and log these operations as Fig. 3 shows. This log contains request time duration, equipment Id, request resource, and service operations. We use this log data as experimental data.

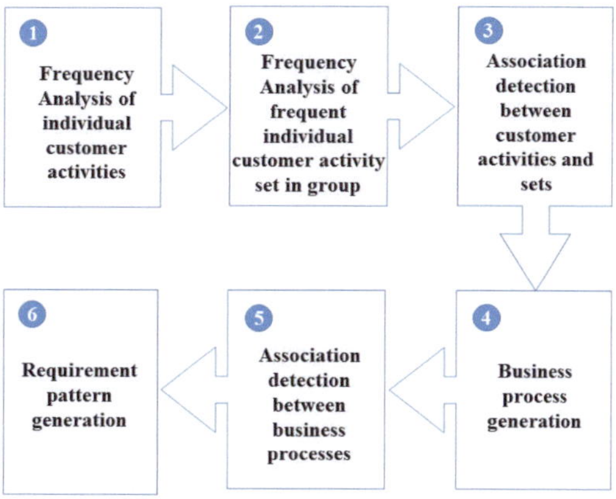

Fig. 2 Steps of requirement pattern elicitation approach

StartTime	EquipID	App	ServiceOperations	EndTime	Redirection
May 7 2015 12:34:20	186*****457	eleme	QueryRestaurant;QueryMenu;OrderFood	May 7 2015 12:38:11	start alipay
May 7 2015 12:38:14	186*****457	alipay	SelectPayMode;SMSVerify;PayConfirm	May 7 2015 12:40:34	Redirect eleme
May 7 2015 12:40:42	186*****457	eleme	ConfirmOrder;ShareInfo	May 7 2015 12:42:08	start weixin
May 7 2015 12:42:22	186*****457	weixin	PubABlog	May 7 2015 12:43:34	Redirect eleme
May 7 2015 12:43:54	186*****457	eleme	CheckOrder	May 7 2015 12:44:04	

Fig. 3 Example of experimental data

Table 2 Example of individual customer activities set list ID

ID	Individual customer activities set AS_{id}
1	A_1, A_2, A_4, A_5, A_7, A_{11}, A_{12}, A_{13}
2	A_1, A_3, A_4, A_7, A_{11}, A_{15}
3	A_2, A_4, A_5, A_7, A_{11}, A_{15}, A_{16}
4	A_5, A_7
5	A_1, A_2, A_4, A_5, A_7, A_{11}, A_{12}, A_{13}
…	…
n	A_1, A_4, A_5, …, A_i, …, A_j, …, A_n

Each line of the log data can be considered as an activity. One user requests one specific service resource at a certain moment. Afterwards, this request might redirect to other service resource, which will trigger another activity. With the consequent time series and the redirection relations, we generate a list of individual customer activities set (as shown in Table 2).

Table 2 gives an example of the individual customer activities set list in one domain. ID is the task performance id. If, one task is performed many times, it has many performance id, such as the ID 1 and 5 in Table 2. Individual customer activities set AS_{id} contains the activities that are used to perform the task by this individual user. $AS_{id} \subseteq A^{D_k}$, $A^{D_k} = \{A_1, \ldots, A_X\}$.

By analyzing the AS_{id} in Table 2, we can find that some of the subset of AS_{id}, $\left\{ \underbrace{(A_i, \ldots, A_j, \ldots, A_k)}_{n} \middle| A_i, A_j, A_k \in AS_{id}, 1 \leq n \leq |AS_{id}| \right\}$, are frequently appear in the individual customer activities set list. For example, (A_4, A_7), (A_5, A_7), (A_4, A_7, A_{11}), and etc. The main task of this step is to find out all the frequent activities and activity sets. Algorithm 1 is the pseudo code of the frequent activities analysis algorithm.

Algorithm 1: The frequent activities analysis algorithm

```
Begin
  m=1
  FS_m = {A_i | A_i ∈ A^{D_k} ∧ σ({A_i}) ≥ n × min sup}
  Repeat
     m++
     TF = FS_{m-1}
     for each item set IS in FS_{m-1}
        for each item set IS* in TF
           SS_1 = SubSet_{m-2}( IS )
           SS_2 = SubSet_{m-2}( IS* )
           if ( (SS_1 == SS_2 ) or (SS_1 = ∅ and SS_2 = ∅ )
                and IS - SS_1 ≠ IS* - SS_2)
              put SS_1 ∪ { IS - SS_1, IS* - SS_2} in C_m
           end if
        end for
     end for
     for each AS_id ( id ≤ n )
        for each item set C' in C_m
           if C' ⊂ AS_id
              σ(c')++
           end if
        end for
     end for
     FS_m = {C' | C' ∈ C_m ∧ σ(C') ≥ n × min sup}
  until FS_m = ∅
  FS = ∪ FS_m
End
```

First of all, we have to define threshold value (min sup) that represents the frequency degree of an activity or activity set, such as 10 percentage. If the activity or activity set appears as a full set or sub set in activities set list *ASL*, $ASL = \{AS_{id} | AS_{id} \subseteq A^{D_k}, 1 \leq id \leq |ASL|\}$, more than $|ASL| \times \min\text{sup}$, then it is a frequent activity or activity set. We use $\sigma(activity)$ or $\sigma(activitySet)$ to represent the appearance times of one activity or activity set. If it is a frequent activity or activity set, it will be saved in a frequent set FS_m, $1 \leq m \leq |activitySet|$. This Algorithm starts from FS_1, then uses $FS_{m-1} \times FS_{m-1}$ method to find all the possible FS_m. If one activity set *IS* in FS_{m-1} only have one element e different from another activity set

IS^* in FS_{m-1}, and if $\sigma(IS^* \cup \{e\}) \geq |ASL| \times \min \sup$, then $IS^* \cup \{e\}$ can be one frequent activity set of FS_m.

4.2 Frequency Analysis of Frequent Individual Customer Activity Set in Group

As mentioned above, the requirement pattern is not only for individual customer. The requirement pattern belongs to a group of customers is more credible and typical. So, the frequent activity or activity set of one individual customer may not suitable for another user. However, some of the requirement activities are commonly requested by most of the users, who have similar experiences, habits or comment interests. Thus, we cluster users into various kinds of user group by analysis their attributes, such as age, sex, and etc., or their behaviors.

After categorizing users into different groups, we will analyze the frequency of frequent individual customer activity sets. So that, we can identify whether these frequent individual customer activity sets are really popular in a specific user group. Meanwhile, because the next step is to find out the association between activity sets, the relationship between activities need to be recorded as well during the frequency analysis. Thus, we construct an activity relationship network with frequency mark by reforming frequent pattern (FP) growth algorithm [18]. Algorithm 2 is the pseudo code of the relationship network construction algorithm of frequent individual customer activity set (FICA) in one user group.

Algorithm 2: The relationship network construction algorithm of FICA in one user group

---

Begin

  $FS^* = \varnothing$

  for each user in one user group

    $FS^* = FS^* \cup FS$

  end for

  $N = |FS^*| \times \min \sup$

  $FS_1^* = \{A_i \mid A_i \in A^{D_k} \wedge \sigma(\{A_i\}) \geq N\}$

  Descending sort $A_i \in FS_1^*$ according to $\sigma(\{A_i\})$

  Create root node of the tree noted as T

  $T = \text{null}$

  for each $AS_{id} \in FS^*$

   Descending sort $A_i \in AS_{id}$ according to $\sigma(\{A_i\})$

    and delete $A_i \notin FS_1^*$ from AS_{id}

    call function insert_tree (AS_{id} , T)

  end for

End

Algorithm 3: The function insert_tree (*AS*, T)

```
Input AS is activity set, T is a tree node
Begin
    sameNode = false
    tempNode = T.childNodeList.next
    Repeat
      if tempNode.name== AS.next.name
          tempNode.count+=1
          sameNode = true
          break
      else
          tempNode = tempNode.next
      end if
    until tempNode is null
    if !sameNode
       create a new tree node N
       N = AS.next
       N.count = 1;
       T. childNodeList.add(N)
       if T is not null
          if T.output==N or T ==N.output
              save dependency relation (Type.De,T, N) or (Type.De,N,T) in Activity relation collection ARC.
          end if
       end if
    end if
    AS = AS.next
    if AS is not null
       call function insert_tree( AS , N)
    end if
End
```

Firstly, the frequent activity set of each individual customer *FS* has to be combined together as FS^*. Then, the threshold value of frequent appearance time is $|FS^*| \times \min \sup$. This threshold value is used to select the real frequent activity A_i in one user group. As the Apriori theory [14, 15] shows, if a set *C* is not frequent, then the superset that contains *C* is not frequent. So, if the $\{A_i\}$ is not frequent, it will be deleted from every AS_{id}. And then, sort $A_i \in AS_{id}$ in descending order according to $\sigma(\{A_i\})$. Afterwards, this algorithm uses a recursive function insert_tree (AS_{id}, T) to build up the FP tree. Considering all the $A_i \in AS_{id}$ is lined up as a path, then if this path has overlap with one path of the existing tree, then the count of overlap nodes increase 1, and the rest $A_i \in AS_{id}$ become the nodes of the new path of the existing tree with one count. During this tree construction, if the neighbor nodes have dependency relationship, one node is the input or output in another, then this relationship has to be stored. This recursive procedure will be executed until there is not AS_{id} in FS^*. Finally, we can obtain the FICA relationship network with frequency mark.

With this network, we can identify the frequent activity sets in one user group. The pseudo code of this identification process is shown in Algorithm 4.

Algorithm 4: The pseudo code of one user group's frequent activity sets identification function find_FreAS(*Tree*, *S*)

Input: *Tree* is the FP-tree constructed by alg. 2 and alg. 3, or conditional FP-tree of one frequent set. *S* is the suffix of one frequent set, but it is null when this function is invoked at the first time.

Output: the list of frequent activity sets in one user group

```
Begin
  if Tree only have one single path P
    get the list CL of the combination of P's nodes
    for each C ∈ CL
      create a frequent activity set AS = C ∪ S
      AS.support = min(n_i.support), n_i ∈ C
      save AS in AS Collection ASC
    end for
  else
      get node list TNL of Tree
      for n_i ∈ TNL (from n_i with min(n_i.support))
          create a frequent activity set AS = n_i ∪ S
          AS.support = n_i.support
          save AS in AS Collection ASC
          find the sub-tree sTree that only with the
          path contains n_i
          change the support of each node into the
          sum of n_i.support if it has a linked path
          with n_i
          delete n_i from the sTree
          if sTree is not null
              call function find_FreAS(sTree, AS )
          end if
      end for
  end if
End
```

---

function find_FreAS (*Tree*, *S*) is a recursive function, which scans the FP-tree constructed by Algorithm 2 and Algorithm 3 from leaf node to the root node to find the frequent activity set.

4.3 Association Detection Between Customer Activity Sets

After identifying the frequent individual customer activity sets in one user group, we have to discover the association between customer activities and sets. So that, we can know how these activities and sets are connected, and whether they need to be connected to form a business process. The activities have the input and output, so it is easy to find the relationship among activities, such as parallel relation, invoked dependency relation, sequential relation, and etc. In this paper, we only consider parallel relation and invoked dependency relation.

When the relationship network of FICA was built up in the previous step, part of the associations between activities have been added to this network as an arrow between nodes. This step will complement the rest relations. Algorithm 5 shows the algorithm of the activity association detection

Algorithm 5: The activity association detection algorithm

```
Begin
  for each frequent activity set AS in ASC
    for each activity A in AS ( A∈ AS )
      Scan rest activity A' in AS
      if A.output==A' or A == A'.output
        save dependency relation (Type.De,A, A') or (Type.De,A',A) in
        Activity relation collection ARC .
        (Type.De,A, A').count++ or
         (Type.De,A',A) .count++
      else
        save dependency relation (Type.Pa,A, A')
        in Activity relation collection ARC .
        (Type.Pa,A, A').count++
      end if
    end for
    if A has no dependency relation with others
      {A} becomes an individual subset of AS
    end if
  end for
End
```

The algorithm will check each frequent activity set *AS* in the *AS* collection *ASC* one by one. This investigation process will scan each activity in *AS*, if it has dependency relation with others, then save this relation in the activity relation collection *ARC* and increase the counting of this relation. If it has not dependency relation, we assume that they have parallel relation. If an activity has no dependency relation with other activities in one *AS*, then it will become one individual subset of this *AS*.

4.4 Business Process Generation

After identifying the associations between activities, we can use these to link activities up, if they have dependency relation. Algorithm 6 shows the pseudo code of the business process generation function. This function is a recursive function, which is used to scan the FP-tree constructed by Algorithm 2 and Algorithm 3. This function is tree preorder traversal function. Before the traversal, a threshold value of association degree between activities has been defined. This value will be used to remove some activity dependency pairs, which are not always connected. During the tree traversal, if one node has the dependency relation with its child node, then its child node will be added into the business process *BP*. If not, the rest nodes on this path will be ignored, because the nodes on this path are ordered by their frequency in descending order. But, their association will still be stored for the pattern generation in the follow-up steps.

Algorithm 6: The pseudo code of business process generation function BP_Gre (*N*, *BP*)

Input: *N* is one node of the FP-tree constructed by alg. 2 and alg. 3.

BP represents one business process. It has been defined in section III as

$$BP = \left\langle AL, E^{AL}, BPDescription \right\rangle \; BP \in BP^{D_k} = \left\{BP_1, \cdots, BP_Y\right\}$$

Predefine the threshold value of association degree (*AD*) between activities

```
Begin
   if N is root node
      node = N.childNodes.next //from left to right
      if BP is not null
         BP^{D_k} .add(BP)
         clear BP's elements into a new BP
      end if
      BP.AL.add(node)
      call function BP_Gre(node, BP)
   end if
   node = N.childNodes.next
   while node is not null
      if (Type,De,N,node) or (Type,De,node,N) is in
         the ARC && the count of this relation is
         bigger than AD
         BP.AL.add(node)
         BP.E^{AL}.add((N,node) or (node,N))
         call function BP_Gre(node, BP)
      end if
   end while
End
```

4.5 *Association Detection Between Business Processes*

In the previous step, we have ignore some activity pairs that have associated link. Even though, these links are not frequent, but they can also represent the specific associations between business processes and sets, which can be used to help the requirement pattern aggregation. Meanwhile, there are also the associated activity pairs with parallel relation. This pairs can be used to predict different business processes that may be required by user in one requirement proposition. Algorithm 7 shows the pseudo code of business process association detection function.

Algorithm 7: The pseudo code of business process association detection function

```
Begin
   Load the business process set (BPS)
   while the BPS is not null
      Get the first business process (BP)
      Get BP's first activity (FA)
      Get BP's end activity (EA)
      Get other BPs's first activities (FAs)
      Get other BPs's end activities (EAs)
      if (Type.De, EAs_i, FA) exist in ARC
          if (Type.De, BP, BPs_i) doesn't exist in the
             BP relation collection BPRS
             Save the (Type.De, BP, Bpsi, (Type.De,
                EAs_i, FA).count) in BPRS
         end if
      else
          if (Type.De, EA, FAsi) exist in ARC
             if (Type.De, BPsi, BP) doesn't exist in
               BPRS
               Save the (Type.De, BP, BPsi, (Type.De, EA, FAsi).count) in
                  BPRS
             end if
          else
             if (Type.Pa, FAsi, FA) or (Type.Pa, FA,
               FAsi) exist in ARC
                if (Type.Pa, BP, BPsi) doesn't exist in
                  BPRS
                  Save the (Type.Pa, BP, BPsi,
                  (Type.Pa, FAsi, FA).count) in BPRS
                end if
             end if
           end if
         end if
         Move the pointer of BPS to the next
     end while
End
```

4.6 Requirement Pattern Generation

As mentioned, the requirement pattern is constituted by a set of business processes, which have certain relations. After the previous steps, the business process relation collection is ready with the association degree. If their association degree (no matter dependency relation or parallel relation) is high enough, then it means that they may always exist in the same customer requirement proposition. Thus, they can be organized together as a requirement pattern. The association type and degree will be used in the further study, the matching of requirement pattern and service pattern. Meanwhile, in this phase, we consider the domain a priori experience. We maintain a priori requirement pattern list $RP^{Apriori} = \left\{RP_1^{Apriori}, \ldots, RP_z^{Apriori}\right\}$. If a business process used to be one business process of one requirement pattern in the $RP^{Apriori}$, then even its association degree is not high enough, we consider it as a member of requirement pattern based on the domain a priori experience. Algorithm 8 shows the pseudo code of requirement pattern generation function.

Algorithm 8: The pseudo code of requirement pattern generation function

---

Predefine the threshold value of association degree (*AD*) between business processes. *RP* represents one requirement pattern. It has been defined in section III as $RP = \left\langle BPL, E^{BPL}, RPDescription\right\rangle$ $RP^{D_k} = \left\{RP_1, \cdots, RP_Z\right\}$ Apriori requirement pattern list $RP^{Apriori}$ $RP^{Apriori} = \left\{RP_1^{Apriori}, \cdots, RP_z^{Apriori}\right\}$

Begin
    scan the business process relation collection with parallel relation BPR^{Pa}
    for each $BPR_i^{Pa} \in BPR^{Pa}$
        if $BPR_i^{Pa}.count \geq AD$ or BPR_i^{Pa} exist in one of the $RP_i^{Apriori} \in RP^{Apriori}$
            $RP.BPL.\text{add}\left(BPR_i^{Pa}.BP_1\right)$
            $RP.BPL.\text{add}\left(BPR_i^{Pa}.BP_2\right)$
            $RP.E^{BPL}.\text{add}\left(\left(BPR_i^{Pa}.BP_1, BPR_i^{Pa}.BP_2\right)\right)$
            scan the business process relation collection with dependency relation BPR^{De} to find the subset ($BPR^{De_{sub}} \subseteq BPR^{De}$) that contains $BPR_i^{Pa}.BP_1$ and $BPR_i^{Pa}.BP_2$
            for each $BPR_i^{De} \in BPR^{De_{sub}}$

---

if $BPR_i^{De}.count \geq AD$ or BPR_i^{De} exist in one of the $RP_i^{Apriori} \in RP^{Apriori}$
$RP.BPL.\text{add}\left(BPR_i^{De}.BP_1\right)$
$RP.BPL.\text{add}\left(BPR_i^{De}.BP_2\right)$
$RP.E^{BPL}.\text{add}\left(\left(BPR_i^{De}.BP_1, BPR_i^{De}.BP_2\right)\right)$
end if
end for
end if
$RP^{D_k}.add\left(RP\right)$
end for
End

After the execution of this function, some of the requirement pattern will still have business processes with parallel relation, then if it is necessary, we can do the set union operation again to combine these requirement patterns into one.

5 Experiment Result Analysis

The experiment has proved that this requirement pattern elicitation approach can accurately elicit the requirement pattern from a big set of requirement activities from different users. However, the experiment result also exposes some inherent weaknesses of this approach.

5.1 Predefined Threshold Value Issue

In this approach, many steps have to predefine the threshold value to dig out the frequent patterns. For example, min sup introduced in step 1, which represents the frequency degree of an activity or activity set. As defined, only when $\sigma(activity)$ or $\sigma(activitySet)$ is more than $|ASL| \times \min\text{sup}$, this activity or activity set is frequent. The experiment result shows that the number of frequent activity sets depends upon different min sup. Figure 4 shows the frequency degree affection to the frequent activity sets selection. The result shows that if the degree is defined too high, then more than half of the activity sets are frequent. It means that the high frequency degree cannot help us to distinguish the real frequent activity or activity set. If the degree is defined too low, then only a few activity sets are frequent. It means that too many activities won't be involved in the requirement pattern, so that these activities are very difficult to be predicted for service pattern matching.

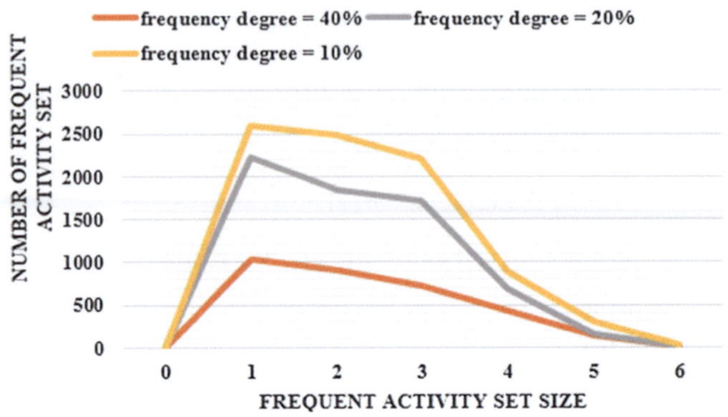

Fig. 4 Frequency degree affection

Meanwhile, in different domain, the customer behaviors are different, which will affect the activity frequency situation. Thus, the frequency degree is not a fixed value, and its definition depends on different domain scenarios.

The analysis result of other predefined threshold values is quite similar with the frequency degree. As the result, the definition of these threshold values must refer the domain prior knowledge.

5.2 Requirement Activity Set List Size Versus Execution Time

Most of the algorithms in this approach are based on the tree traversal algorithm and sequence alignment algorithm. Some of these are recursive functions. Thus, the execution time of the requirement pattern elicitation application will increase quite fast with the growth of analysis object set size. Figure 5 shows the analysis application execution time of corresponding requirement activity set list size.

In this experiment, the maximum data set size is 8000. The execution time is more than 9 s. Even it is the time including the I/O execution time, it is still too long. Also, by analyzing the increasing trend of this curve, we can image that if the size becomes 10,000, 50,000 or more, the execution time will extremely high. Thus, the task of customer clustering or similar data clustering must be refined, so that the size of the analysis object set can be reasonable.

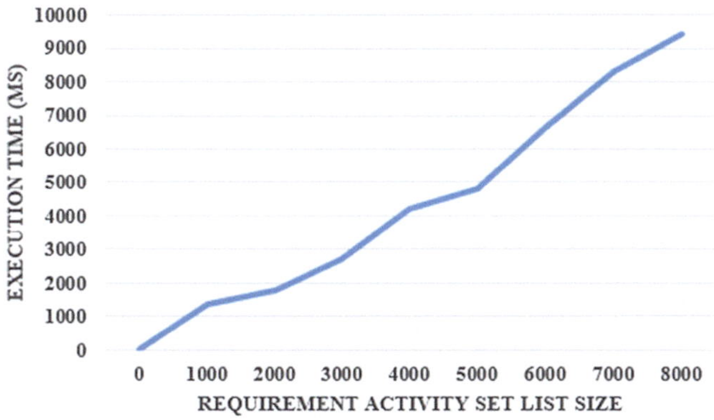

Fig. 5 Requirement activity set list size versus execution time

5.3 Overall Analysis Versus Extreme Situation

This approach is an overall analysis approach. It does not consider the extreme situations. For example, some of the activities are required by some customers many times, so that their frequency degrees are high. However, they are exceptional behaviors of some specific customers, and they cannot be considered as a pattern for one user group. Also, some of the activities are required by most of the customers once a day. If the number of customers or analyzing days are not big enough, then the frequency degree of these activities may not reach the threshold value. But, actually, they must be considered as a pattern.

Anyway, without the consideration of the extreme situation is a drawback of this approach. We will pay more attention on this and improve the algorithms in the further study.

6 Conclusions and Future Work

As mentioned in previous section, this approach can accurately elicit the requirement pattern from a big set of requirement activities from different customers. However, it still has some drawbacks need to be fixed up. Also, as mentioned, the requirement patterns are always having a precondition, which is the context constrains. So, the associations between business processes and sets is not the only standard for the requirement pattern aggregation, the context constrains must also be fully complied with. Thus, one of the future works of this research is to involve context constrains in the requirement pattern aggregation step. The algorithm of this step will be complemented, so that we can predict more accurate requirement pattern.

Recently, we consider the domain a priori experience in the final step, when we generate the requirement pattern. In the further study, we will try to refer to the domain a priori experience in the previous steps, and compare the result with the current result. Involving the domain a priori experience earlier, may increase the complexity of the algorithm, but may also increase the accuracy of the requirement pattern generation result. More experiments must be performed in the further study.

Another group of our research center is working on the service pattern elicitation approach. So, another future work of this research is propose an approach to match the requirement patterns with service patterns accurately. With this matching approach, we can obtain the feedback that can indicate the accuracy of the requirement pattern generation result. We can use this feedback to improve the requirement pattern elicitation approach iteratively. Also, by involving the Nonfunctional requirement and QoS into the matching approach, we can final implement the RE2SEP paradigm to achieve intelligent customer oriented service recommendation.

References

1. Armbrust, M., Fox, A., Griffith, R., et al. (2010). Above the clouds: a Berkeley view of cloud computing. *Communications of the ACM, 53*(4), 50–58.
2. Xu, X., Sheng, Q. Z., Zhang, L.-J., Fan, Y., & Dustdar, S. (2015). From big data to big service: a visionary view. *IEEE Computer, 48*(7), 112–115.
3. Xu, X. (2014). A new service engineering paradigm for big service in big data environment, Keynote speech in ICSS' 2014. In *International Conference on Service Sciences, Wuxi, China*, May 22–23, 2014.
4. Royce, W. W. (1970). *Managing the development of large software systems: Concepts and techniques*. Technical Papers of Western Electronic Show and Convention (WesCon), Los Angeles, USA, August 25–28, 1970.
5. Sommerville, I. (2004). *Software engineering* (7th ed.). London: Pearson Education Ltd.
6. Boehm, B. W. (1988). A spiral model of software development and enhancement. *IEEE Computer, 21*(5), 61–72.
7. McIlroy, M. D. (1968). Mass produced software components. In *Software Engineering: Report of a Conference Sponsored by the NATO Science Committee, Garmisch, Germany* (p. 79). Scientific Affairs Division, NATO. October 7–11, 1968.
8. Rumbaugh, J., Blaha, M., Premerlani, W., et al. (1991). *Object-oriented modeling and design* (Vol. 199, No. 1). Englewood Cliffs (NJ): Prentice Hall.
9. OMG. OMG pursues new strategic direction to build on success of past efforts: Model driven architecture ™. http://www.omg.org/news/releases/pr2001/2001-03-08a.htm
10. Bieberstein, N., et al. (2008). *Executing SOA: A practical guide for the service-oriented architect (Paperback)*. Indianapolis: IBM Press. (ISBN-13: 978-0-13-235374-8).
11. Michael, B. (2008). *Introduction to service-oriented modeling, service-oriented modeling: Service Analysis, design, and architecture*. Hoboken: Wiley. (ISBN 978-0-470-14111-3).
12. Xu, X. F., Mo, T., & Wang, Z. J. (2007). SMDA: A service model driven architecture. In *Proceedings of the 3th International Conference on Interoperability for Enterprise Software and Applications, IESA 2007* (pp. 291–302). Funchal, Portugal, Berlin: Springer, March 27–30, 2007.

13. Sheng, Q. Z., & Benatallah, B. (2005). ContextUML: A UML-based modeling language for model-driven development of context-aware web services. In *International Conference on Mobile Business—ICMB* (pp. 206–212).
14. Agrawal, R., Imielinski, T., & Swami, A. (1993). Mining association rules between sets of items in large database. In *Proceedings of ACM SIGMOD International Conference on Management of Data, Washigton, DC* (pp. 207–216).
15. Agrawal, R., Aggarwal, C. C., & Prasad, V. V. V. (2001). A tree projection algorithm for generation of frequent itemsets. *Jounal of Parallel and Distributed Computing (Special Issue on High Performance Data Mining), 61*(3), 350–371.
16. Cai, C. H., Fu, A., Cheng, C. H., & Kwong, W. W. (1998). Mining association rules with weighted Items. In *Proceedings of IEEE International Database Engineering and Applications Symposium, Cardiff, Wales* (pp. 68–77).
17. Silverstein, C., Brin, S., & Motwani, R. (1998). Beyond market baskets: Generalizing association rules to dependence rules. *Data Mining and Knowledge Discovery, 2*(1), 39–68.
18. Han, J., Pei, J., & Lin, Y. (2000, May). Mining frequent patterns without candidate generation. In *Proceedings of ACM-SIGMOD International Conference on Management of Data (SIGMOD'00), Dallas, TX* (pp. 1–12).
19. Lucchese, C., Orlando, S., & Perego, R. (2006). Fast and memory efficient mining of frequent closed itemsets. *IEEE Transactions on Knowledge and Data Engineering, 18*(1), 21–36.
20. Shen, B., Yao, M., Wu, Z., & Gao, Y. (2010). Mining dynamic association rules with comments. *Knowledge and Information Systems, 1*, 73–98.
21. Palmer, C., Harding, J. A., Swarnkar, R., et al. (2013). Generating rules from data mining for collaboration moderator services. *Journal of Intelligent Manufacturing, 24*(2), 313–330.
22. Lim, A. H. L., & Lee, C. S. (2010). Processing online analytics with classification and association rule mining. *Knowledge-Based Systems, 23*(3), 248–255.

Semantic Data Integration Approach for the Vision of a Digital Factory

Marco Franke, Konstantin Klein, Karl A. Hribernik and Klaus-Dieter Thoben

Abstract The vision of Digital Factories is "…to reduce the need for physical prototyping when designing future factories, and support the management of the entire lifecycle of products…" Gorecky in Session A: PLM digital factory simulation optimization for the smart factory, [3]. For that purpose the exchange of information over the boundaries of processes and applications is a mandatory step to achieve the required interoperability. This paper proposes on basis of Digital Factory's requirements an approach to achieve the interoperability in their processes. For that purpose, the proposed approach is implemented as an extension of XML Wrapper. This wrapper is part of the Specific Enabler SEMed. Finally the extended SEMed is evaluated according to the Digital Factory's requirements.

Keywords Semantic web · Ontologies · Digital factory in future internet · Interoperability

1 Introduction

Megatrends like globalization, an increasing number of product variants, shorter product life cycles, and the fasten integration of new technologies into products require that companies move its production and logistic processes to more agility and flexibility. These megatrends result e.g., through current trends of customers who demands for individualized and intelligent products with product updates and customization options during the middle of live of a product. These customer requirements are located in most of today's available products. In consequence,

M. Franke (✉) · K. Klein · K.A. Hribernik · K.-D. Thoben
BIBA—Bremer Institut für Produktion und Logistik GmbH,
Hochschulring 20, 28359 Bremen, Germany
e-mail: fma@biba.unibremen.de

K.-D. Thoben
Faculty of Production Engineering,
University of Bremen, Bibliothekstraße 1, 28359 Bremen, Germany

© Springer International Publishing Switzerland 2016
K. Mertins et al. (eds.), *Enterprise Interoperability VII*,
Proceedings of the I-ESA Conferences 8, DOI 10.1007/978-3-319-30957-6_6

these trends have shorten the product lifecycles and have resulted in a diversity of product variants. The possibility to individualize the products by the customer have resulted in smaller production batches and faster changes in the production processes. At the same time, rising costs for resources like energy and raw material costs and the demand for the highest product quality are challenges for companies on the global market.

The quality of the product and its containing services are transparent through marketplaces like Amazon and social media channels like Facebook and Twitter. Product's weakness are preserved through the internet capabilities.

The above mentioned challenges require a shift from conventional strategies to more flexibility and adaptability to achieve the interoperability between information flows and processes. Especially, the relevant processes of the design phase are not integrated into operative manufacturing processes. The knowledge exchange between both phases are still a challenging task. This weakness could be solved by modular, collaborative and flexible based approaches. The different domains require to share information among each other not data any longer. Bellinger defines information as "data that has been given meaning by way of relational connection" [1]. In the domain of production the context is defined by the specific processes which would specify both the meaning and the kind of exchangeable information. In Fig. 1, a vision is presented by Gorecky [2] in which the digitalization, improved communication and collaboration processes of the early phases of design reduces the effort of the over whole product development significantly. The continuous synchronization between all kinds of engineers, especially for material, mechanical and software developments is a mandatory step. The impact of a complete and seamless integration would be the integration of the physical world related information with digital representation as mock-ups. Furthermore the digitalized information requires a corresponding knowledge management to be applicable. In consequence, the need for expensive and time consuming physical prototypes will be reduced by Digital Factories, because all information of the entire lifecycle of the production environment and its products could be integrated seamlessly.

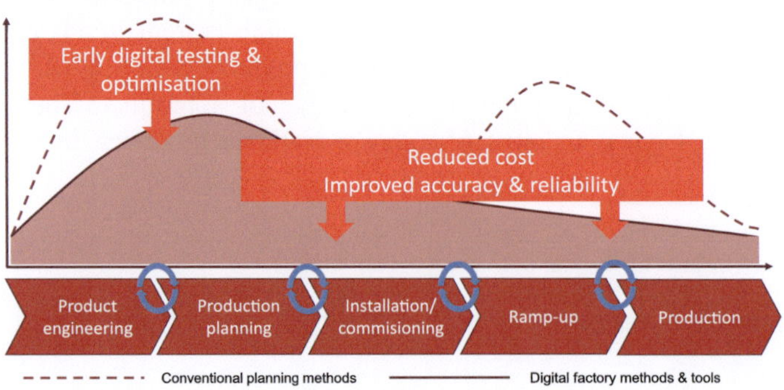

Fig. 1 The digital factories vision, cited from [2]

This paper focuses on the integration of Future Internet technologies to implement the vision of the Digital Factory. For that purpose, the aim of this paper is proposing a data integration middleware based on available Generic and Specific Enabler of Future Internet PPP for achieving a robust and seamless communication. As an outcome, the proposed approach shall enable the vision of Digital Factories in real production environments. The paper is structured as follows: First, industrial requirements of two use cases are presented. Both the motivation and the requirements are from previous published work [3]. On basis of these requirements, the problem statement is presented in detail. Subsequently the approach and the implementation are presented. Finally a conclusion and outlook are given.

2 Use Cases

The research project FITMAN [4] developed plattforms including Generic Enabler, Specific Enabler and Proprietary Components. These platforms were created for 10 industry-led use case trials in the application domains of Smart, Digital and Virtual Factories. The Trials of AIDIMA and Consulgal are located in the Digital Factory and their requirements were the foundation for this paper. A set of functional requirements of both Trials have been published in [3]. In the following, a subset of these requirements are listed/cited which focus on the data integration functionality.

The requirements in Table 1 arise of the existing heterogeneity of data sources in companies in general and therefore also in the Digital Factory. The intention of this paper is to achieve the interoperability for all covered heterogamous data sources in

Table 1 Functional requirements of digital factory, cited from [3]

Req.-ID	Description
SASG-01	Enable an extendable and application specific semantic description
SASG-02	Enable a strict separation between data and information view through adding a semantic description of the data beyond a data source
SASG-03	Enable a simplified and dynamic method for adding data sources
SASG-04	Enable the aggregation of information over the boundaries of data sources
SASC-01	Standardized access to process information
SASC-02	Fast and efficient information sharing between stakeholders
SASC-03	Digitization of previously physical information carriers
SASF-01	Analysts shall be able to introduce and configure information sources where weak signals will be identified
SASF-02	The information sources of different analyses shall be kept private
SASF-03	Analysts shall be able to search references by search words
SASF-04	Obtained a pondered list customer or potential customer requirements
SASF-05	Different users should be identified by the system
SASF-06	Make possible the attachment of new blogs or social sites

the Trials. Hereby, the interoperability shall not be reached through programmatic solutions but rather through a configurable semantic data integration solution. In the following, the problem description according to the increasing heterogeneity is presented in detail.

3 Problem Description

The data of a company is stored not only in one data source but in a couple of data sources. The set of data sources are not homogenous according to the kind of data source, the syntax, the semantics and the provided logical views. This heterogenous mentioned Hannus [5] in 1996 and summerized his impression in Fig. 2.

Since then, the number of applied IT systems have increased through several facts like price reduction, continous software developments, trends like embedded systems and IoT. Today, for each company, the storage of product lifecycle related data is spread over varying and heterogenous IT systems. Reeve [6] mentioned that: "*…Most organizations of middle to large size have hundreds or, more probable, thousands of applications, each with its own various database and other data stores. Whether the data stores are from traditional technologies or document management systems, it is critical to the usefulness of these applications to the organization that they share information between them…*".

The implementation of the Digital Factory requires the opportunity to pass an information over the boundaries of different data sources which implies the transformation of an information of the language L_i of the source data source$_i$ into the language L_t of the target data source$_t$. Here, a language defines only the syntax of

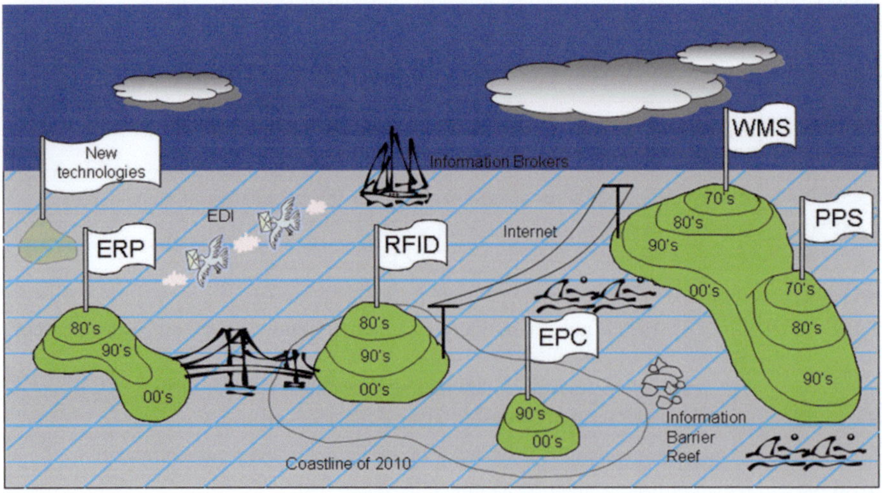

Fig. 2 Bridging Islands (based on [5])

the data source. The transformation process must take into consideration the specifics of both languages which means the transformation of an element of L_i into an element of L_t considering the semantics of elements$_{i,t}$. During the transformation process, common data integration conflicts according to the syntax and semantics have to be considered which are presented in detail in [7].

The consideration of the semantics of L_i and of L_t is only possible if the contained semantic models$_{i,t}$ are explicit available as part of a data source. If the semantics is only implicit available in the data model/data schema of the data source, the extraction of the semantics of the data source schema or the enrichment of the semantics is not possible as part of the data integration process. To achieve the interoperability between data sources, the interpretation of data beyond a data source is necessary. The unambiguity interpretation can be achieved on different levels. Oren et al. [8] defined different levels of understanding, which are

- Lexical understanding
- Syntactical understanding
- Morphological understanding
- Semantic understanding
- Pragmatic understanding.

The lexical, syntactical and morphological understanding enables the recognition of the structure of the information and the grouping of relevant entities together, but the meaning is still unclear. The semantic understanding is the key to understand the meaning of the data and enables first the application of common data integration approaches to achieve the interoperability. The enforcement of data integration approaches in daily business requires both automated data integration processes and the maintenance of the established data sources within the processes of Digital Factory. Up to now, Microsoft Office especially Excel, PLM, ERP tools which enable data export in proprietary data formats, CSV or XML files are common. In consequence, most of the common data sources in the Digital Factory do not enable an unambiguity interpretation out of the box. Table 2 demonstrates the differences between common data sources and data sources containing a complete semantics for semantic understanding.

The above listed kinds of data sources holds e.g., relevant information for the Digital Factory. The access to this information is usually restricted through authentication and the provision of logical views which abstracts of the internal data structures. Web services are a common approach to provide a public interface to other stakeholders, offering a WSDL for describing the semantics of the data. The interpretation of the WSDL enables the lexical, syntactical and morphological understanding but not the necessary semantic understanding to achieve interoperability between data sources. The semantic web service which is an extension of web service enables the annotation of ontology concepts to enhance the semantics. The annotation capabilities does not achieve the necessary level for the unique semantic understanding [9].

Table 2 Kind of data sources containing explicit semantic model

Kind of data source	Semantic understanding is possible for syntax of the language	Semantic understanding is possible for modelled information in the language	Structural coverage of semantics for information modelling	Common data source
CSV	Y	N	Nothing	Y
Database (relational)	Y	N	Nothing	Y
Plain text	Y	N	Nothing	Y
Ontology represented as files or in RDF database	Y	N	Hierarchical, taxonomical object oriented data structures including primitive data types and instance specific data types	N
Web service	Y	Y, the WSDL contains schema	Hierarchical object oriented data structures including primitive data types und simple relationships between data structures	Y
XML files in a repository	Y	Y, if the XSD schema is also available	Hierarchical object oriented data structures including primitive data types	

Current data integration approaches enables the linkage of a semantic model to a data source to close the gap for the semantic understanding. A semantic model describes the data of the data source formally and provides therefore a light-weight self-description of the data source. Common approaches uses multiple ontologies to achieve the semantic understanding on integration level. In such an approach each data sources has its own ontology such as in OBSERVER [10]. Wrappers, containing the self-description of the data source, are used to transform the data of a data source into a data source specific ontology. The transformation of data into ontologies through wrappers has been shown successfully for CSV files, XML files and MySQL databases.

A wrapper implementation as an extension for a web service, namely a web service wrapper, was proposed by Franke et al. [9]. The proposed approach uses the concept of a wrapper and a semantic mediator defined in [11, 12] to enable the mapping of a XML tag of a web service result, which correlates to a value of a data or object property of an ontology. This approach implements a global as view (GAV) approach over multiple data source specific ontologies in which each ontology concept is a merged concept according to the available data properties and axioms in the data source specific ontologies. The GAV integration approach "*…is effective whenever the data integration system is based on a set of sources that is stable*" [13].

The analysis of the use case specific properties of the data sources identified that semantic mediator approaches are appropriate in general. The weakness of existing transformation mechanism lies in the fact that a value for a data or object property of an ontology is assigned from the value which is a property or the smallest distinct part in the data source. For example, in the case of a XML file the smallest distinct part is the value of a XML tag or in the case of a database it is the value of a column. The definition of the smallest distinct part does not satisfy the requirements of the use cases, in which one value often codifies more than one information. The frequency of this possibility was very high in a specific Trial which is contained in FITMAN [4] but not covered by [3]. For example, a value of a data source could be a barcode or a phone number including more than one information. As a consequence, the smallest distinct part for the transformation mechanism has to be redefined. A string of a XML tag or another extracted string of another kind of data source cannot be the smallest distinct part. The smallest part should be any token of a string which satisfy specific integrity conditions. Apart from that, the data integration approach using multiple ontologies and an inferred merged ontology for a GAV approach satisfy the above mentioned functional requirements for the semantic data integration in the Digital Factory.

4 Approach

The presented general wrapper approach of [11, 12], the specialization of this approach for web services of [9] and the integration of wrappers as part of a mediator oriented data integration solution like proposed in [11] shall be extended to additional mapping and transformation capabilities. Here by, the objective is to set the smallest distinct part from a data source value to a subset of characters of this value. A transformation definition shall also be capable to handle the subset of a data source value as a corresponding object. The above mentioned semantic mediator has been become a Specific Enabler in FITMAN and is available as the open source tool SEMed in the FITMAN Catalogue [14]. The proposed and available extension redefine the smallest possible part of the object of a transformation rule as a subset of a value of the logical source view. For that purpose, all data source values will be handled as strings interpreted as a sequence of letters. A letter is a character codified in a Universal Character Set like ASCII or UTF-X.

The extended wrapper concept of [11, 12] defines a transformation in which each transformation rule maps a data property of the ontology to a data source specific value. Hereby, the possible axiom for functional data property is ignored on the wrapper level and will be handled on the mediator level. In the case of a wrapper representing XML files as data source or a wrapper for web services, the data source specific smallest part of a XML element is a primary type such as an integer or a string. The existing structure of a transformation definition and the

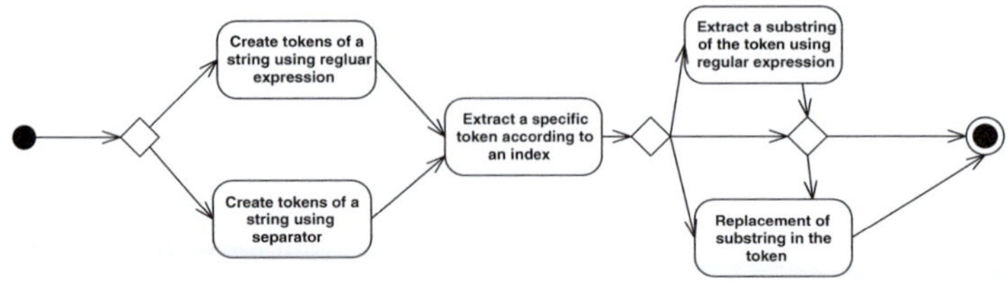

Fig. 3 Capabilities of the string post processing unit

scope of a transformation rule shall be kept to guarantee the compatibility and comparability of existing wrappers [14]. For that purpose, a post processing unit is added on top. That means, that a transformation rule is still defined on basis of the smallest distinct part of the data sources and the corresponding value is forwarded to the post processing which enables the extraction of tokens. The string post processing unit shall be capable of extracting a specific token at a specific position according to given constrains. The proposed functionality of the required string post processing unit is shown in Fig. 3 as an activity diagram.

The string post processing unit of Fig. 3 enables the creation of tokens according to a simple separator or to a regular expression. Subsequently it allows the extraction of a token to a specific index. Finally, an additional execution of a regular expression or a replacement operation is optional. These possible post processing steps are included in a transformation definition as optional transformation rules. The current available implementation of a XML Wrapper represents its transformation definitions in a XML file. In consequence, the transformation rules due to the string processing capabilities are provided as an extension of the XSD schema, which is shown in Fig. 4.

```
<xsd:complexType name="TokenExtraction">
    <xsd:sequence>
    <xsd:element name="Seperator" type="xsd:string" minOccurs="0" maxOccurs="1"/>
    <xsd:element name="RegularExpression" type="xsd:string" minOccurs="0" maxOccurs="1"/>
    <xsd:element name="index" type="xsd:int" minOccurs="1" maxOccurs="1"/>
    </xsd:sequence>
    </xsd:complexType>

<xsd:complexType name="Replacements"><xsd:annotation></xsd:annotation>
    <xsd:sequence>
    <xsd:element name="Replacement" type="Replacement" minOccurs="1" maxOccurs="unbounded"/>
    </xsd:sequence>
</xsd:complexType>

<xsd:complexType name="Replacement"><xsd:annotation></xsd:annotation>
    <xsd:sequence>
    <xsd:element name="source" type="xsd:string" minOccurs="1" maxOccurs="1"/>
    <xsd:element name="target" type="xsd:string" minOccurs="0" maxOccurs="1"/>
    </xsd:sequence>
</xsd:complexType>
```

Fig. 4 XSD schema for string post processing

5 Conclusions and Outlook

Most of the requirements of both use cases have been satisfied through the usage of SEMed in which the mediator based data integration approach of [11, 12] was implemented. The integration of a string post processing unit was added to enrich the wrapper's capabilities. The Table 3 summerizes how the use case specific requirements have been satisfied through this approach.

The string post processing unit has been included in the XML Wrapper of SEMed and has been used in the Trials of the FITMAN project. The evaluation assigned the applicability of SEMed for the adressed business processes in the Digital Factory. The proposed flexibility of SEMed requires the creation of information models, transformation definitions and additional configuration files for each data source. The evaluation pointed out that the complexity of the configuration is so high, that the creation of the configuration needs initial training of computer scientists to be applicable. Further research shall address mostly this issue. The authors vision is that the solution should be configurable and useable by both engineers and information scientists without requiring initial trainings. For that purpose, the usability and support tools for the configuration creation lies in the focus of next research activities.

Table 3 Requirement coverage of SEMed

Req.-ID	Covered through this approach
SASG-01	Each wrapper configuration provides an information model for a data source
SASG-02	The semantic description of the data beyond a data source is represented as an OWL ontology and is part of a wrapper configuration
SASG-03	Data Sources can be added through a wrapper configuration
SASG-04	The mediator, namely SEMed, offers the functionality to aggregate information of all connected data sources
SASC-01	The mediator, namely SEMed, offers a web based access and a query language for a standardized access
SASC-02	The query language of SEMed could be used to enable the specific formulation of information which should be exchanged between stakeholders
SASC-03	Not covered through the approach because data is required in a digitized form
SASF-01	The data sources for extracting weak signals are social media channels which contain unstructured text. The post processing unit could be used to extract specific information in XML exports of such channels. The corresponding Trial haven't contained such a scenario
SASF-02	The authentication is implemented through the data sources and a wrappers uses it only
SASF-03	This requirement is not satisfied
SASF-04	This requirement is not satisfied
SASF-05	The authentication is implemented through the data sources and a wrappers uses it only
SASF-06	This requirement is not satisfied

Acknowledgments This work has been partly funded by the European Commission through the Project FITMAN: Future Internet Technologies for MANufacturing industries (Grant Agreement No. 604674). The authors wish to acknowledge the Commission for their support. We also wish to acknowledge our gratitude and appreciation to all the FITMAN Project partners for their contribution during the development of various ideas and concepts presented in this paper.

References

1. Bellinger, G., Castro, D., & Mills, A. (2004). Data, information, knowledge, and wisdom.
2. Gorecky, D., & Session A. PLM digital factory simulation optimization for the smart factory. In *Joint Dissemination Conference FoF-H2020.*
3. Klein, K., Franke, M., Hribernik, K., Coscia, E., Balzert, S., & Sutter, J. (2014). Potentials of future internet technologies for digital factories. In *IEEE/ACS 11th International Conference on Computer Systems and Applications* (*AICCSA*) (pp. 734–741). IEEE.
4. FITMAN. Adopting future internet technologies in manufacturing industries. http://www.fitman-fi.eu/ (Accessed October 26, 2010).
5. Hannus, M. (1996). Islands of automation in construction. In Ž. Turk (Ed.), *Construction on the information highway* (number 198). CIB publication, University of Ljubljana.
6. Reeve, A. (2013). *Managing data in motion data integration best practise techniques and technologies*. USA: Elsevier Inc.
7. Wache, H. (2003). Semantische mediation für heterogene informationsquellen, KI, Bd. 17, Nr. 4, 56.
8. Oren, T., Ghassam-Aghaee, N., & Yilmaz, L. (2007). An ontology-based dictionary of understanding as a basis for software agents with understanding abilities. In *Proceedings 2007 Spring Simulation Multiconference*. IEEE Press.
9. Franke, M., Klein, K., Hribernik, K., Lappe, D., Veigt, M., & Thoben, K. D. (2014). Semantic web service wrappers as a foundation for interoperability in closed-loop product lifecycle management. *Procedia CIRP, 22*, 225–230.
10. Mena, E., Kashyap V., Sheth, A., & Illarramendi, A. (1996). Observer: An approach for query processing in global information systems based on interoperability between pre-existing ontologies. In *Proceedings 1st IFCIS International Conference on Cooperative Information Systems* (*CoopIS'96*), Brussels.
11. Klein, K., Franke, M., Hribernik, K. A., & Thoben, K. D. (2014). Identification of interface information for a virtual data integration. In *Enterprise interoperability* (Vol. VI, pp. 89–99). Springer International Publishing.
12. Franke, M., Pirvu, B.-C., Lappe, D., et al. (2014). Interaction mechanism of humans in a cyber-physical environment. In *Proceedings of the 4th International Conference on Dynamics in Logistics* (LDIC-14), 4th, February, Bremen, Germany, LDIC. 10–14.
13. Lenzerini, M. (2002). Data integration: A theoretical perspective. In *Proceedings of the Twenty-first ACM SIGMOD-SIGACT-SIGART Symposium on Principles of Database Systems* (pp. 233–246). ACM.
14. FITMAN. Specific enablers. http://catalogue.fitman.atosresearch.eu/enablers (Accessed October 26, 2010).

Negotiation Coordination Model for Supporting Enterprise Interoperability

A. Cretan, C. Coutinho, B. Bratu and R. Jardim-Goncalves

Abstract The scientific evolution of negotiation in collaborative working environments allowed companies to benefit from new interoperability standards. The proliferation of SMEs led to a highly competitive environment, in which the various partners rely on interoperability and collaboration to be efficient. This paper highlights the role of negotiation in solving interoperability issues by proposing a distributive coordination model in order to manage multiple parallel negotiations. The research results will be validating within the European research project H2020 C2NET.

Keywords Enterprise interoperability · Negotiation · Coordination rules · Model-driven · Product design

A. Cretan
"Nicolae Titulescu" University, 185 Calea Văcăreşti, District 4,
040051 Bucharest, Romania
e-mail: badina20@yahoo.com

C. Coutinho (✉)
Caixa Mágica Software, Rua Soeiro Pereira Gomes, Lote 1-4 B,
1600-196 Lisbon, Portugal
e-mail: carlos.coutinho@caixamagica.pt

B. Bratu
Atos Big Data and Security—R&D, Rue Jean Jaurès B.P. 68,
78340 Les Clayes-Sous-Bois, France
e-mail: bbratu@yahoo.com

R. Jardim-Goncalves
CTS, Faculdade de Ciências e Tecnologia, Universidade Nova de Lisboa,
UNINOVA, Lisbon, Portugal
e-mail: rg@uninova.pt

© Springer International Publishing Switzerland 2016

K. Mertins et al. (eds.), *Enterprise Interoperability VII*,
Proceedings of the I-ESA Conferences 8, DOI 10.1007/978-3-319-30957-6_7

1 Introduction

In order to survive in a global economy, enterprises, especially Small and Medium Enterprises (SMEs), must collaborate by exchanging information. In this respect, Enterprise Interoperability (EI) is defined as the capacity of an enterprise to interact and to exchange information with others within a collaborative networked environment [1]. In this environment, any change in any network partner affects the others, leading to system interoperability breaking. In this context, sustainable EI (SEI) is defined as the ability of maintaining interoperability along the enterprise systems and applications [2].

This paper highlights the role of negotiation in resolving these discrepancies, and proposes a generic coordination negotiation model by describing coherent sets of rules that manage multiple bilateral negotiations to support the interoperability within the collaborative working environments.

2 Related Work

In order to support sustainable interoperability, the proposed approach states that one important aspect is the designing of intelligent software components. These components are able to support the communication, coordination and collaboration activities at all levels within the networked environment. This paper tackles two main issues related to low-level business decision support and interoperability breaking among different structures of the company. Regarding this, our approach proposes the use of intelligent agents for decision support at business level. For example, it is presented in [3] an interesting approach by creating different agents with Expert Systems. Previous papers [4], have used Artificial Intelligence (AI) and algorithms in the supervised learning process, achieved through Restricted Bolzmann Machines (RBM) and employed as latent factor analysis application [5].

However, these procedures providing flexible ways of making inferences have not been truly exploited in Multi-Agent Systems (MAS). In this regard, this paper proposes MAS models for supporting the coordination of negotiation process within the dynamic environment. In addition, our approach does not promote the entire substitution of human decision, but proposes the agent technology helping to communicate results among the different software systems. In this context, rule-based systems are the main AI technique that has been added to software agents. As first objective, the proposed negotiation coordination model based on MAS serves to break boundaries within a negotiation environment represented by the contracting authority, several contractors and subcontractors. In second, the proposed solution should be as generic as possible in order to be successfully employed in heterogeneous business environments addressing dynamical changes in the product offers or the business processes, workflows, policies and rules of an enterprise or group of enterprises.

3 Negotiation System with MAS

There are three main steps to the negotiation process: Preparation; Refinement of the job under negotiation; and Closure. The first step, Preparation, establishes the *Negotiation Object* (i.e., the task that will be negotiated) and the *Negotiation Framework* (i.e., the manner in which the Negotiation Object will be negotiated). In the second step, Refinement, both parties exchange proposals and counter proposals with the goal of meeting their constraints. The third step, Closure, finalizes the negotiation.

We have described in detail the architecture of the negotiation system in our previous papers [6]. In order to describe the complex types of negotiation scenarios, we have proposed in our previous work [7] seven different services: *Outsrc; Insrc; Block; Split; Broker*; *SwapIn/SwapOut; Transport.*

Our coordination approach proposes two different classes of services: (i) *Coordination services in closed environment*—refer to the services that manage the coordination constraints among multiple valid proposals only based on information extracted from a single negotiation; (ii) *Coordination services in an open environment*—refer to the services that manage the coordination constraints among valid proposal extracted from several or all ongoing negotiations into the system. The different coordination services in open or closed environment highlight the main features implemented in this negotiation process: distributive and parallelism.

In [8] we have proposed a formal model to settle and to manage the coordination rules of one or more negotiations, which can take place in parallel.

In this paper we are proposing a negotiation method based on coherent sets of coordination rules that can be easily instantiated and triggered on top of a communication middleware level. Then using those rules, we will present an example of a negotiation tactic as a coherent set of coordination rules.

4 Coordination Rules

Before presenting the rules, in the next section will detail our proposed constraint model and several definitions.

4.1 Fundamental Concepts

The fundamentals of the negotiation model are given by the following basic concepts:

A *Negotiation Model* is defined as a quintuple $M = <\mathcal{T}, \mathcal{P}, \mathcal{N}, \mathcal{R}, \mathcal{O}>$ where: $\mathcal{T}$ denotes *the time of the system*, $\mathcal{P}$ denotes *the set of participants* in the negotiation framework, $\mathcal{N}$ denotes *the set of negotiations* that take place within the negotiation framework, $\mathcal{R}$ denotes *the set of policies of coordination* of the negotiations (a coordination policy is a set of rules establishing dependencies between several negotiations) and $\mathcal{O}$ denotes *the common ontology* that consists of the set of definitions of the attributes that are used in a negotiation.

A *negotiation* is described at a time instance through a set of negotiation sequences. Let $\mathcal{S} = \{\text{si} \mid \text{i} \in \mathbb{N}\}$ denote the set of *negotiation sequences*, such that $\forall$si, sj $\in \mathcal{S}$, i $\neq$ j implies si $\neq$ sj. A *negotiation sequence* si $\in \mathcal{S}$ such that si $\in$ N(t) is a succession of negotiation graphs that describe the negotiation N from the moment of its initiation and up to the time instance *t*. The function *view()* returns the participant, the negotiation and the coordination policy described by a negotiation sequence: *view*: $\mathcal{S} \rightarrow \mathcal{P} \times \mathcal{N} \times \mathcal{R}$.

The negotiation graph created at a given time instance, *G = (A,E)* where *A* is the *set of nodes* and *E* is the *set of edges*, is an oriented graph in which the nodes describe the negotiation proposals that are present at that time instance and the edges express the precedence relationship between the negotiation proposals.

The *Status* (*Status* $\in$ *{initiated, undefined, success, failure}*) is the possible state of a negotiation, with *initiated* defining the sequence in which the negotiation has just been initiated; *undefined* defining the sequence with ongoing negotiation proposals; and *success* and *failure* are defining the sequence in which an agreement has been reached or the negotiation has been stopped with a denial.

Issues is the set of attributes with associated values that describe the proposals made in a negotiation.

The functions *status* and *issues* return, respectively, the state of a negotiation proposal and the set of the negotiated attributes.

We define *Role* as the *set of participant roles* such that *Role = {initiator, guest}*; with *initiator* being the participant initiating a negotiation N and *guest* being the participant invited in the negotiation N.

The functions *role()*: $\mathcal{T} \times \mathcal{P} \times \mathcal{N} \rightarrow$ *Role* and *role_s*: $\mathcal{T} \times \mathcal{S} \rightarrow$ *Role* returning the role of the participant p involved in negotiation N or in a particular sequence s, with the property that a participant has only one role in a negotiation and this role does not change in time.

4.2 Constraints Definition Model

We use these coordination rules as basic rules to describe the complex links among negotiations. We consider the coordination rule as the implementation of a dependency relation among several sequences of negotiation. A coordination rule has the following structure:

<Rule_Definition>':'[<Parameter_Definition>]* ';'<Graphs_conditions> <Condition><Relation><Result>

- The first part of the coordination rule (**<Rule_Definition>**) is used to define the name of the rule and its parameters—ex.: *name_rule(v1,v2,..,vn)*, with $v_i \in \mathcal{T} \cup S$.

- In **<Parameter_Definition>** each variable is related to its field of possible values—ex.: *synchronize*$(T_1,T_2,s1,s2)$: $T_1 \in \mathcal{T}$, $T_2 \in \mathcal{T}$.
- **<Graphs_conditions>** sets the conditions related to graphs and, in particular, related to the nodes of the negotiation sequences involved in coordination.
- The second part of the coordination rules is composed of a left part named *condition* **<Condition>** and a right part named *conclusion* **<Result>** and a relationship between both named (**<Relation>**).

The proposed model establishes two types of relationships between condition and conclusion: (i) *hard relationship* "→•" and (ii) *soft relationship* "→".

Hard dependency relations (denoted → •) ensures the fact that if there is a time instance t_1 where conditions are met, then the conclusion of the relation will be satisfied to the next time instance ($t_1 + 1$).

Soft dependency relations (denoted →) ensures the fact that if there is a time instance t_2 where the conclusion is met, then the conditions have been met at the time instance $t_1 < t_2$ and they have been remained true until the instance time $t_2 - 1$.

In other words, even if at some point the conditions of the *soft relationship* are satisfied, there is no guarantee that the result of the relation will be obtained. The conditions of a soft relation are necessary but not sufficient to obtain the result.

Our negotiation-centric set of rules can express constraints between the execution time of negotiations and their corresponding states (status dependencies), between the tasks and the attributes negotiated (attribute dependencies), and between the participants involved in the negotiation process (role dependencies). As an example, the *status dependences* establish the constraints between two or more states of the negotiation sequences involved in the same negotiation or in different negotiations.

We have defined the function *status(t,s,a,s)* with values in the set *Status* $\in$*{initiated, undefined, success, failure}; status*: $E_{\Phi 4} \rightarrow$ *Status* where $E_{\Phi 4}$ is the graph of Cartesian product $\mathcal{T} \times \mathcal{S} \times \mathcal{Ph} \times \mathcal{S}$ that meets the relation Φ_4 such that:

$\forall(t,s_i,ph,s_j) \in \mathcal{T} \times \mathcal{S} \times \mathcal{Ph} \times \mathcal{S}$, *exists* $\Phi_4(t,s_i,ph,s_j)$ *if and only if* $s_i, s_j \in \mathcal{S}$ *and* ph *initiated in* s_i

For example: $status(t,s_0,2,s_2)$ returns the state of the negotiation proposal in the negotiation proposal 2 initiated in sequence s_0 and visible in s_2 at the time instance t.

5 Coordination Pattern

At a certain point, for a participant involved in several negotiations, the infrastructure should handle many coordination rules, which can be very different. The proposed coordination model is not defined as a centralized process, managed by a single module, but it is distributed on several coordination modules. Therefore, we do not need to describe all the coordination rules and all possible actions, but we want to set out coherent sets of coordination rules managed by the coordination process that may have an isolated execution.

Coordination Pattern is a set of coordination rules applicable within a given context and according to the chosen negotiation tactic. The coordination rules are expressed both by *Program Formula* and by the sequences involved in defining the coordination rules:

PattCoor = (TriggerSet, RulesSet, ProgramFormulaSet)

TriggerSet is the set of trigger conditions which must be met at the same moment of time, in order to apply the coordination pattern. These are descriptive conditions of negotiation sequences involved in a set of coherent coordination rules.

TriggerSet is described by three types of expressions:

(<Exp(t,N,p) >) ';' (<Exp(s)>)* ';' (<Exp(t,s)>)**

where

- *<Exp(t,N,p)>* are expressions that identify negotiations and participants as being the object of one or more dependence relations. These expressions are functions defined on the Cartesian product $\mathcal{T} \times \mathcal{P} \times \mathcal{N}$. These functions establish the involvement and role of the participants in the negotiations (ex. *role() function*);
- *<Exp(s)>* are expressions that identify punctually the sequences that will be involved in the dependence relations. These expressions are functions by type *view()* that establish the link between the sets $\mathcal{P}, \mathcal{N}, \mathcal{R}$ and the set $\mathcal{S}$;
- *<Exp(t,s)>* are expressions that establish the conditions on the characteristics of negotiation visible in sequences previously identified to trigger execution of the coordination rules. According to our model, the characteristics of negotiation are returned by the functions *status(), issues()* şi *role_s ()*.

RulesSet is the set of coordination rules that the coordination pattern is committed to synchronize.

First, the coordination rules are set and globally represented (visible in one or multiple negotiations and their corresponding sequences). Then, using the proposed negotiation model, these rules are splitted into coordination policies locally represented (visible in a single sequence), in order to be handled by a single negotiation sequence.

The implementation of a coordination pattern will correspond to a negotiation tactic that can be activated for a set of proposals in a negotiation (i.e., negotiation sequence), for the entire negotiation (i.e., negotiation graph) or for multiple negotiations (i.e., dependent negotiation graphs). We have proposed in [8] a modelling solution based on IAM but other rules based coordination frameworks can be employed (e.g., JESS or Drools).

Next we will detail an example of negotiation tactics using our coordination rules model.

5.1 Coordination in a Closed Environment

The two main characteristics of the coordination in a closed environment are the following: (i) The defined model refers only to dependences among bilateral negotiations of a single negotiation (i.e., one negotiation initiator with multiple negotiation participants); (ii) The evolution of a negotiation is carried out without taking into account other negotiations involving the same participant.

The proposed model will manage the coordination of exchanged proposals among partners on the attributes of the negotiation object considered in the concurrent bilateral negotiations.

As an example, a tactic stating that a task has to be outsourced as a block shall be described using our coordination rules model in a closed environment.

5.1.1 Negotiation as a Block

The block tactic is used in the negotiations where the task must be executed in its totality by a single participant of the negotiation process. The interactions take place between the enterprise that initiated the negotiation and all the other enterprises invited in the negotiation.

The following scenario provides the constraints of the negotiation process. A manager of a SME (participant P1) initiates a negotiation with the goal of establishing a contract regarding the execution of the entire outsourced task by a single participant. The negotiation ends when the participant P1 reaches an agreement with one of the partners (e.g., participant P2) regarding the set of attributes that describes the task being negotiated. At the same time, participant P1 ends all bilateral negotiations with the other partners.

These constraints can be described by the coordination pattern detailed below:

TriggerSet

The conditions of coordination pattern refer mainly to the role of the enterprises involved in negotiation. The coordination pattern manages the constraints on the participant p1 proposing a task within the collaborative working environments. Thus, he has the role of *initiator.*

$(p_1 \in participants(t,N))\ (role(t,\ p_1,N) = initiator);view(s_1) = (p_1,\ N,\ R_1);$
$(\exists\ a \in s_1(t) : status(t,s_1,a,s_1) = undefined)$

RulesSet

For the negotiation N, this coordination pattern manages dependences between different bilateral negotiations on the current task. These dependences refer to the status of negotiations. The participant P1 can independently manage each bilateral negotiation proposals. In other words, if the participant P1 reaches an agreement within a bilateral negotiation, then he must stop all bilateral negotiations for the same task. Therefore, the coordination should manage dependences when the proposal is accepted. These dependences between bilateral negotiations are established by the coordination rule *(**competition**)*.

If the task is contracted by a single participant the coordination module should ensure that the execution of tasks was entirely accepted.

By defining rule *(**block**)*, the coordination pattern ensures that the negotiation considers only the proposals on overall task.

competition$(T_1, T_2, s1, Si)$ $T_1 \in \mathcal{T}$ $T_2 \in \mathcal{T}$ $Si \in negotiation(T1,N)-\{s1\}$;

$\exists a \in s1(T_1)$; $\forall b \in s1(T_2)$ cu $b \neq a$; $\exists a' \in s1(T_2)$ cu $a' = a$

$status(T_1,s1,a,s1) = success \wedge status(T_1,Si,a,Si) \wedge issues(T_1,s1,a,s1) = issues(T_1, Si,a,Si) \wedge role\_s(T_1,Si) = guest \rightarrow \bullet\ status(T_2,s1,b,s1) = failure$;

In other words, if in the sequence s1 of the participant p1 there is a negotiation proposal with the status *success* and, also, the sequence *si* of the participant pi (guest) has the same status *success* and, if the sets of attributes (Issues) are equal, then the negotiation N stops all the active proposals present in s1.

block$(T_1, T_2, s1, Si)$: $T_1 \in \mathcal{T}$ $T_2 \in \mathcal{T}$ $Si \in negotiation(T1,N)-\{s1\}$;

$\exists a \in s1(T_1)$; $\exists a' \in s1(T_2)$ cu $a' = a$ $\neg(status(T_1,s1,a,s1) = failure) \wedge \neg(issues(T_1,s1,a,s1).size = issues(T_1,Si,a,Si).size) \wedge role\_s(T_1,Si) = guest \rightarrow status(T_2,s1, a',s1).failure$;

In other words, the negotiation stops in the proposals where the size of the task is not the one specified by the participant p1.

These global rules can be interpreted using coordination policies composed of visible local rules. These policies model the behavior of a particular type of negotiation service (named *Block*) that negotiates for the participant p1. Next we are presenting the policy allowing a participant p1 to integrate a new negotiation sequence in an ongoing negotiation with a tactic Block. Other policies can be defined similarly for the guest participants' tactics.

policy for sequence *sk*

In order to comply with the rules defined by coordination pattern, this new sequence must be able to observe and act upon exchanged proposals between the participant p1 and other participants (guests). Therefore, this sequence should manage the active proposals that participant p1 shares through sequence *s*1, with other participants:

$\exists\ sk \in N(t1) : sk \subset$ s1

The following two rules represent the transcription of global rules defined at the beginning of the coordination pattern, according to local rules rules visible in sequence *sk*.

competition$(T_1, T_2, sk, s1, Si)$: $T_1 \in \mathcal{T}$ $T_2 \in \mathcal{T}$ $Si \in N(T_1)-\{s1,sk\}$; $\forall a \in sk(T_1) \wedge \forall b \in sk(T_2)$ cu $b \neq a \wedge \exists a' \in sk(T_2)$ cu $a' = a$

$status(T_1,sk,a,s1) = success \wedge status(T_1,sk,a,Si) = success \wedge issues(T_1,sk,a, s1) = issues(T_1,sk,a,Si) \wedge role\_s\ (T_1,Si) = guest \rightarrow \bullet\ status\ (T_2,sk,b,sk) = failure$

block$(T_1, T_2, sk, s1, Si)$: $T_1 \in \mathcal{T}$ $T_2 \in \mathcal{T}$ $Si \in N(T_1)-\{s1,sk\}$; $\forall a \in sk(T_1)$ $\exists a' \in sk(T_2)$ cu $a' = a$

$\neg(status(T_1,sk,a,sk) = failure) \wedge \neg(issues(T_1,sk,a,s1).size = issues(T_1,sk,a, Si).size) \wedge role\_s\ (T_1,Si) = guest \rightarrow status(T_2,sk,a',sk) = failure$

ProgramFormulaSet

Further, the policies presented are translated through different *Program Formula*, in methods modelling the structure and the content of the corresponding negotiation graphs.

The corresponding *Programs Formula* are described in our previous work [8].

Therefore, using the proposed coordination model we can manage the coordination among several bilateral negotiations on a task that must be fully accepted by a single contractor.

6 Application to the Manufacturing Segment

The outcomes of this research are being applied to the European Project C2NET. The goal of this project is the creation of cloud-enabled tools for supporting the supply network optimization of manufacturing and logistic assets, based on collaborative demand, production and delivery plans.

Particularly for the development of the Data Collection Framework in C2NET, which is responsible for a seamless integration among the developed components, modules and the cloud, the project proposes to complement the data collection framework with methods for interoperability decisions and steps that lead to adequate business process activities between them. To enforce the proper interoperability between systems, supported by a maturity environment, the set of negotiation mechanisms proposed in this paper are being implemented by technologies including Multi-Agent Systems (MAS) and the inference of reasoning rules. Another application of the negotiation services here developed will naturally be in the C2NET collaboration tools suite, for generic negotiation purposes. With respect to this, a possible scenario for collaborative working environment is the interaction between the plants and their suppliers. Each plant has its supply chain, which is used to satisfy any needs of the factory, and according to the requirements for each need, multiple suppliers may propose solutions that can be used, and which can make use of the negotiation system.

In the implemented scenario using the proposed framework, the manufacturing domains are interconnected by a set of services that support the development and maintenance of the interoperable enterprise collaborative environment.

Whenever the changes occur in the interoperable space, the negotiations will take place in order to find the most suitable solution mutually accepted by all parties.

The C2NET environment is the proof-of-concept that is under development by the authors to prove the framework.

7 Conclusions and Future Work

This paper proposes a generic coordination model based on a coherent set of rules and patterns that can be applied in order to model a rapid solution delivery in response to changes in a business process environment. The proposed solution allows handling several negotiations in parallel satisfying the possible dependencies among them. Managing parallel and dependent negotiations based on the users' constrains and negotiation tactics empowers the users to change their business workflows, policies and to cope with a continuous changing business processes. In the current work we have described the coordination of negotiations in a closed environment where the coordination is achieved only through bilateral negotiations that compose the same negotiation. It is presented an example of B2B interactions with the goal to outsource the entire task to a single participant. A negotiation process may end with a contract and in that case the supply schedule management and the well going of the contracted task are both parts of the outsourcing process. In the sequence of our research we will complete our model with the coordination of negotiation in an open environment that allow the coordination of constraints among several different negotiations in parallel.

The proposed model allows to coordinate negotiations of systems and applications' changes towards interoperability in enterprise interactions, supporting a sustainable interoperability networking environment along its life cycle.

Acknowledgments This work has been partly funded by the European Commission through the Project H2020 C2NET: Cloud Collaborative Manufacturing Networks (Grant Agreement No. 636909).

References

1. Panetto, H., & Cecil, J. (2013). Information systems for enterprise integration, interoperability and networking: Theory and applications. *Enterprise Information Systems, 7*(1), 1–6.
2. Jardim-Goncalves, R., Sarraipa, J., Agostinho, C., & Panetto, H. (2011). Knowledge framework for intelligent manufacturing systems. *Journal of Intelligent Manufacturing, 22*(5), 725–735.
3. Kadar, M. (2013, July 16–19). Assessement of enterprise interoperability maturity level through generative and recognition models. In *Proceedings of the 2013 International Conference on Economics and Business Administration (EBA 2013)*, Rhodes Island, Greece, ISBN 978-1-61804-200-2.
4. Wooldridge, M., & Jennings, N. R. (1995). Intelligent agents: Theory and practice. *The Knowledge Engineering review, 10*(2), 115–152.
5. Sycara, K., & Dai, T. (2010). Agent reasoning in negotiation. In *Advances in Group Decision and Negotiation* (Vol. 4, pp. 437–451). Handbook of group decision and negotiation, Part 4.
6. Cretan, A., Coutinho, C., Bratu, B., & Jardim-Goncalves, R. (2012). NEGOSEIO: A framework for negotiations toward sustainable enterprise interoperability. *IFAC Journal Annual Reviews in Control, 36*(2), 291–299.

7. Coutinho, C., Cretan, A., & Jardim-Goncalves, R. (2012, June 18–20). Cloud-based negotiation for sustainable enterprise interoperability. In *Proceedings of the 18th International ICE Conference on Engineering, Technology and Innovation* (ICE'12), Munich, Germany.
8. Coutinho, C., Cretan, A., & Jardim-Goncalves, R. (2013, October). Sustainable interoperability on space mission feasibility studies. *Computers in Industry—Special Section on Interoperability and Future Internet for Next-Generation Enterprise*, *64*(8), 925–937.

Part III
Semantics for Enterprise Interoperability

The Systems Development Life Cycle to Facilitate Progression Towards Semantic and Organizational Interoperability for Healthcare System

Daliborka Mačinković and Nenad Aničić

Abstract Model of healthcare system interoperability should provide accurate interpretation of medical information and the exchange of data and services through modern technologies and infrastructure for improvement of the healthcare system. The existence of numerous standards in medicine and their adoption facilitated the establishment of interoperability, but a complete semantic and organizational interoperability is still not achieved. In this paper, interoperability is considered within the methodological framework of the systems development life cycle on semantic and organizational level of interoperability. Life cycle of semantic web services and semantic technology through the proposed phases indicates the solution of misunderstandings in the exchange of information and improvement of services in the health system. Life cycle of business processes interoperability through the proposed phases facilitate understanding of common business processes, which increases the efficiency of the healthcare system. An example of the introduction of new procedures into the health system was presented through life cycle of the common stream of proposed phases as one methodological proceeding.

Keywords Semantic interoperability · Organizational interoperability · Common business processes · Semantic web services · Ontology

1 Introduction

Establishment of full interoperability in complex domains such as Healthcare systems is a challenging task. Solving interoperability is the subject of numerous research projects and initiatives such as ATHENA, INTEROP NoE, C4IF, EIF 2.0.

D. Mačinković (✉) · N. Aničić
Faculty of Organizational Sciences, University of Belgrade, Jove Ilića 154, Belgrade, Serbia
e-mail: daliborka.macinkovic@teol.net

N. Aničić
e-mail: nenad.anicic@fon.bg.ac.rs

© Springer International Publishing Switzerland 2016
K. Mertins et al. (eds.), *Enterprise Interoperability VII*,
Proceedings of the I-ESA Conferences 8, DOI 10.1007/978-3-319-30957-6_8

Interoperability can be perceived through interoperable dimensions such as barriers: conceptual, technological and organizational; approaches: integrated, unified, federated [1] and layers of interoperability: technical, syntactic, semantic, organizational [2]. Interoperability on technical and syntactic layers is fully developed. According to the research in the paper [2], for semantic IOP (interoperability) there are concepts and methods available, but which are not yet standardized, and for organizational IOP it is by far less obvious what has to be standardized, who could develop and establish appropriate standards, and what is necessary for their operation and maintenance.

This paper discusses interoperability on the semantic and organizational level within the methodological framework of systems development life cycle. Aspects related to the specifics of the healthcare system are added at each phase.

The paper is organized into five parts. After the first part which represents the introduction, the second part points out the challenges of interoperability, using the known dimensions of interoperability. It goes through the ways of achieving interoperability and barriers for achieving full interoperability, as well as ways for improvements that are specific for the healthcare system. The third part deals with the semantic interoperability through the proposed phases of the life cycle of web services and semantic technologies. The fourth part discusses the organizational interoperability through the proposed life cycle of business processes interoperability. The fifth part gives the example of the new procedures in the health system obtained through the Life cycles of semantic and organizational interoperability. The conclusion contains the basic achievements and future plans.

2 The Challenges of Interoperability

Ways for solving interoperability problems are application standards, interoperability framework, a framework for system architecture, service orientation, semantic technologies based on ontologies, common modeling and business process management.

The framework for establishing interoperability is an agreed approach for organizations that want to work together and it includes a set of principles, guidelines, best practices, vocabulary and concepts.

In order to achieve effective communication between the organizations you need to clearly define the framework for the architecture according to the following domains: Business architecture, Services architecture, Technology architecture, Information architecture.

The paper [3] considers three types of measurements interoperability: potentiality, compatibility and performance. Measuring interoperability allows a company to know its strengths and weaknesses in interoperability and to prioritize actions to improve the interoperability.

The paper [4], from the aspect of measuring the interoperability compatibility, points to the possibility of extracting the phases and activities in the methodology of structured approach to interoperability: Definition of objectives and needs, Analysis of existing system, Select and combine solutions, Implementation and testing.

The complexity of health systems is facilitated by the introduction of some international standards for the exchange of medical data such as HL7, DICOM, CEN/TC251 and other. Their complexity and different interpretations, as well as the existence of different versions lead to errors in the work and that is still an obstacle to achieving full interoperability [5].

Health documents which are in accordance with EDI, DICOM, ISO, HL7 and other standards can be automatically converted into any other standard, if mapping mechanisms are defined. Increasing efficiency of mapping in achieving interoperability can be done by enriching their structures with semantics [6].

Improvements that have been discussed in current works in health care are: applying the principle of re-engineering of business processes, removing duplication of administrative tasks by orchestration of activities and to allow those who are the most effective to perform these tasks [7, 8], the aspect of knowledge management to excellence that are best for control of continuous management, improvements in quality and cost, access to online health information, clinical systems for decision support including expert systems and software agents can also contribute to reducing medical errors [9], integration of different functionalities of hospitals [10], during the creation of value for the patient, it is necessary to considered every business process layer as the layer of value for the service [11].

As support in solving interoperability, this paper observes semantic and organizational levels of interoperability within the framework of the systems development life cycle (SDLC) by defining phases of planning, analysis, design, implementation and maintenance. Phases follow the interoperability of business processes and semantic web services and semantic technologies shown in Table 1 and they are intended to help ensure each step that will lead closer to the goal of achieving full interoperability. Due to specificity that occur in individual phases, they are observed as separate phases of the life cycle of semantic web services and phases of life cycle of business processes interoperability for healthcare system.

3 Semantic Interoperability in Healthcare

Projects in the field of health systems such as Cocoon [12], Artemis [13], PPEPR [14], HLH [15], Saphire [16], Shape [17] have been launched with the aim of developing infrastructure to improve the exchange of information in medicine for resolving semantic interoperability. Semantic interoperability should provide automatic, precise and unambiguous interpretation of data and business processes which are to be exchanged between different systems, mechanical understanding of the objectives of the user and system capabilities to be provided in order to avoid mistakes and misunderstandings. Semantic Web services combine Web services communication technology with the intelligent processing of ontology-based

Table 1 Semantic and organizational interoperability within the framework of the systems development life cycle

Layers of Interoperability	Technical IOP	Syntactic IOP	Semantic Interoperability	Organizational Interoperability
	Signals	Data	Information	Processes
	fully developed		practical implementation problems for specific domains	lacking clear concept for specific domains

Systems development life cycle (SDLC)	Life Cycle of Semantic and Organizational Interoperability		Technology
	Life Cycle of Semantic Web Services	Life Cycle of Business Processes Interoperability	
Planning	• Determine needs, intentions or goals • Determine capability and readiness • Establish Agreement for standards, communication, collaboration	• Identify common project • Determine the objective, strategy, changes • Readiness and capability of organizations for cooperation • Reach an agreement for cooperation, standards, tools, frameworks	
Analysis	• Identify of the users / providers of the web services • Analysis of semantic linking with other ontologies • Analysis of users' goals • Analysis of functionality of web services • Analysis of mediation - the target user and functionality the web services	• Identify business processes for common actions • Identify stakeholders • Analysis of state " is-as "of business processes • Analysis of redundant business processes	Semantic Annotation Data structure Common Processes
Design	• Design of domains ontology • Design of linking with other ontologies • Design of users goals • Design of capability, conditions, behaviour WS • The modelling of the composition of the WS	• Design of " to-be 'model • Simulation • Identify improvement • Aligning of the model with the architecture of the system	Domain ontology Architectural models Common repository
Implementation	• Implementation component of services • Discovering WS • Mediation users goals and capability WS • Composition of WS	•Introduce technologies for implementation of processes •Implement the 'to be' scenario	Intelligent Agents Linked Open Data Semantic Web, BPEL, BPR,SWS, Semantic Web of Things
Maintenance	• Monitoring • Results of services • Optimization-reallocation of resources	•Monitoring •Compare actually KPI •Identify opportunities •Optimizing - redistribute resources	

metadata to achieve highly integrated enterprise application integration scenarios, according to [18].

Increasing spread of the use of technology is encouraging the development of interoperability solutions with these technologies. However, application of semantic solutions, as individual solutions, is still evident. This paper also looks at the possibility of wider application of semantic technology in the health system.

3.1 Life Cycle of Semantic Web Services and Semantic Technology

Semantic web services and semantic technology are observed through the phases of Planning, Analysis, Design, Implementation and Maintenance.

In the Planning phase it is necessary to define the needs, intentions, goals and challenges for using the services. It is necessary to achieve agreements for standards, communication and collaboration.

In the Analysis phase it is necessary to identify the participants in interoperability as users and providers of web services. Medical information systems should be able to communicate with complex medical terms unambiguously, which requires a deep analysis of the structure and the concept of medical terminology in this phase. Linked Open Data technologies [19] should allow the presentation and linking of knowledge in the field of healthcare. They represent the possibility of linking health information on the Web using a similar set of information on the two resources, over the same domain using the semantic web technologies. The use of sensors-based domain ontologies as a Semantic Web of Things (SWoT) [20], for new smart services in healthcare is another analysis of the possibilities of modern technology. The main part of the analysis is the analysis of functionality of the web service which should be semantically enriched. According to the paper [21] the need for semantic web services for eHealth, as well as improving web services, is particularly emphasized.

The domain ontology should be defined in the Design phase. Capability, conditions and behaviour of web services can be semantically described where it can more accurately describe the services in relation to the WSDL description. It is possible to use the transformation mechanisms that will facilitate semantic description. The basic elements of web service capabilities are axioms: precondition, assumption, postcondition and effects as well as the terms and conditions under which they are running. The terms of use are represented by parameters or logical expressions such as accuracy, quality of service, performance, reliability, scalability and security of web services. Interface-behavioural characteristics describe the way in which the functionality of the web service is realized.

In the Implementation phase, the service components should be deployed. Discovering web services is seeking for candidates with the best service in accordance with the priorities and conditions of the user. Mediation can eliminate misunderstandings between capability of web services and users goals. Web services composition enables combining of several web services to address customer requirements to services.

The Maintenance phase includes monitoring and optimization. At this stage, information about the quality and success of execution services are provided. Existing functionality of services can be improved with the help of the information. Optimization gives the possibility for reallocation component of services for more successful execution of web services.

4 Organizational Interoperability in Healthcare

Health systems are looking to improve the efficiency of the health care system, better quality services and standardized procedures in treatment. Understanding of common business process in healthcare system means understanding of the links that exist between health organizations, the degree of matching in business processes, the flow of information through these processes and technologies which should facilitate cooperation. According to the OECD [22], common business processes are: Those business processes that exist in different organisations yet have, in essence, the same goals and outputs, thereby creating the possibility for the arrangements to conduct these business processes to be optimised and delivered in a more efficient and standardised manner.

The Business Process Interoperability Framework—BPIF [23], in accordance with a common architecture, should enable organizations to work together based on the principles, policies, tools, standards and guidelines. The challenge is to fix the coordination and collaboration across organizational cooperation, delivery of necessary services and quick response to emergency situations. The request is to reform and redesign business processes, define the strategy as the cooperation of people, processes and systems for delivering more efficient health services. The life cycle of business processes interoperability, which is described in the next section, looks at the possibility of improving organizational interoperability.

4.1 Life Cycle of Business Processes Interoperability in Healthcare

Business processes interoperability can be observed in the framework of the systems development life cycle as Planning, Analysis, Design, Implementation and Maintenance. The business processes life cycle is an iterative process, which allows review of individual phases [24].

The analysis phase precedes the Planning phase, which should identify the mutual project, determine the objective, strategy, program and organizational changes, critical points of cooperation, readiness and capability of organizations and participants for cooperation and change. It is necessary to reach an agreement for cooperation, responsibilities, standards, frameworks, tools and mechanisms for communication.

In the Analysis phase it is necessary to identify business processes for mutual action that will contribute to more efficient and effective work of the health system. It is necessary to identify stakeholders in the business processes, and identify clear boundaries of the process. The analysis of "as-is" current situation for performing of business process serves as the basis for determination of redundant business processes and specific business processes.

In the Design phase using the simulation of the desired business process "to be" should point out the impact of changes to business processes and introduce improvements in the business processes. The key performance indicators KPI, as a

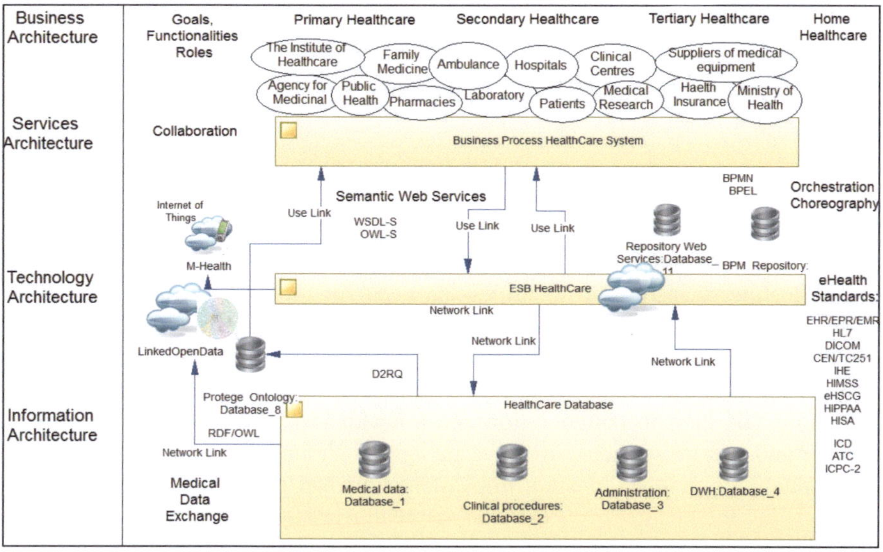

Fig. 1 Healthcare system architecture

new standard of performance, should be established based on the results of simulation. It is necessary to align the proposed model with architecture of the system. The architecture includes all participants in the healthcare system, business processes, information, modern technologies, standards and relationships as shown in the Fig. 1. The architecture of health systems facilitates access to joint health services and allows flexibility with frequent changes in business processes. The effects of changes in outcomes, improving services and technologies can be traced through architecture.

In the Implementation phase, 'to be' scenario should be implemented. It should include the users, owners and participants in the process and they should provide the necessary support for the implementation of changes. Implementation of new processes should ensure goals and objectives. It is possible to continually monitor and track performance of activities, duration, communication costs, measure the performance of new business processes (KPI) and satisfaction of users due to the constant process of improvement. The review of processes needs to be done in the case of falling level of expected values. The purpose of monitoring is to look at the impact of business processes on the efficiency and effectiveness of the whole organization. Optimizing should include redistribution of resources in order to achieve maximum results, which should be identified.

5 The Introduction of New Procedure in the Healthcare System Through Life Cycle of Semantic and Organizational Interoperability

This work contains one example of introduction of the new procedures in the health system, as a new procedure in the administration and monitoring of patients. The example is seen trough the Life cycle of semantic and organizational interoperability and shown in the Fig. 2 in BPMN notation.

The healthcare institution identifies the need for the introduction of new procedures for the administration and monitoring of the patient during treatment, which can be achieved by collaboration in the system. It is necessary to align new procedure with common business processes and involve it into the system. Health institutions take part in collaboration and a health institution can play the role of provider and user of services. Middleware system should allow the phases of a life cycle, planning, analysis, design, implementation and maintenance, to be carried out following the business processes in the healthcare system. Life cycle phases of semantic web services and life cycle of business processes interoperability go in the common way to provide interoperability.

In the Planning phase, the required cooperation between the institutions is identified which is necessary for the realization of a business process. It is determined by the current state of readiness and capabilities of institutions for cooperation, checking and approval of contracts. The necessary harmonization of business processes with strategies and goals of the system is being executed. The institution should give permission to standards, communication and collaboration.

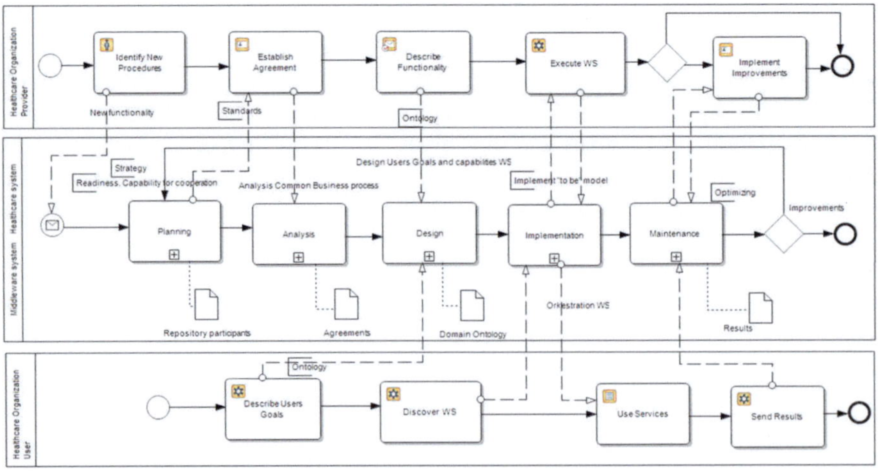

Fig. 2 Life cycle of semantic and organizational interoperability for new procedure in healthcare system

In the Analysis phase the following steps are carried out: analysis of the participants of business processes, defining of users and providers of services and determining the boundaries of the process. Analysis of functionality, users' goals and possible misunderstandings is performed. Business processes were harmonized with other processes and analysis of the system for common action system was performed.

In the Design phase, using the domain ontology, the following steps are carried out: describing the capabilities, setting out the conditions and determination of behaviour of the service. By simulation of "to-be" model, KPIs are determined and improvements are identified. The new model is aligned with architecture of the system. The user describes his objectives that he wants to achieve by collaboration.

In the Implementation phase "to-be" model is implemented. Based on the description of users goals discovers services and solve possible mismatches with mediation. The services composition can includes necessary web services for the new process.

The results which sent about executed services can be constantly monitored in the Maintenance phase. The value of service is determined by the quality of output and quality of service. When the health institution can not fulfill its strategic objectives, it is necessary to re-plan activities to fit business processes into the strategy. If the business process does not fit into the strategy, the total value will be reduced. Identified improvements can be implemented as a new business processes thus achieving the life cycle of continuous monitoring and improvement.

6 Conclusions and Future Work

Solving interoperability has been observed on the semantic and organizational level of interoperability within the methodological framework of the systems development life cycle. Each phase has indicated specificities of the healthcare system. The work is focused on the resolution of ambiguity in the semantic and organizational level, while it is less focused on technical and syntactic levels, which are fully developed. The proposed life cycles of semantic web services and life cycle of business process interoperability had a goal to facilitate progression towards semantic and organizational interoperability for Healthcare system.

The above presented example had a task to present inclusion of the new business processes in the health system within common stream of the Life cycle of semantic web services and Life cycle of business processes interoperability, which were jointed in one methodological proceeding—Life cycle of semantic and organizational interoperability. In the presented sequence, the new functionality has been semantically described by using domain ontology. According to the phases of organizational level, the new procedure has been included in the collaboration taking care of the joint strategy and coordination with the proposed Healthcare system architecture and necessary technologies. Users of service, using domain ontologies, have described their goals and found the required service avoiding misunderstandings by mediation. The services composition allowed calling necessary web services for executing the business process. The results, that are sent after the process is finished, have pointed to the possibility of continuous improvement, re-design and optimization of the system.

Understanding common business processes and semantic capabilities, through the Life cycle of semantic and organizational interoperability, may indicate the relationships that exist between health organizations, the degree of matching in business processes, the flow of information through these processes and technologies to facilitate and enhance collaboration.

The topic of the next work will be related to the integration and extension of proposed Life cycle of semantic and organizational interoperability.

References

1. Chen, D., & Shorter, D. (2008, March 25–28). Framework for manufacturing process interoperability—CEN/ISO 11354. In *Standardisation Workshop in Conjunction to I-ESA 2008,* Berlin.
2. Herbert, K., & Ralf, C. (2009, January). Three dimensions of organizational interoperability. *European Journal of ePractice.* www.epracticejournal.eu. ISSN: 1988–625X.
3. Chen, D., Vallespir, B., & Daclin, N. (2008). An approach for enterprise interoperability measurement. In *Proceedings of the International Workshop on Model Driven IS Engineering: Enterprise, User and System Models (MoDISE-EUS'08),* Montpellier, France.
4. Chen, D., & Daclin, N. (2007). Barrier driven methodology for enterprise interoperability. In *Proceeding of Establishing the foundation of Collaborative Networks,* PROVE2007 (pp. 10–12), Guimarães.
5. Varlamis, I. (2007). A flexible model for the delivery of multi-facet information in patient–centric healthcare information systems. *The Electronic Journal for E-Commerce Tools & Applications* (special issue in interoperability and security in medical information systems).
6. Bicer, V., Kilic, O., Dogac, A., & Laleci, G. B. (2005). Archetype-based semantic interoperability of web service messages in the health care domain. *International Journal of Semantic Web and Information Systems (IJSWIS).*
7. Bliem, M., & Hassanein, K. (2004). E-health: Applying business process reengineering principles to healthcare in Canada. *International Journal of Electronic Business.*
8. Hammer, M. (1990). Reengineering work: Don't automate, obliterate. *Harvard Business Review,* 104–112.
9. Sanmartin, C., Houle, C., Berthelot, J. M., & White, K. (2002, July). *Access to health care services statistics Canada.* Health Analysis and Measurement Group: Statistics Canada.
10. Alnuaimi, O. A., Cronan, T. P., Douglas, D. E., & Limayem, M. (2011). Healthcare professionals' reactions to health enterprise system implementations. In *44th Hawaii International Conference on System Sciences,* Hawaii.
11. Anshari, M., & Almunawar, M. N. (2011). Evaluating CRM implementation in healthcare organization. In *International Conference on Economics and Business Information IPEDR* (Vol. 9, pp. 30–34), Bangkok, Thailand.
12. Della Valle, E., & Cerizza, D. (2005). COCOON glue: A prototype of WSMO discovery engine for the healthcare field. In *Proceedings of the WIW 2005 Workshop on WSMO Implementations.*
13. Dogac, A., Laleci, G. B., Kirbas, S., Kabak, Y., Sinir, S., & Yildiz, A. (2005). Artemis: Deploying semantically enriched web services in the healthcare domain. *Information Systems Journas* (special issue on semantic web and services).

14. Sahay, R., Akhtar, V., & Fox, R. (2008). PPEPR: Plug and play electronic patient records. In *Proceedings of the 2008 ACM Symposium on Applied Computing (SAC)*, Fortaleza, Ceara.
15. HLH [Online]. http://hl7.seecs.nust.edu.pk/index.php?id=2. Accessed May 2015.
16. Laleci, G. B., Dogac, A., Akcay, B., Olduz, M., Yuksel, M., Orhan, U., et al. (2006, October). SAPHIRE: A semantic web service based clinical guideline deployment infrastructure exploiting IHE XDS. In *eChallenges Conference*, Barcelona, Spain.
17. SHAPE Projekat FP7 EU. http://www.shape-project.eu/. Accessed July 2015.
18. Studer, R., Grimm, S., & Abecker, A. (Eds.) (2007). *Semantic web services: Concepts, technologies, and applications.* Springer: Heidelberg.
19. Linked Open Data [online]. http://linkeddata.org/. Accessed July 2015.
20. Gyrard, A., Bonnet, C., & Boudaoud, K. (2014, June 25–26). Domain knowledge interoperability to build the semantic web of things. In *W3C Workshop on the Web of Things*, Berlin, Germany.
21. Della Valle, E., Cerizza, D., Bicer, V., Kabak, Y., Laleci, G. B., & Lausen, H. (2005). The need for semantic web service in the ehealth. In: *W3C Workshop on Frameworks for Semantics in Web Services.*
22. OECD. (2005). E-Government for Better Government, OECD E-Government studies. http://www.oecd.org/gov/public-innovation/e-governmentforbettergovernment.htm
23. The Australian Government Business Process Interoperability Framework, AGIMO. (2007). [online] http://www.agimo.gov.au/government/australian_government_architecture. Accessed June 2015.
24. Weske, M. (2007). *Business process management: Concepts, languages, architectures*. Berlin: Springer.

Performance Oriented Decision Making to Guide Web Service Lifecycle

Tehreem Masood, Chantal Bonner Cherifi, Néjib Moalla and Muhammad Fahad

Abstract Service composition networks are becoming increasingly important and more critical for the enterprise collaborations. The sustainability of service based systems is driven with the support of performance metrics analysis for decision making. Some of the existing works only focused performance at messaging level. Ontology based performance monitoring techniques did not provide the recommendation model for better quality of service. In this paper we propose a semantic service recommendation framework (SSRF) based on Service Networks Monitoring (SNM) concepts. This framework includes creation of Service Network Monitoring Ontology (SNMOnt), extracting traces of technical indicators for services and storing decision data in the existing database. The decision support covers the unavailability of service and recommendation of services based on technical indicators. Our framework is meant to support the creation of shared domain knowledge ensuring an optimal decision making system for web service lifecycle.

Keywords Service network monitoring · Decision support · Ontology · Performance · Technical indicators

1 Introduction

Service-Oriented Architecture (SOA) is an emerging cross-disciplinary paradigm for the distributed and service oriented computing [1]. One of the challenges in SOA is to provide an efficient way to discover compatible Web services and compose them into business process. It is important to create dynamically scalable business processes and guarantee their performance, availability, reliability and

T. Masood (✉) · C.B. Cherifi · N. Moalla · M. Fahad
Decision and Information for Production Systems (DISP), Université Lumière Lyon 2, Lyon, France
e-mail: tehreem.masood@univ-lyon2.fr

© Springer International Publishing Switzerland 2016
K. Mertins et al. (eds.), *Enterprise Interoperability VII*,
Proceedings of the I-ESA Conferences 8, DOI 10.1007/978-3-319-30957-6_9

throughput to be used by internal and external partners. Selecting the most significant web service according to a consumer requirement is a challenging task. Technical indicators needs to be considered when there are several functionally similar web services.

Semantic web uses the notion of ontology's for the creation and elicitation of domain knowledge [2]. Ontology's represent formal specifications about the component of systems and their relationships in a machine understandable and processable manner [3]. They have played an important role in both semantic web applications and knowledge engineering systems [4]. Several tasks such as information storage, processing, retrieval, decision making etc. are done on the basis of ontologies by such systems.

The problem of providing seamless decision making capabilities based on the new project requirements of users is very important. There are different levels of performance in the service network monitoring like binding level, service level, server level and business process level. Business process level is used for the performance analysis at the composition of service [5]. Binding level covers the performance at the protocols level [6]. Service is composed of different operations. Performance can be monitored at both the service and different operations level. Server level covers the performance of resources available. Current research support specific activities, but do not fulfil the requirements of all the activities in the Service-Based System's lifecycle. Different technologies are available that can be effectively utilized for better performance of services. Now-a-days, WSO2 application server is widely being used for the deployment of web services. It has sophisticated infrastructure to capture statistics at operation, service and server level [7].

Our objective in this research is to provide decision support combining service/process technical indicators in order to generate validation arguments for the expansion of the service environment. Therefore in this paper, we propose SSRF based on the SNM. We provide decision support that covers the unavailability of services and recommendation of services based on technical indicators. The proposed model integrates the following building blocks. The first block is the identification and classification of performance metrics from Information Technology Infrastructure Library (ITIL) [8]. Second block is to structure the performance based technical indicators for atomic service, business process, server and binding. For this purpose we design a sophisticated SNMOnt as a base infrastructure. It aggregates the main concepts and relationships between them. QoS requirements, service domain concepts, technical indicators and performance levels are the major concepts. Third block is the decision support based on some inference rules like availability, reliability, throughput, memory available and delay.

The remaining of the paper is organized as follows: Sect. 2 includes related work of different frameworks. Section 3 discusses our proposed approach. We conclude the paper in the last section.

2 Related Works

There are different categories of techniques available that can be effectively utilized in order to provide better performance of service-oriented networks. Therefore related work section has two dimensions. First, we study techniques related to the performance issues of web services. Second, we include ontology based QoS monitoring techniques. Third, we provide analysis of techniques of both sections.

2.1 Performance Based Web Services Techniques

In this category we will discuss some techniques that have been proposed to ensure performance. While analyzing these techniques, it has been observed that all these techniques only focused the binding level performance.

Tari et al. proposes a benchmark of different SOAP bindings in wireless environments. They conduct experiments in loopback, wireless network and mobile device modes. The experimental results show that UDP binding has the lower overhead due to not establishing connections before datagram's transmission. It does not discourse reliability. Therefore response time decreases and throughput decreases. On the other side, HTTP binding receives high protocol overhead from TCP. The reason is the slow connection establishments, tear-down processes and acknowledgement of packet mechanism [9].

Tari et al. proposes a similarity-based SOAP multicast protocol (SMP) which overcomes the network load by the help of reduction in generated traffic size. This technique uses syntactic similarity matching of SOAP messages. In particular, the SMP reuses payload values of SOAP messages and common templates. It sends one copy containing common part of messages to various clients. Their technique relies on the WSDL description of a SOAP based webservice while finding the similarity of response messages. They reuse the duplicated parts of messages that are extremely similar for different clients [10].

Tekli et al. conduct a review of XML processing for SOAP-based webservices. They analyse various techniques for enhancing performance in XML serialization, parsing and deserialization. They propose a model for analyzing the delay distribution. This model analyses the time required to parse an XML input message, operation execution and for building SOAP output message. The results of the service testing on the execution time of operation are used for analysing the distributions of waiting time [11].

2.2 Ontology Based QoS Monitoring Techniques

We have narrow down our related work with respect to ontology because of its various advantages. Several tasks such as information storage, processing, retrieval, decision making etc. are done on the basis of ontologies by such systems. In this section we list ontology based research that focuses on quality of services.

Lin et al. proposes an ontology based QoS-Aware support for semantic web services. They have used ontology to describe Quality of service metrics. They have composed their ontology into upper and lower level property. They have not considered the input/output operations. They have not included real time values for the performance analysis to reach towards optimal web service [12].

Moraes et al. designed an ontology named MonONTO for proposing recommendation for the advanced internet applications users. They have considered information concerning the application type, traffic generated and user profile along with network performance metrics. Their expert system monitors the performance of advanced internet applications. Their ontology provides high-level information to the consumers about the service level agreements of the network. They use a fixed list of network parameters. Therefore, it does not deal with the heterogeneity and extensibility issues. Implementations of web services have not been done by them. Additionally, it does not deal with quality of service mapping [13].

Benaboud et al. proposed Semantic Web Service Discovery Based on Agents and Ontologies considering the fuzzy constraints. Their framework is modeled by adding semantics of QoS attributes with web service profiles. It describes the design and implementation of a web service matchmaking agent. Their agent uses an OWL-S based ontology. They uses OWL reasoner to compare ontology based service descriptions. They use fuzzy constraints increases the efficiency of the web service discovery approach by providing the customers the web services which are not actually satisfying the input QoS constraints, but are close to the QoS constraints specification [14].

2.3 Analysis

The experimental results of the first technique of Sect. 2.1 show that UDP binding has very less overhead due to not establishing connections before datagram's transmission. It does not discourse reliability. Therefore response time decreases and throughput decreases. On the other side, HTTP binding receives high protocol overhead from TCP. Results for the second technique of Sect. 2.1 shows that SMP has medium response time, throughput and network traffic. The third technique of Sect. 2.1 is a survey of XML processing for SOAP-based web services. All these techniques are related to the performance of web service at binding level which we can utilize in proposed framework. The focus of our research work is at the infrastructure level covering service, server, binding and business process level.

Section 2.2 covers the techniques that are related to ontology based QoS monitoring. First technique of this section has not considered the input/output operations. Second technique of this section is related to the ontology of performance at the network traffic level. The third ontology based technique is specific to the monitoring of agent based systems.

To be more effective in terms of performance, Qos based approaches are needed to present the systems in an accurate or realistic way. Also these approaches should be dynamic and adaptable to adjust parameters at run-time based on measured data. Therefore we propose a semantic service recommendation framework that infers decision support at the infrastructure level. Infrastructure level covers the service, server, binding and business process levels.

3 Proposed Approach

This section presents SSRF to recommend the services based on technical indicators requested by the consumer. Figure 1 illustrates SSRF. First of all, user send request to consume existing service. Second step has two parts. First part is to get traces or events from application server. First we deploy services and BPEL in the WSO2 server. Server stores values of technical indicators in some database. Second part includes the design and instantiation of SNMOnt by storing the values of technical indicators. It combines the major concepts and relationships between them. Third step is the decision support. Decision support constitutes some inference rules. Inference rules includes availability, reliability, throughput, memory availablc and delay. Based on these rules, proposed SSRF generates some recommendations like reuse existing service, service is not available and physical resource is not available.

3.1 Service Network Monitoring Ontology (SNMOnt)

We define an ontology named as SNMOnt in order to store the traces of events gathered from the execution of service/business process in WSO2 server. We use some concepts from the literature as well as define some new concepts. Figure 2 presents SNMOnt.

Service provider, service consumer and service host are the concepts that have already been used in the literature. Technical indicators like response time and delay are also used [15]. But the technical indicators that we show in our ontology are specific to WSO2 server. Further we classify different performance levels. Performance Level concept uses the level where service network can be monitored. It has various sub concepts such as binding level, service level, business process level, and server level.

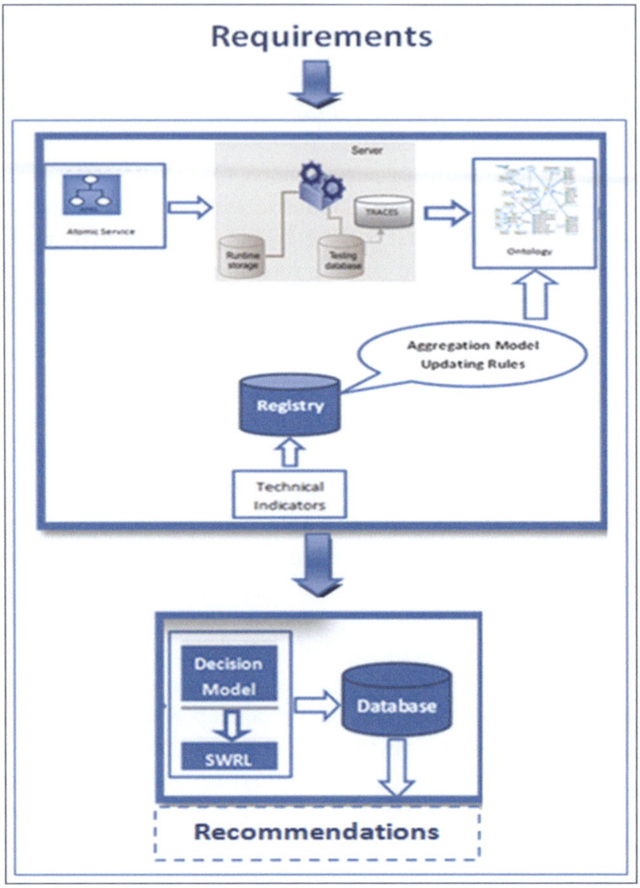

Fig. 1 Semantic service recommendation framework (SSRF)

Each of the service and its operations are monitored. Response Time captures the response time of a service/process. It has three sub concepts to record Maximum, Minimum and Average response time. Request Count shows the number of invocation of a service. Response Count shows the number of replies for an invocation of a service. Fault Count shows the number of invocations the service has not replied. Deploy Time shows when the service is deployed at the server. Up Time shows the time period the service is available since its deployment. Down Time shows the time period of un-availability of a service since its deployment. Delay shows the average response time of a service. Loss shows that the service is un-available.

Data properties that are used to link concepts are shown in Table 1. Consumed_By relationship captures the relationship of each service consumed by its consumers. Has_Indicators relationship describes that performance of each

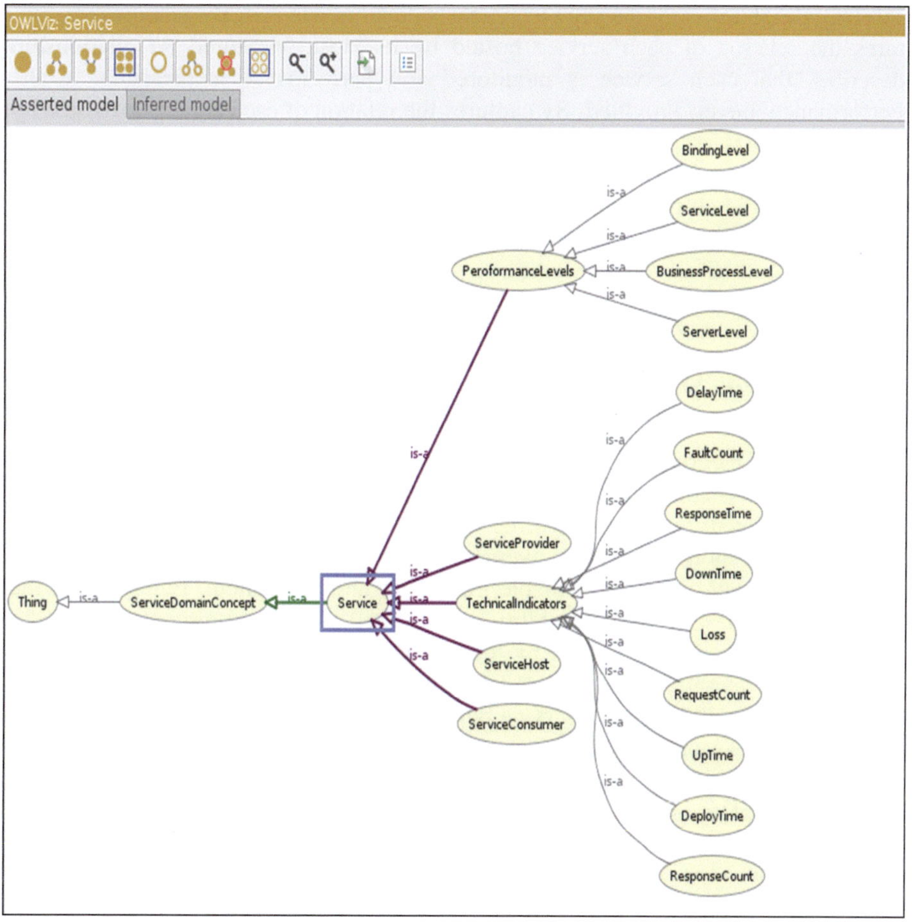

Fig. 2 Service network monitoring ontology (SNMOnt)

Table 1 Object properties of SNMOnt

Concepts	Relationships
Service consumer	Consumed By
Service	Consumes
Service	Has Indicators
Service host	Hosted By
Performance levels	Measured At
Service	Measures
Service provider	Provided By

service is monitored via various technical indicators. Hosted_By relationship captures the relation of each service hosted by its Host. Measured_At relationship describes that each service is monitored and performance is measured at the Performance_Level. Provided_By captures the relation of each service provided by its provider.

We design our SNMOnt in Protégé 5 and uses jess reasoner to verify its consistency to avoid all types of ontological error [16]. Protégè structure allows to add new plug-ins. It helps to visualize concept trees and ontology flow diagrams. OWL code can automatically be generated.

3.2 Decision Support

We use SNMOnt concepts values and generate inference rules. SNMOnt concepts are total response count, total request count, service deployed time, service up time, memory allocated, memory usage, no of successful invocations, total no of invocations, total response count and average response time. Inference rules are for the availability, delay, memory available, reliability and throughput. Following is the description of these rules. Availability is calculated as the total response count of a service/business process divided by total request count of a service/business process. If the requested service is available then we can reuse this service. Delay is measured as the difference of service deployed time and service up time. Memory available is measured as the difference of memory usage and memory allocated. Reliability is calculated as number of successful service operation invocation divided by total number of service operation invocations. Throughput is calculated as the total response count divided by average response time. Table 2 shows these inference rules that compose the knowledge organization originated from the relationships between SNMOnt concepts.

Table 2 Inference rules using SWRL

Technical indicators	SWRL rules
Calculate_Availability	swrl:divide(Total_Response_Count,Total_Request_Count,?x)-> Availability(?x)
Calculate_Delay	sqwrl:difference(Service_Deployed_Time, Service_Up_Time,?x)->Delay(?x)
Calculate_Memory_Available	sqwrl:difference(Memory_Allocated,Memory_Usage,?x)->Available_Memory(?x)
Calculate_Reliability	swrl:divide (no_of_successful_invocations,total_no_of_invocations, ?x)-> Reliability(?x)
Calculate_Throuput	swrl:divide(Total_Response_Count, Average_Response_Time,?x)->Throughput(?x)

We use aggregate QoS functions (like divide, difference) that can be easily defined and managed. Based on these rules we recommend that "the selected service or business process is available or not", "reuse existing service","delay of a deployed service or business process", "memory is available or not to support the new deployment", "reliability of a service or business process in terms of service operation invocations" and "throughput of a service or business process". We store decision data in existing database to deploy decision. We define these rules using Semantic Web Rule Language (SWRL) [17] in Protégé. Future recommendations will include "service composition is not possible" and "security compliance problem in the deployment".

4 Conclusion

In this paper, we focus on how to accomplish and improve enterprise business and IT together. We propose a framework named as SSRF based on the SNM. This framework includes the extraction of traces of technical indicators for services or business process, design of ontology named as SNMOnt, instantiating ontology by values of technical indicators, providing decision support based on inference rules and storing decision data in the existing database. Proposed framework generates recommendations like service availability, reuse existing service, delay of a deployed service, memory is available or not to support the new deployment and reliability in terms of service operation. The use of the framework aims at increasing particularly the alignment of the requirements and of the objectives of the enterprise business and IT in decision making. We will enhance our work to cover the recommendation of service composition and security compliance problem. We will implement our work by using real time case study.

References

1. Gottschalk, K., Graham, S., Kreger, H., & Snell. J. (2002). Introduction to web services architecture. IBM Systems Journal, pp 170–177.
2. Antoniou, G., & Harmelen, F. V. (2004). A semantic web primer. Cambridge: MIT Press. ISBN 0-262-01210-3.
3. Fahad, M., & Qadir, M. A. (2008). A framework for ontology evaluation. *ICCS Supplement 354*, 149–158.
4. Gomez-Perez, A., Lopez, M. F., & Garcia, O. C. (2001). Ontological engineering: With examples from the areas of knowledge management, e-Commerce and the semantic web. Berlin: Springer. ISBN 1-85253-55j-3.
5. Documents Associated with Business Process Model and Notation (BPMN) Version 2.0. Release date January 2011. http://www.omg.org/spec/BPMN/2.0/PDF
6. Simple Object Access Protocol (SOAP) 1.2, Part 0, Primer (2007). World Wide Web Consortium. http://www.w3.org/TR/soap12-part0/
7. http://wso2.com/products/application-server/

8. Donna, K. (2014). *A guide to service desk concepts: Service desk and the IT infrastructure library* (4th ed.). Boston, MA, United States: Course Technology Press, ©2014. ISBN-10: 1285063457, ISBN-13: 9781285063454.
9. Tari, Z., Phan, A. K. A., Jayasinghe. M., & Abhaya. V. G. (2011). *Benchmarking soap binding. On the performance of web services* (pp. 35–58), Berlin: Springer.
10. Tari Z., Phan, A. K. A., Jayasinghe, M., & Abhaya, V. G. (2011). *The use of similarity & multicast protocols to improve performance. On the performance of web services* (pp. 59–104). Berlin: Springer.
11. Tekli, J., Damiani, E., Chbeir, R., & Gianini, G. (2011). SOAP processing performance and enhancement. *IEEE Transactions on Services Computing, 5*, 387–403.
12. Lin, L., Kai, S., & Sen, S. (2008). *Ontology-based QoS-aware support for semantic web services*. Technical report, Beijing University of Posts and Telecommunications.
13. Moraes, P., Sampaio, L., Monteiro, J., & Portnoi, M. (2008). MonONTO: A domain ontology for network monitoring and recommendation for advanced internet applications users. In: *Network Operations and Management Symposium Workshops*, IEEE 2008.
14. Benaboud, R., Maamri, R., & Sahnoun, Z. (2012). Semantic web service discovery based on agents and ontologies. *International Journal of Innovation, Management and Technology, 3* (4), 467–472.
15. Chhun, S., Cherifi, C., Moalla, N., & Ouzrout, Y. (2015). A multi-criteria service selection algorithm for business process requirements. CoRR abs/1505.03998.
16. Protégè. Stanford Medical Informatics (2007). Available at: http://protege.stanford.edu/
17. W3C (2004). SWRL: semantic web rules language. http://www.w3.org/Submission/SWRL/

Profiling Based on Music and Physiological State

João Gião, João Sarraipa, Fonseca Francisco-Xavier, Fernando Ferreira, Ricardo Jardim-Goncalves and Milan Zdravkovic

Abstract Technology is driven by the objective of improving our lives, attending our needs and supporting our work and daily activities. Such guidelines are important to lead technological developments in many areas, whereas the most important result of such enabler is the enhancement of the users' wellbeing. It is questionable how such technologies make us feel better. One can argue whether environmental conditions can be modified in order to enhance people's wellbeing, and in what direction. One of the adopted methods in this work explores that thought, on whether the usage of a person's physiological state can wield adequate sensorial stimulation to be usefully used thereafter. Another question considered in this work is whether it is possible to use such collected data to build a user's musical playlists that tries to match a user's physiological and psychological state with the stimuli evoked by the music that he or she is listening.

Keywords Ontologies · Knowledge representation · Physiological states

J. Gião · J. Sarraipa (✉) · F. Francisco-Xavier · F. Ferreira · R. Jardim-Goncalves
CTS, UNINOVA, DEE, Faculdade de Ciências e Tecnologia,
Universidade Nova de Lisboa, 2829-516 Caparica, Portugal
e-mail: jfss@uninova.pt

J. Gião
e-mail: jcd.silva@campus.fct.unl.pt

F. Francisco-Xavier
e-mail: fx.fonseca@campus.fct.unl.pt

F. Ferreira
e-mail: flf@uninova.pt

R. Jardim-Goncalves
e-mail: rg@uninova.pt

M. Zdravkovic
Faculty of Mechanical Engineering, University of Nis,
ul. Aleksandra Medvedeva 14, 18000 Nis, Serbia
e-mail: milan.zdravkovic@gmail.com

© Springer International Publishing Switzerland 2016
K. Mertins et al. (eds.), *Enterprise Interoperability VII*,
Proceedings of the I-ESA Conferences 8, DOI 10.1007/978-3-319-30957-6_10

1 Introduction

Scientists and consumer product companies try to use technology to aid people's daily life. Through the creation and improvement of new technologies, the comfort of people is becoming better every day. This wellbeing can be improved with a feature that is present, most of the times, in our lives, the music. The interaction of music with our affective state could bring huge benefits, since it can positively impact people's performance in many ways (e.g., increasing the concentration, or the performance while developing tasks). In this paper authors want to present the usability of technology and music and its effects on people's emotions. We depart from the fact that each person has a personal reaction to external stimuli, and in particular, the reaction to the same music can vary significantly, according to present affective state and surrounding environment. For that reason, we may infer the usefulness of a learning system that, using previously retained information, would generate new contents and new musical sequences. Thus it becomes relevant to start by measuring the effects of selected music on a user's physiological state so that later we can assess the transition to another state, which according to literature, seems to be possible to distinguish happy from sad music within as little as half a second from the beginning of the music [1]. That will be taken as a baseline for the user's emotional state at that departing point of the enrolment. Since data analysis of physiological measurements by itself would be a complex and would require very specialized knowledge, an alternative approach will be developed as described in the next sections.

In Sect. 2 a comprehensive study on the state of the art on physiological measurements is performed. It includes how to perform physiological measurements. This section finishes with emotion representation approaches and how to infer emotions from physiological measurements. Section 3 suggests a framework that captures physiological data and reacts to the deducted emotions processed by the framework. The architecture used to build the framework is described in this section along with all the sub-components of the framework. The following sections describe three scenarios used to test the devised framework, and the document finishes with a section that draws conclusions and future work. The study reported in this chapter was carried out with human subjects.

2 Background Observation

2.1 Physiological Measurements

Physiological monitoring systems have being developed for different research reasons. They exist, mostly, as medical devices with the purpose of verifying a patient's health. They can be classified into three groups: (i) electrical signals generated by the patient (e.g., electroencephalography, electrocardiography and electromyography); (ii) conversion of a measurable variable into an electrical signal (e.g., thermocouple); and (iii) passing energy directly into the patient and measure its effects on the patient (e.g., thoracic bio impedance cardiac output monitor) [2]. Beyond capturing data for medical knowledge consolidation and support pro patient's best clinical treatment, physiological measurements can also be used to infer emotional states. Voice and facial expression can be deliberately changed occasionally, but automatic nerve physiological activities cannot be intentionally controlled. Since automatic nerve physiological activities are true emotional indicators, they are very used in literature [3–5].

2.1.1 The Processing of One Physiological Measurement

In order to understand physiological processes with physiological measurements, in the human body, more than one biosensor is used to simultaneously get different parameters. The data acquired by an electronic physiological measurement is usually obtained in the form of analogue electronic data proportional to the variable being measured. Since acquired data has low amplitude and may have non-negligible quantities of "noise", the signal is amplified and filtered electronically. This noise results from unwanted signals or interferences, commonly due to induction from external currents and magnetic fields. After filtered and amplified, the signal is converted to digital by an analogue-to-digital converter (ADC), being then ready to be processed [2].

2.1.2 Existing Solutions

EEG readings are used to understand the ongoing brain activity as in [6]. A machine-learning algorithm is used with the purpose of categorizing EEG dynamics according to self-reported emotional state during music listening. It results that DASM12, a spectral power asymmetry across multiple frequency bands, is an important metric for characterizing brain dynamics in response to emotional states of joy, anger, sadness and pleasure. With EEG physiological measurement, an accuracy of 82.29 ± 3.06 % is achieved, across 26 subjects to identify four emotional-states (joy, anger, sadness and pleasure). According to music subject, the EEG dynamics is characterized during music listening with fewer electrodes, which makes measurement a little less invasive.

A set of multimodal sensor data with EMG, ECG, Electro Dermal Activity (EDA), BVP, Peripheral Temperature (SKT) and RESP is used in [7]. To improve data analysis, by maximizing the signal and reduce noise multiple filtering techniques are used, as a low-pass filter for SKT and RESP physiological signals, the Hamilton algorithm for ECG and a band-pass filter (1–18 Hz) for BVP signal.

In another study physiological measurements had different signals while listening to opera [8]. Cardiovascular, skin conductance, respiration and blood pressure measurements were made and observations reveal a decrease of RR (Respiratory Rate), IBI (cardiac Inter Beat Intervals) and SCL (Skin Conductance Level) while listening to music.

2.2 *Emotions*

In order to understand emotions, as they are relevant for the present study, it is necessary to define the similar definition of affective domain by differentiate emotion, mood and core affect. Since such definitions are similar, many researchers tend to use an individualistic approach of the definitions.

Core Affect is a neurophysiological state that is consciously accessible as a simple *feeling*. Some examples are feeling of wellbeing (good or bad), pleasure and displeasure, tension and relaxation, energy and tiredness. Core affective feelings vary in intensity, and arguably everybody is always in some state of core affect, even if neutral [9]. Mood is subtler, lasts longer, is less intensive and endures over longer timescales. The mood only exists through the interaction between people [10]. There is a connection between mood and emotion. For instance, when a person has a negative effect (e.g., anger), that person also tend to have elevated levels of negative moods, like fear, sadness and/or guilt. Emotion is defined as a complex set of interactions among subjective and objective factors, mediated by neural/hormonal systems, which can (a) give rise to affective experiences such as feelings of arousal and pleasure/displeasure; (b) generate cognitive processes such as emotionally relevant perceptual effects, appraisals, labelling processes; (c) activate widespread physiological adjustments to the arousing conditions; and (d) lead to behaviour that is often, but not always, expressive, goal directed, and adaptive [9–11].

2.2.1 How to Represent Emotions

Emotion is a difficult concept to universally agree on. Therefore, in order to develop methods to determine what is the emotion of a given person, researchers created models, some of those are presented in Fig. 1.

The OCC model (Fig. 1a) was developed to be implemented on a computer and to be used in an Artificial Intelligence application, for the purpose of tracking emotions. This tracking system starts, not with emotions themselves, but with how people perceive the world. The supposition is that there are three major aspects or changes in the world, which are events, agents and objects. The base of their assumption is that emotions represent valence reactions of these perceptions of the

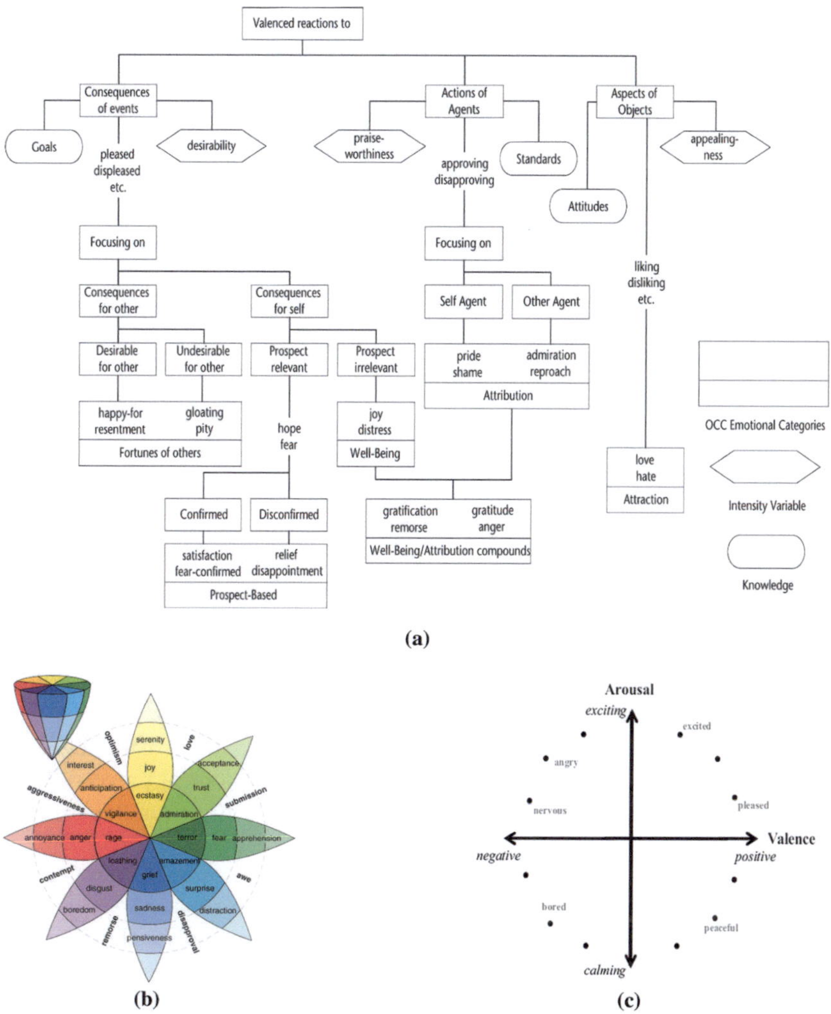

Fig. 1 Emotional models. **a** The OCC model [12]. **b** The wheel of emotions model [13]. **c** The arousal-valence space model [14]

world. This model hierarchically described 22 emotion type specifications. Though relating to music general descriptions represent emotions induced by music, such as soft, neutral and aggression and happy, neutral and sad [15].

The Wheel of Emotions (Fig. 1b) is a three-dimensional circumplex model that describes relations between emotions. Eight primary emotions can mix with each other and form another emotion. The intensity of the emotion increase as the emotion moves towards the wheel's centre. This wheel allows similar emotions grouped together or opposites 180° apart.

The Arousal-Valence Space model (Fig. 1c) is widely used in music emotion detection or recognition [15]. Valence describes how positive or negative an emotion is classified, and ranges from unpleasant to pleasant feelings. Arousal refers to how excited or how apathetic an emotion is classified, and ranges from sleepiness or boredom to excitement. This results from evidence on how affective states are dependent from each other [16]. Dimensional emotion theory believes that emotion should be analysed in a psychological dimensional space, as it helps to represent a wide range of emotions not necessarily depicted by a particular emotion descriptor [15].

2.2.2 Emotional Deduction

It is arguably not possible to measure the emotional state of a person, because it cannot be measured. There is, however, the possibility to deduct the emotional state based on physiological measures. However some inaccuracies can happen, as it is a very hard task to uniquely map physiological patterns (e.g., time, context, space, culture) onto specific emotional states. Thus such physiological patterns correspond uniquely to one individual and to a single situation. Then to record physiological measurements the user needs to biosensors and sensing using surface electrodes need to be attached to the body, which can be disturbed by motion. Finally, the use of various biosensors at the same time, with specific characteristics becomes a complex multivariable task and requires knowledge of biological processes related to neuropsychological functions, which makes hard to identify emotions based only on collected signals.

Nevertheless, the usage of biosensors provides some advantages as it is possible to continuously gather information about user's affective states. Since automatic nervous system (ANS) actions are mostly involuntary, generally they cannot be triggered intentionally, ANS physiological sensors can claim the user is telling the truth or not. For instance, it is unusual that a person smiles and have a negative emotion [7].

3 Framework

The proposed framework is composed by five parts: (i) users, (ii) profile monitoring, (iii) knowledge, (iv) module for analyse and propose the best solution and (v) external stimulus. The framework's objective is to support human oriented knowledge ground through analysing the current user's profile.

Figure 2 shows a general overview of this framework as well as the correct flow of the system's events. The system described by this framework has to handle from one monitor system to a huge amount of them in (ii), because the framework can require more data to complement the user's profile to get better results. On the other hand, they need to be capable of deliver the output to (i), depending on the actual user's profile and the effects of that output on the user. The (iii) is responsible for storing all the information provided from (ii) (to enable learning from the previously acquired data) and must search in one ontology, in order to get all the necessary information to user's affective state evaluation.

With this information about the user, the module (iv) can deduce the best solution for user. The system must also be capable to take advantage from previously stored data in order to refine its results to benefit the user. For this reason, it is important to store the current user's profile, the actions proposed to the user and the profile after the actions made. According to the results on module (iv), the user is presented by external stimulus on module (v), which has the capability to change the user affective state.

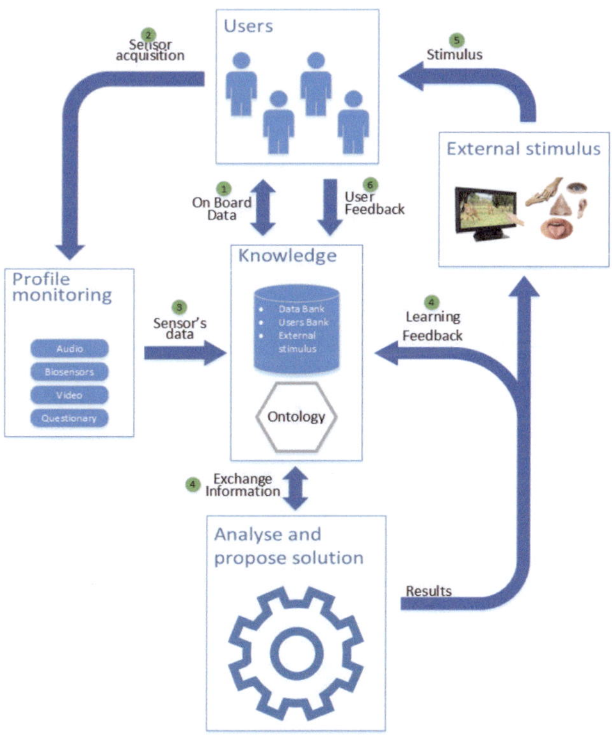

Fig. 2 Conceptual framework for user knowledge

3.1 System Architecture

The developed system is based on the framework proposed above. Such system main objective is to suggest music accordingly to the identified emotional states so that a comfortable feeling can be induced to its user. The system architecture is represented in Fig. 3, which is composed by seven modules: (i) physiological measurements (data capturing), (ii) database (with MySQL query language), (iii) system's engine, (iv) ontology, (v) data analysis (with the MATLAB software), (vi) choose the appropriate music and (vii) audio stimulus. The module (i) is where it is possible to constantly receive data from physiological sensors. The data acquired from physiological sensors in module (i) is saved in a database in the module (ii). The module (iii) is responsible for receiving information from the user, regarding its initial/final emotional states, and for sending the most appropriate music to module (vii) based on profiles saved in ontology (iv). These profiles are created from each user's pattern, which accomplishes data analysis (v). This data analysis encloses specific filters applied above the physiological measurements to obtain specific information as identification or recognition of a specific emotion.

The list of users, the initial/final emotional state, the music list, as well the one chosen is saved in the ontology using the Java framework Apache Jena [17], as it allows the management of the ontology. As it can specify beforehand the necessary queries, and be efficient accessing the information to obtain, as an example, the

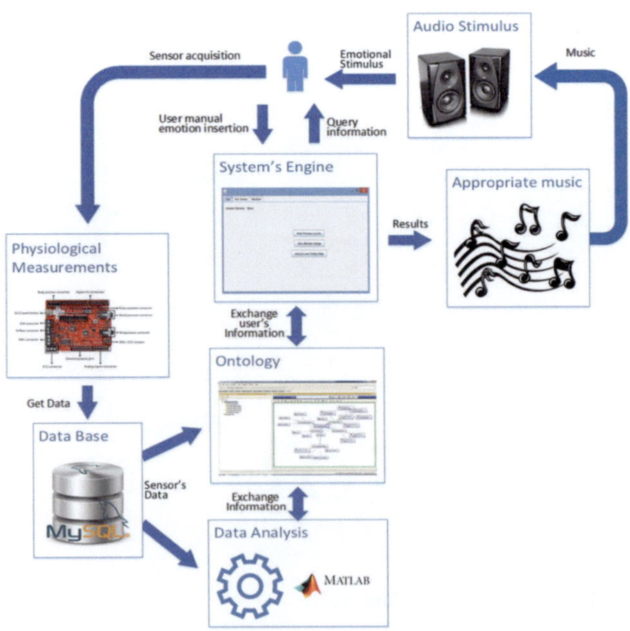

Fig. 3 System architecture

appropriate music for a specific profile it was used SPARQL language. Finally, due to its countless useful tools as signal filters, Matlab was used for data signal analysis.

3.2 *Music Deduction Using an Ontology*

3.2.1 Ontology Model

The ontology model in this research has some main classes that are represented in Fig. 4, which are:

- User: represents personal information about the user. It could also contain all the emotional records performed by the user.
- Emotional Records: contains each event data, thus it links all the information about the emotional records of a user, as the initial and final emotion, and the information of the music that the he was listening.
- Music: contains all the information of each music.
- Physiological Measurements: contains the physiological measurements of the user, while listening a specific music.
- Feeling State: defines the different feeling states of the user.
- Emotion and Mood: has the description of different emotions and moods.

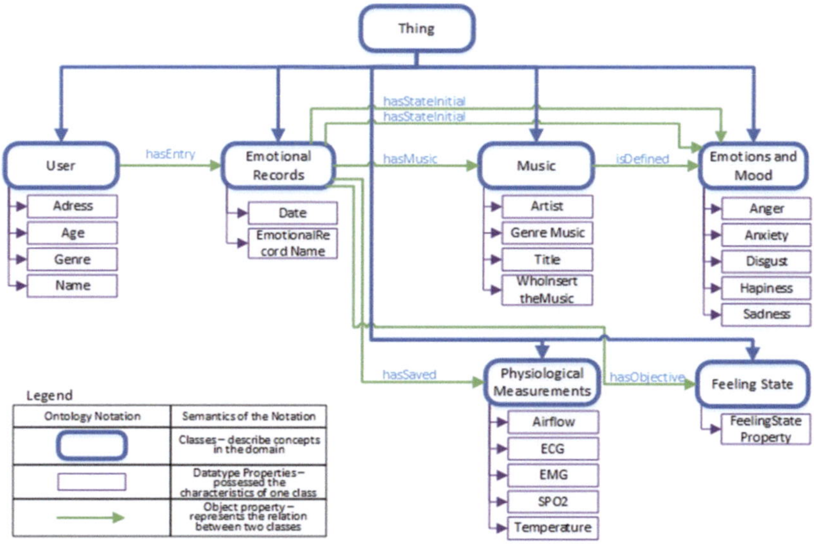

Fig. 4 Music ontology

4 Use Cases

This system was developed in order to support three different scenarios with the objective of suggesting the most suitable music accordingly to the current assessed feeling state in order to feel better (i.e., feeling good). To perform the validation of the system, invitation was made to members of our team, master and PhD students, which filled informed consent forms prior to any measurements being made. Thus, this study was carried out with human subjects.

In a first scenario (Fig. 5a), the user introduces its initial emotional state, and accordingly to it, the system will suggest the most appropriate music. This suggestion is made accordingly to records existent in the database related to past use patterns, which comes from various emotion the user defined he was feeling while listening different music.

In the second scenario (Fig. 5b), it is gathered physiological measurements from the user to automatically determine his feeling state. Based on data analysis over such measurements, the system suggests the most appropriate music also based in past pattern information existent in the ontology. This scenario represents a conceptual experiment using sensors that can be used unobtrusively as soon as its technology integrates such sensors in small wearables.

Figure 6 presents one example of physiological measurements used with this system. Additionally, in Fig. 7 it is presented three different values for Pulse measurements, from which through MATLAB the system identified the possible emotion state of the person. In the first plot the person was nervous, and the values were higher, from 72 to 76 BPM. In the second plot, the person was indifferent, and that resulted on BPM values from 60 to 64 BPM. In the third plot, the person was relaxed, and for that reason, the values from Pulse were between 51 and 58 BPM.

The proposed system can be used in different applications, as in evaluating attention in students or trainees. It could help to identify the type of music should be integrated in eLearning materials to potentiate trainees attention. Other similar

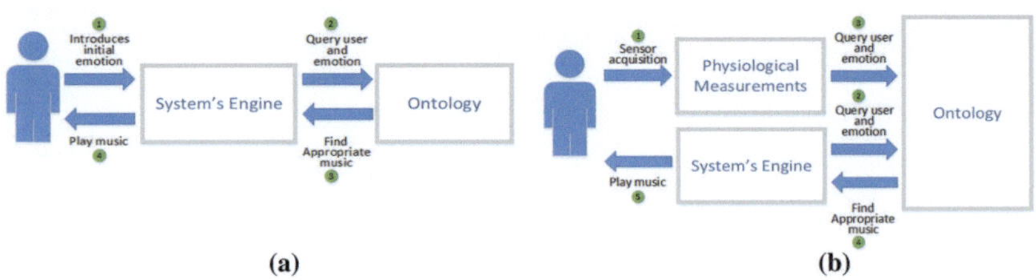

Fig. 5 Scenarios. **a** Scenario 1. **b** Scenario 2

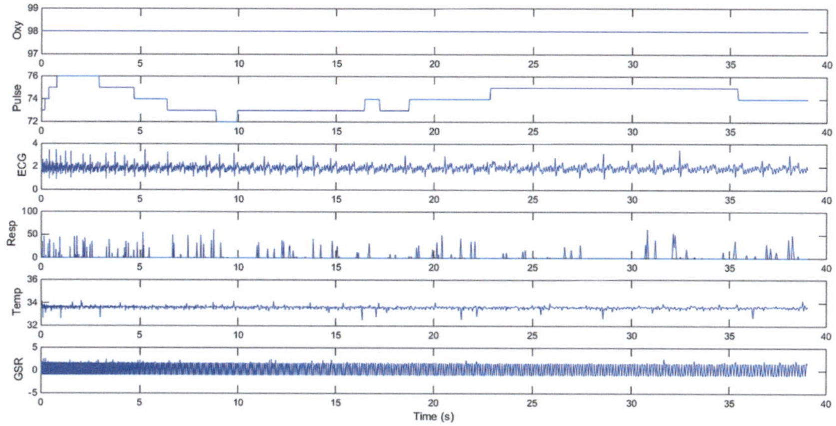

Fig. 6 Physiological measurements

applications could be applied to industry as in trying offering better working environments to workers in such a way to increase their efficiency accomplishing their tasks. However, the main purpose of this system is not to be a product but a prototype to analysts/researchers to facilitate their emotional and physiological analysis, which results/conclusions would potentiate better environments or feelings to people.

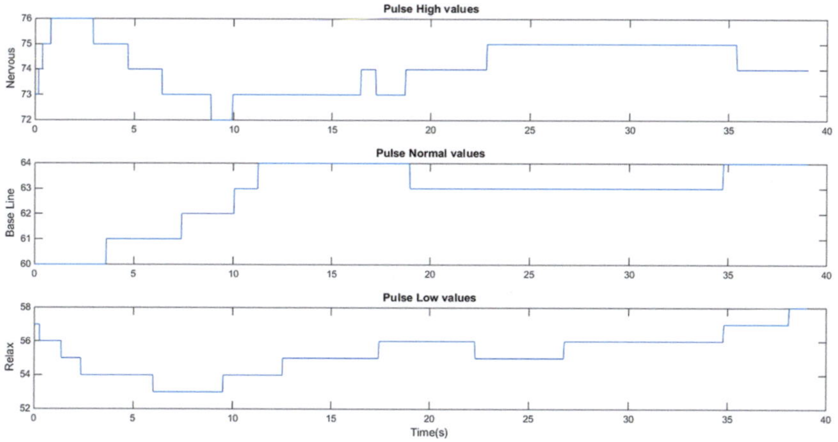

Fig. 7 Pulse different measurements

5 Concluding Remarks and Future Work

The presented research work demonstrates that human affective states are very complex, but in what regard to musical effects, interesting interactions can be achieved. This early work generated results that lead us to conclude that it is possible to change affective states based on profiling and user preferences.

The proposed framework establishes a person's profile with early physiological assessment and later proposes music selections for targeted affective states. That is based on learning through previously records, saved on the ontology model that will empower a selection of music that aims to improve the user's affective state. The usage of this framework, with a somehow invasive approach, it is possible to determine the best music selection, from a pool, according to a person's inferred affective state.

The system can support alternative scenarios, as mentioned in this document, but those was not explored yet due to the requirements of complex data analysis. A possible third scenario could be based on all the registered users in the system, which may use a statistical probability of all these users patterns information in the database to make recommendations. Thus, when the user wants to feel comfortable, the system uses personal information (e.g., age, address and genre), from which can identify the user profile related to geographic or gender tendencies but also taking in consideration the music's characteristics (e.g., beats per minute, rhythm, tonality). These scenario characteristics facilitate interoperable integrations between this system and independent music players, like Spotify, Youtube, Itunes or Soundcloud. The implementation of those alternative scenarios will be explored in future work.

Acknowledgments The research leading to this result has received funding from the European Union 7th Framework Programme (FP7-ICT) under grant agreement of the OSMOSE—OSMOsis applications for the Sensing Enterprise, project ID nr. 610905 (http://www.osmose-project.eu/). It has also received funding from the Portuguese-Serbian bilateral research initiative specifically related to the project entitled: Context-aware Smart Cyber—Physical Ecosystems.

References

1. Vieillard, S., Peretz, I., Gosselin, N., Khalfa, S., Gagnon, L., & Bouchard, B. (2008). Happy, sad, scary and peaceful musical excerpts for research on emotions. *Cognition and Emotion, 22*, 720–752. doi:10.1080/02699930701503567
2. Dingley, J. (2011). Processing, storage and display of physiological measurements. *Anaesth Intensive Care Med, 12*, 426–429. doi:10.1016/j.mpaic.2011.06.010
3. Picard, R. W., Vyzas, E., & Healey, J. (2001). Toward machine emotional intelligence: Analysis of affective physiological state. *IEEE Transactions on Pattern Analysis and Machine Intelligence, 23*, 1175–1191. doi:10.1109/34.954607
4. Takahashi, K., Namikawa, S. Y., & Hashimoto, M. (2012). Computational emotion recognition using multimodal physiological signals: Elicited using Japanese kanji words. In *35th International Conference Telecommunication on Signal Process (TSP)* 2012 (pp. 615–20), Prague. doi:10.1109/TSP.2012.6256370

5. Wen, W., Liu, G., Cheng, N., Wei, J., Shangguan, P., & Huang, W. (2014). Emotion recognition based on multi-variant correlation of physiological signals. *IEEE Transactions on Affective Computing, 5*, 132–140.
6. Lin, Y. P., Wang, C. H., Jung, T. P., Wu, T. L., Jeng, S. K., Duann, J. R., et al. (2010). EEG-Based emotion recognition in music listening. *IEEE Transactions on Biomedical Engineering, 57*, 1798–1806. doi:10.1109/TBME.2010.2048568
7. Canento, F., Fred, A., Silva, H., Gamboa, H., & Lourenço, A. (2011). Multimodal biosignal sensor data handling for emotion recognition. In *Sensors* (pp. 647–650). Limerick: IEEE. doi:10.1109/ICSENS.2011.6127029
8. Balteş, F. R., Avram, J., Miclea, M., & Miu, A. C. (2011). Emotions induced by operatic music: Psychophysiological effects of music, plot, and acting: A scientist's tribute to Maria Callas. *Brain and Cognition, 76*, 146–157. doi:10.1016/j.bandc.2011.01.012
9. Russell, J. A., & Barrett, L. F. (1999). Core affect, prototypical emotional episodes, and other things called emotion: Dissecting the elephant. *Journal of Personality and Social Psychology, 76*, 805–819. doi:10.1037/0022-3514.76.5.805
10. Feidakis, M., Daradoumis, T., & Caballé, S. (2012). A multi-fold time approach to address emotions in live and virtualized collaborative learning. In *2012 Sixth International Conference on Complex, Intelligent and Software Intensive Systems (CISIS)* (pp. 881–886). Palermo: IEEE. doi:10.1109/CISIS.2012.197
11. Kleinginna, P. R., & Kleinginna, A. M. (1981). A categorized list of emotion definitions, with suggestions for a consensual definition. *Motivation and Emotion, 5*, 345–379. doi:10.1007/BF00993889
12. Ortony, A., Clore, G.L., & Collins A. (1990). *The Cognitive Structure of Emotions*. Cambridge Univ. Press
13. Ivan Akira. *Plutchik's eight primary emotions and how to use them (Part 1 of 2)*. https://dragonscanbebeaten.wordpress.com/2010/06/04/plutchiks-eight-primary-emotions-and-how-to-use-them-part-1/. Accessed October 1, 2015.
14. Yang, Y., Lin, Y., Su, Y., Chen, H.H. (2008). A regression approach to music emotion recognition. *IEEE trans audio, speech, and language processing* , *16*, 448–457. doi:10.1109/TASL.2007.911513
15. Leung, C.H.C., Deng, J. (2015). Emotion-Based music retrieval. *Encycl. Inf. Sci. Technol*, IGI Global, *3*, 143–53. doi:10.4018/978-1-60566-026-4
16. Nicolaou, M.A., Gunes, H., Pantic, M. (2011). Continuous prediction of spontaneous affect from multiple cues and modalities in valence-arousal space. *IEEE trans affect comput, 2*, 92–105. doi:10.1109/T-AFFC.2011.9
17. Apache Jena (2015). https://jena.apache.org/index.html. Accessed October 1, 2015

Part IV
Architectures and Frameworks for Interoperability

Automated Process Model Generation for Internet of Things Oriented Enterprise Systems

D. Repta, I.S. Sacala, M.A. Moisescu, I. Dumitrache and A.M. Stanescu

Abstract Several paradigms have emerged in the last few years which have the ability to reshape the architecture of the Enterprise Systems. These refer to "Internet of Things and Cyber-Physical Systems" with proposed forms of implementation such as "Sensing Enterprise" and "Intelligent Cyber Enterprise". The main focus of this paper is related to enterprise systems required to support the implementation of next generation of enterprise architectures. The research presented in this paper aims to solve the problem of recovering process models from sensor observations given a reasonable amount of background information. As such, the main objective is the development of a framework for building systems capable of solving this problem. These systems, using data received from an existing sensor network, will present to the end-users a process model, in a common graphical notation, depicting the observed dynamics.

Keywords Enterprise systems · Process modelling · Process model generation

1 Introduction

The main focus of this paper concerns the IT systems required to support the implementation of next generation of enterprise architectures.

In broad lines, the defining features envisioned for the future enterprise architectures are a reaction to the realities of the increasingly connected global economy. In order to stay competitive in the modern markets, the enterprises need to be able to quickly adapt to the changing environment and take advantage of the new business opportunities brought by the ever increasing rate of technological innovation. This has led to the appearance of concepts such as "enterprise agility" and a

D. Repta · I.S. Sacala (✉) · M.A. Moisescu · I. Dumitrache · A.M. Stanescu
Faculty of Automatic Control and Computers, University Politehnica of Bucharest, Splaiul Independentei 313, Bucharest, Romania
e-mail: ioan.sacala@acse.pub.ro

© Springer International Publishing Switzerland 2016

K. Mertins et al. (eds.), *Enterprise Interoperability VII*,
Proceedings of the I-ESA Conferences 8, DOI 10.1007/978-3-319-30957-6_11

push for automation at the upper levels of the enterprise through the application of techniques developed in the fields of artificial intelligence and machine learning.

In the last decades, several paradigms that pertain to the architecture of the future enterprise systems have been developed and investigated. New concepts such as "Internet of Things", "Cyber-Physical Systems", "Sensing Enterprise" and "Intelligent Cyber Enterprise" have been the subject of research projects around the world and most notably in the EU-backed "FInES" research cluster.

This study follows this trend by investigating the possibility of using techniques developed in the field of process mining—which concerns with the extraction of knowledge from the observed behavior of business process—in cases in which the processes use not only virtual resources but also physical ones. Specifically, this investigation will analyze the feasibility of using data/observations collected directly from sensing components to recover a process model that will prove useful for a business analyst.

2 Process Modelling in the Context of Future Enterprise Systems

The research presented in this paper aims to solve the problem of recovering process models from sensor observations given a reasonable amount of background information. As such, the main objective is the development of a framework for building systems capable of solving this problem. These systems, using data received from an existing sensor network, will present to the end-users a process model, in a common graphical notation, depicting the observed dynamics.

In order to tackle this complex problem in an efficient manner, the following sub-problems have been considered:

1. Event identification.
2. Process instance (or case) identification.
3. Activity recognition.
4. Process model discovery.

The first sub-problem refers to the identification of the relevant events (state changes) for a certain process, from the set of collected sensor observations. This step is required because several distinct process types may be executed at the same time in the observed environment. For example, if the users are interested in analyzing a supply chain process, the events referring to various devices from the production line should be ignored.

Because the process model that needs to be recovered from the sensor observations is expected to represent a recurring behavior, its instances (or cases according to the BPM terminology) need to be isolated. The result of this step will be a (multi-)set of event sequences that represent the effects of each instance (execution) of the target process. As a side-note, this data structure represents the "event log", which represents the input of the "standard" process discovery problem, as defined in the field of process mining.

It is however obvious that the sequences of events directly related to sensor observations will have little value for a business analyst as a process model extracted directly from this data set will be very complex and incomprehensible. This issue has been identified in the field of process mining as one of the main causes of the "spaghetti model" problem [1]. However, the problem of achieving the level of abstraction expected by the end user is still being investigated and, as it will be shown in the following sections, may necessitate using results from other research areas as activity recognition.

Finally, after iteratively aggregating events, sensor observations and low-level activities up to the desired level of abstraction, the resulting sequences of activities, each corresponding to a process instance (case), can be processed using existing process discovery algorithms developed in the field of process mining.

From a process mining perspective, this work is concerned with the data and information processing steps required in order to enable the correct and efficient application of process discovery techniques. These steps are especially important in the case of physical processes, whose behavior is observed using deployed sensors. Although similar investigations have already been made in the field of process mining [2–4], this work takes a wider perspective on this problem, by integrating results from other research areas such as semantic technologies and activity recognition.

3 The Development of an Automated Process Model Generation Framework

The development of a framework for process generation is presented in the following section. This development phase has proceeded in an incremental manner starting from a baseline solution. The description of this baseline solution. in the context of the example process further described, will highlight the scope and complexity of the sub-problems addressed in this work.

In order to better explain the various aspects of the process discovery from sensor observations problem, a small example is introduced in this section consisting of a supply chain process.

The process involves a single object—a crate or a pallet—marked with an RFID Tag that passes in front of a set of RFID Readers as it exits the production line and is transported through various warehouses to the final retailer. The transported object has a sensor installed on it that periodically measures its temperature. The measurements will be collected using a wireless connection or be stored on the sensing device and downloaded once the object arrives in one of the storage areas. The temperature observations will be used at some point during the process to make a decision regarding the future destination of the crate—if the temperature is below a certain threshold, the object will be shipped to the retail and if not, the object will be sent to a location in order to be disposed of.

For this first version of the process discovery system, two important design choices have been made:

- the implementation effort should be minimum, using only available software components and process discovery techniques;
- Semantic Web technologies will be used to encode the application's domain—both the working information (sensor observations and events) and the background information. This choice is similar to that used in various IoT Middleware architecture [5, 6] and is motivated by the need to efficiently encode a complex system domain.

In order to achieve this objective, a generic system architecture that incorporates results and techniques from the Semantic Web and process mining research fields is proposed.

As mentioned earlier, the input data for this system will be a set of observations collected from the available sensors. From these observations the system will need to detect relevant events (state changes). This event identification step will be followed by a process case identification step that will partition the set of relevant events into distinct instances. On the identified event sequences, representing process instances, a number of "activity recognition" steps are performed, in which higher level activities are identified. For example, two consecutive events referring to RFID readers in different locations might indicate that a "Transport" activity was performed. Finally, the resulting sequences of activities can be compiled into a process log and used as the input of a process discovery procedure.

An architectural overview of the proposed system is presented in Fig. 1. The system operates at three distinct levels. At the data level the system manages and transforms sensor observations from their domain-/application-dependent format

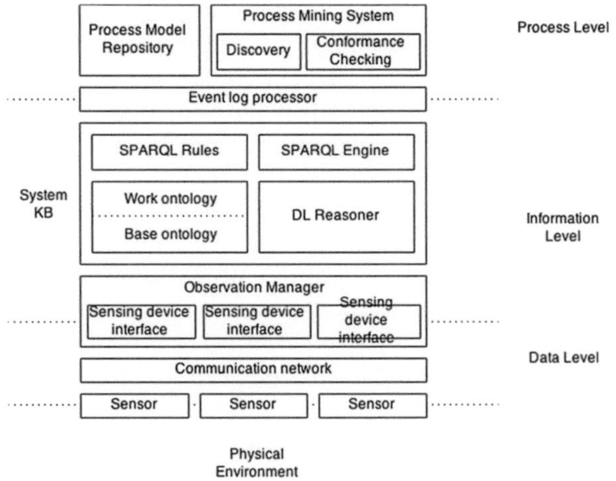

Fig. 1 Architectural overview of automated process model generation enterprise system

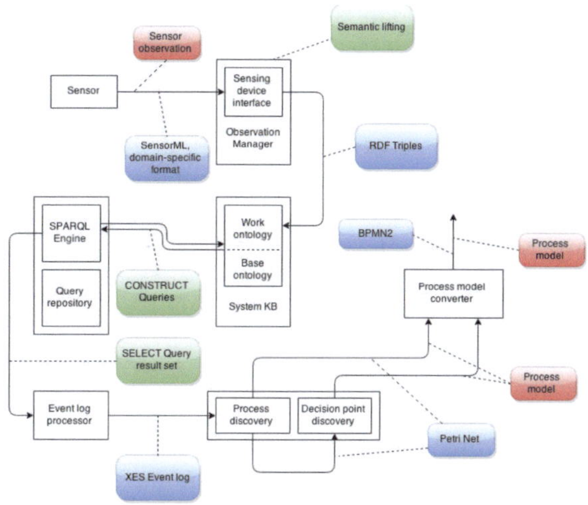

Fig. 2 Enterprise system information flow diagram

into RDF serialized assertions referring to the concepts from the system's base ontology, an extension of the SSN ontology [7]. The information level contains the SPARQL Query Engine and DL Reasoner that will be used in the "Process case identification" and "Activity recognition" steps. Finally, the components from process layer discover the process models from the sequences of activities extracted at the lower level.

As depicted in the information flow diagram (Fig. 2), the main processing steps occurring in the baseline system are:

1. Semantic lifting of the sensor observations.
2. Action detection using CONSTRUCT SPARQL queries.
3. Process case identification using CONSTRUCT SPARQL queries.
4. Process model extraction—involving 3 sub-steps: Petri Net discovery, Decision point mining and Petri Net to BPMN2 conversion.

4 Automated Process Model Generation Framework Implementation

4.1 Semantic Lifting of Sensor Observation

This function is accomplished by the Observation Manager component that acts as a container for Sensing Device Interfaces. It is assumed that the Observation Manager can communicate with the underlying sensor network and receives data packets containing the observations. A further assumption regarding the Observation

Manager refers to its ability to identify the source of each packet of information and dispatch it to the Sensing Device Interface corresponding to its source's type. Finally, the (appropriate) Sensing Device Interface is responsible with the semantic lifting of the incoming raw sensor data into a set of RDF triples that will be added into the system's work ontology. Both the Observation Manager and the contained Sensing Device Interfaces may need to query the ontology in order to retrieve the location of the Sensing Device Interface (URI) or the concepts to which the new triples must refer to.

4.2 Action Detection Using CONSTRUCT SPARQL Queries

Using an approach similar to the ones presented in [8], SPARQL Construct queries are used to create new individuals in the system's work ontology corresponding to activities and process instances.

To showcase this method of identifying activities from registered events, the simple supply chain scenario introduced in the previous section is used.

In this case, the system needs to detect two types of activities:

- the end of production and packing of the shipped item,
- transportation activities.

The activity recognition is performed through a set of SPARQL CONSTRUCT queries, one for each activity type. The queries select a sequence of events from the system's work ontology and register a new "Action" individual in the ontology for each detected sequence.

4.3 Process Case Identification Using CONSTRUCT SPARQL Queries

In the baseline system, the process case identification is performed in a similar manner as the action detection task. This task uses the previously identified Action individuals in the ontology to create Process individuals and associate the actions to them. In order to achieve this, each query is based on the unique traits of each process instance. For example, in the supply chain example, a process instance can be uniquely identified using the RFID code referred by each of its actions. Given the nature of the investigated scenario, each process instance refers to only one object. Furthermore, all Action instances from the work ontology are known to pertain to only one of the instances of the investigated process, so it's just a matter of partitioning the set of action by the object that they refer to. In the general case however, identifying distinct processes instances is a much harder problem, as they can rarely be identified by one actor or involved object and the available action definitions might correspond to instances of different types of processes.

The investigation of the general case will be an important issue in future research. A promising approach involves the discovery and usage of event correlation rules to partition the set of collected events into process execution instances as described in [9–11]. It is expected that these methods will benefit from the way in which information is encoded in the system, using Semantic Web technologies.

4.4 Process Model Extraction

The final task involves the extraction of the process model for the sequences of activities referred by each of the identified process instances.

In this step of process model extraction, the baseline system will perform the tasks of process discovery and process extension.

The process discovery task uses as input an event log which is a multi-set (bag) of event sequences expressing the observed behavior of an investigated process. Each event sequence corresponds to an instance (or case) of the process.

The name of the input data structure used by the process discovery task—"event log"—is a bit of a misnomer given the scope of the current analysis. In this paper, the term "event" is used to refer to a state change in the system under observation, such as the change in the internal temperature of a crate, which can be inferred from the collected sensor observations. On the other hand, in the field of process mining, the elements of "event logs" usually refer to high level activities like the start or completion of a transport activity from the production facility to warehouse A.

This problem of using the appropriate level of abstraction for the events/actions depicted in the final process model is well known in the process mining community and has been investigated in a number of papers [12–15] using clustering and substring detection techniques. It should be noted that in the current investigation, this issue is addressed by the activity recognition sub-problem.

Returning to the baseline system, the process model extraction step involves the following sub-tasks: process model discovery, decision point analysis and BPMN2 conversion

4.4.1 Process Model Discovery

Over the last decade, the research into the field of process mining has led to the development of several important process discovery techniques/algorithms. A review of the existing techniques is beyond the scope of this work but is available in a number of papers [16].

For the baseline system, the two-step approach described in [17, 18] is used. An implementation of this method is available in the ProM Suite [19].

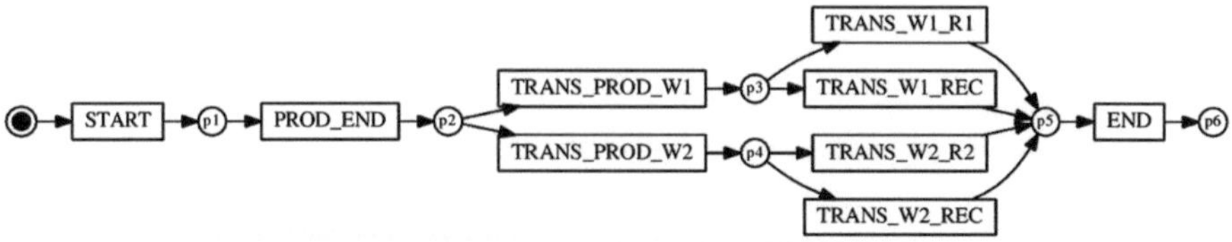

Fig. 3 Petri Net model of the example process

The method is based on the theory of regions that provides a method of building compact representations as Petri Nets of transition systems. The resulting compact model will exhibit the same behavior as the input transition system.

This process discovery method first builds a transition system using the information from the event log and then converts it into a compact Petri Net representation.

In the second phase of the process discovery approach, a Petri Net that exhibits the same behavior as the transition system is built. This phase is based on the results of the research into the synthesis problem—the problem of synthesizing a Petri Net whose behavior coincides with a specified behavior [20].

In order to use the ProM implementation of this process discovery method, the Process OWL individuals created in the previous step must be extracted from the work ontology using a SPARQL query.

A simple component from the baseline system is used to convert the result set into an event log in the XES format [21] that can be imported in the ProM suite.

After applying the process discovery technique, a preliminary Petri Net process model is discovered—Fig. 3.

It should be noted that the process discovery method is not entirely automatic, as the user must tune the settings that control the transition system generation step.

4.4.2 Decision Point Analysis

This subtask involves the discovery of the logical expressions that control the process flow in conflict places. For the baseline system, the approach described in [22] and later revised in [23] was chosen. This technique is already implemented in the ProM suite.

The event log generated in the "Process model discovery" sub-step contained, for each activity related event, an attribute corresponding to the last temperature observation. In this step, the value of this attribute is used to discover the formula based on which a transport decision is taken. The decision-point mining technique instantiates a decision tree based classification problem for each conflict place. The result of applying this technique is a "Petri Net with data", in which some of the transitions have associated "guard conditions" that determine if the transition can be fired.

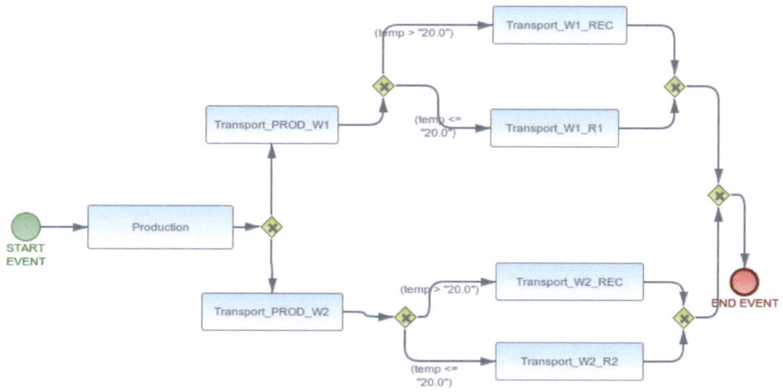

Fig. 4 BPMN2 model of the example process

4.4.3 BPMN2 Conversion

The last sub-step involves the conversion of the "Petri Net with data" process model into a BPMN2 process diagram. The BPMN2 notation has been chosen due to its popularity in the BPM community.

The conversion of the Petri net into a BPMN2 diagram is made using the ProM Suite plugin described in [24]. The resulting BPMN2 diagram for the supply chain scenario is depicted in Fig. 4, where the guard conditions are displayed as labels on the arcs connecting the activities.

4.4.4 Base Solution Analysis

Although the implementation of the baseline solutions reiterates the utility of using linked data representation (for sensor observation) for developing domain-independent systems, it however has several shortcomings, the most important being the fact that the system doesn't use the available domain/expert knowledge in the event/process/activity recognition steps, given that these steps rely on hand-written queries.

5 Conclusions and Further Work

As mentioned in the previous section, the baseline system has some shortcomings, mainly the fact that the problems of action recognition and process identification require that the operators explicitly encode the process case and the activity model in the SPARQL queries issued in the system. The system should be able to at least suggest the correlation rules for process identification and the sets of events referring to the same activity, if a formal activity model is unavailable or the system cannot infer one from the background knowledge.

References

1. van der Aalst, W. M., & Günth, C. W. (2007, July). Finding structure in unstructured processes: The case for process mining. In *Seventh International Conference on Application of Concurrency to System Design, 2007. ACSD 2007* (pp. 3–12). IEEE.
2. Rozinat, A., de Jong, I. S. M., Günther, C. W., & van der Aalst, W. M. P. (2009). Process mining applied to the test process of wafer steppers in ASML. IEEE Transactions on Systems, Man and Cybernetics.
3. Rozinat, A., Zickler, S., Veloso, M., van der Aalst, W. M., & McMillen, C. (2009). Analyzing multi-agent activity logs using process mining techniques. In *Distributed autonomous robotic systems 8* (pp. 251–260). Berlin, Heidelberg: Springer.
4. Jaroucheh, Z., Liu, X., & Smith, S. (2011). Recognize contextual situation in pervasive environments using process mining techniques. *Journal of Ambient Intelligence and Humanized Computing, 2*(1), 53–69.
5. Pfisterer, D., Römer, K., Bimschas, D., Kleine, O., Mietz, R., Truong, C. … Richardson, R. (2011). SPITFIRE: Toward a semantic web of things. *IEEE Communications Magazine, 49* (11), 40–48.
6. Barnaghi, P., Wang, W., Henson, C., & Taylor, K. (2012). Semantics for the internet of things: Early progress and back to the future. *International Journal on Semantic Web and Information Systems (IJSWIS), 8*(1), 1–21.
7. Compton, M., Barnaghi, P., Bermudez, L., GarcíA-Castro, R., Corcho, O., Cox, S. … Taylor, K. (2012). The SSN ontology of the W3C semantic sensor network incubator group. *Web Semantics: Science, Services and Agents on the World Wide Web, 17*, 25–32.
8. Meditskos, G., Dasiopoulou, S., Efstathiou, V., & Kompatsiaris, I. (2013, March). Sp-act: A hybrid framework for complex activity recognition combining owl and sparql rules. In *2013 IEEE International Conference on Pervasive Computing and Communications Workshops (PERCOM Workshops)* (pp. 25–30). IEEE.
9. Rozsnyai, S., Slominski, A., & Lakshmanan, G. T. (2011, July). Discovering event correlation rules for semi-structured business processes. In *Proceedings of the 5th ACM International Conference on Distributed Event-based System* (pp. 75–86). ACM.
10. Motahari-Nezhad, H. R., Saint-Paul, R., Casati, F., & Benatallah, B. (2011). Event correlation for process discovery from web service interaction logs. *The VLDB Journal—The International Journal on Very Large DataBases, 20*(3), 417–444. Springer-Verlag New York, Inc. Secaucus, NJ, USA. doi:10.1007/s00778-010-0203-9.
11. Pérez-Castillo, R., Weber, B., de Guzmán, I. G. R., & Piattini, M. (2012). Improving event correlation for non-process aware information systems. In *ENASE* (pp. 33–42).
12. van Dongen, B. F., & Adriansyah, A. (2010, January). Process mining: Fuzzy clustering and performance visualization. In *Business Process Management Workshops* (pp. 158–169). Berlin, Heidelberg: Springer.
13. Günther, C. W., Rozinat, A., & Van Der Aalst, W. M. (2010, January). Activity mining by global trace segmentation. In *Business Process Management Workshops* (pp. 128–139). Berlin, Heidelberg: Springer.
14. Bose, R. J. C., Verbeek, E. H., & van der Aalst, W. M. (2012). Discovering hierarchical process models using prom. In *IS Olympics: Information systems in a diverse world* (pp. 33–48). Berlin, Heidelberg: Springer.
15. Li, J., Bose, R. J. C., & van der Aalst, W. M. (2011, January). Mining context-dependent and interactive business process maps using execution patterns. In *Business Process Management Workshops* (pp. 109–121). Berlin, Heidelberg: Springer.
16. Tiwari, A., Turner, C. J., & Majeed, B. (2008). A review of business process mining: State-of-the-art and future trends. *Business Process Management Journal, 14*(1), 5–22.
17. Van der Aalst, W. M., Rubin, V., Verbeek, H. M. W., van Dongen, B. F., Kindler, E., & Günther, C. W. (2010). Process mining: A two-step approach to balance between underfitting and overfitting. *Software & Systems Modeling, 9*(1), 87–111.

18. van der Aalst, W. M., Rubin, V., van Dongen, B. F., Kindler, E., & Günther, C. W. (2006). *Process mining: A two-step approach using transition systems and regions* (BPM Center Report BPM-06-30, BPMcenter. org, 6).
19. Van der Aalst, W. M., van Dongen, B. F., Günther, C. W., Rozinat, A., Verbeek, E., & Weijters, T. (2009, January). ProM: The process mining toolkit. In *Proceedings of the Business Process Management Demonstration Track (BPMDemos 2009).* Ulm, Germany, September 8, 2009.
20. Cortadella, J., Kishinevsky, M., Lavagno, L., & Yakovlev, A. (1998). Deriving Petri nets from finite transition systems. *IEEE Transactions on Computers, 47*(8), 859–882.
21. Verbeek, H. M. W., Buijs, J. C., Van Dongen, B. F., & Van Der Aalst, W. M. (2011). Xes, xesame, and prom 6. In *Information systems evolution* (pp. 60–75). Berlin, Heidelberg: Springer.
22. Rozinat, A., & van der Aalst, W. M. (2006). *Decision mining in ProM* (pp. 420–425). Berlin, Heidelberg: Springer.
23. de Leoni, M., & van der Aalst, W. M. (2013, March). Data-aware process mining: discovering decisions in processes using alignments. In *Proceedings of the 28th Annual ACM Symposium on Applied Computing* (pp. 1454–1461). ACM.
24. Kalenkova, A. A., de Leoni, M., & van der Aalst, W. M. (2014). Discovering, Analyzing and Enhancing BPMN Models Using ProM⋆. In *Business Process Management-12th International Conference, BPM* (pp. 7–11).

Use of Big Data for Continuous Interoperability in Crisis Management

Audrey Fertier, Sébastien Truptil, Aurélie Montarnal, Anne-Marie Barthe-Delanoë and Frédérick Bénaben

Abstract Crisis management often involves several stakeholders, who need to be synchronized, despite their heterogeneity or their lack of time, without having any predefined solution for supporting their coordination or data sharing. To help them, crisis cells are set up at the moment the news of crisis is released and actually have only the time to centralize decisions. The purpose of the French ANR (French National Agency for Research) GéNéPi project is to permit interoperability, through supporting coordination and data sharing: inside the decisional level, between hierarchical echelons, and inside the operational level, between the stakeholders. A MIS (Mediation Information System) is provided by this project and encompasses: (i) the information analysis, on the basis of data collected during the preparation phase, (ii) the creation of one collaborative process and (iii) the agility of the solution by comparing two situation models on run-time. Such a system needs to deal with Big Data issues in order to be able to evaluate these models. The paper presents the context of the research and the improvements brought by the MIS. It also expresses the possibility of and the need for automate the data retrieval, thanks to Big Data methods, so that the MIS could evaluate the field situation in real time.

Keywords Interoperability · Crisis management · Collaboration · Big data

A. Fertier (✉) · S. Truptil · A. Montarnal · F. Bénaben
Centre Génie Industriel, Université de Toulouse—Mines Albi, Campus Jarlard, Albi 81000, France
e-mail: audrey.fertier@mines-albi.fr

A.-M. Barthe-Delanoë
Université de Toulouse, INPT-ENSIACET, 4 Allée Emile Monso, Toulouse Cedex 4 31432, France

A.-M. Barthe-Delanoë
LGC (Laboratoire de Génie Chimique), CNRS, Toulouse Cedex 04 31432, France

© Springer International Publishing Switzerland 2016
K. Mertins et al. (eds.), *Enterprise Interoperability VII*,
Proceedings of the I-ESA Conferences 8, DOI 10.1007/978-3-319-30957-6_12

1 Introduction

Every country can be affected by a crisis situation that impacts civilians, national economy or health conditions. In France, to manage crisis, government, military authorities and enterprises are counting on both the experience of their actors and their emergency plans. One of the issues identified by these organizations is the necessity to establish and maintain both communication and coordination among the different stakeholders involved in the crisis response. Difficulties arise because of vocabulary issues or interferences between intervention processes. In addition, the increase of sensors, along with the emergence of new data sources, proportionately drowns the stakeholders under field information, while these are necessary for the decision-makers to give instructions, maintain the coordination of their teams and anticipate the future.

To address this paradox, the French-ANR GéNéPi project, started in October 2014, aims to provide a Mediation Information System in charge of the continuous coordination: (i) among the stakeholders on the field operations and (ii) among the decision-makers between the different crisis cells.

The remainder of this paper is organised as follows: the second section gives definitions of interoperability and crisis management to introduce the major issues faced by the stakeholders to collaborate. These are illustrated by a crisis case study. Then, an overview of the solution and its architecture is proposed in section three.

2 Towards a Better Interoperability in Crisis Management

2.1 Two Elements of Language Specific to Crisis Management

In a general way, a crisis [1] is a 'set of circumstances in which individuals, institutions or societies face threats beyond the norms of routine day-to-day functioning but the significance and impact of these circumstances will vary according to individual perceptions'. The second part of this definition takes into account the subjective aspect of a crisis. For instance, the road services, the fire brigade and the health services will not respond in the same way to the consequences of a gas explosion: their experience, their emergency plans, their rules and their doctrines will influence their choices.

Crisis management is generally divided in four phases [2]:

The **prevention** phase aims to decrease the occurring probability of a trigger event and aims to improve the system in order to avoid the crisis situation;

The **preparation** phase aims to develop the processes planned for the response phase and to adapt the organizations, if need be, in order to be prepared when the trigger event will occur;
The **response** phase starts with the crisis. Quickly after the first diagnosis, a response process is implemented by using the tasks described in the preparation phase and by adapting them to the characteristics of the on-going crisis;
The **recovery** phase aims to return to a suitable situation.

The main objective of the GéNéPi project is to support the coordination of both the stakeholders and the decision-makers when a crisis occurs. Therefore, the following sections of this paper will focus on the response phase, although the results of the preparation and prevention phases sustain the proposed solution.

2.2 Characteristics and Example of a French Crisis Response Management

In France, several kinds of actors are involved in the crisis response and can be observed at different hierarchical levels. Figure 1 illustrates this configuration.

The local level includes the stakeholders working on the field, like the fire brigade or the road services. Despite heterogeneous activities and specific vocabulary, their decision-makers have to collaborate in order to deal efficiently with the crisis situation. Besides, interconnections of backbones network are only serving to highlight this necessity to coordinate.

A **crisis cell** is set up at the county level. It represents a set of decision-makers, directly connected to their teams on the field so that they can gather all the relevant information they need to respond instructions. In order to be more precise, the decision-makers would ask for more data, more information, if their capacity to deal with a given amount of data were to change. Therefore, an intermediate is necessary to transform the massive amount of available data into viewable information for the decision-makers.

Another crisis cell is set up at the zonal level (respectively the national level) when the person in charge is overwhelmed or when the crisis' consequences impact more than one county (respectively one defense zone). The crisis cells of lower levels remain even if higher-level ones are created. Thus, one can focus on the response progress in real time, while the two others focus on the anticipation of how to handle the future consequences. The decision-makers of these higher levels want also to have access to more data sources in order to make their anticipation truer.

In this context, two types of links, with their own issues, have been defined [3]:

The **horizontal distribution** represents the necessity to communicate inside the operational level and between the operational level and the decisional level. The objective is to avoid interferences among the processes of the different stakeholders;

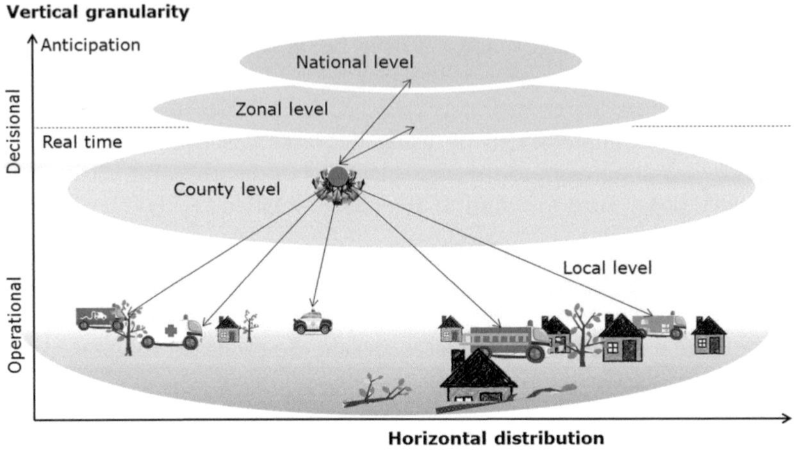

Fig. 1 Organization of crisis response in France

Fig. 2 Representation of the middle Loire and its five vals

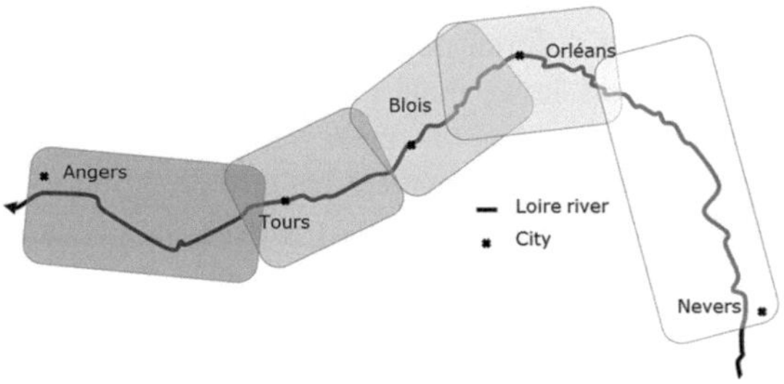

The **vertical granularity** represents the necessity to communicate among the different crisis cells in order to push forward information more and more concise or instructions more and more precise.

Such an organization can be illustrated using a typical French natural disaster: a Loire flood. The Loire is the longest river in France. It flows from South to North and West to East, thus cutting half the length of France. The use case concerned a 1/50 per year flood, with 2 % chance of happening each year. It focuses on the middle Loire area between Nevers (35,000 inhabitants) and Angers (149,000 inhabitants). This region is divided into five 'vals' (a French term designating an area protected from small floods by dikes), as shown in Fig. 2.

Six counties and 2 defense zones are impacted by the Loire flooding. This represents [4] 9 crisis cells and 23 different kinds of stakeholders that have to communicate (vertical granularity issues) and coordinate (horizontal distribution issues).

According to official simulations, a 1/50 per year flood lasts for seven days and four days after the beginning of the flood, the entire middle Loire area is on alert. In addition, four of the five vals have to be partially evacuated at the same time. On this day, the stakeholders have to cover almost 400 km: to succeed at managing such an area, the computerization of both the data retrieval and the collaborative process elaboration is needed. The decision-makers ask for new support to help them on the two dimensions previously defined: the horizontal distribution and the vertical granularity. One of the solutions is to offer them an overall view, freely specifiable, as a visualization of all the relevant information deduced from data retrieval combined with Big Data techniques.

To sum up, the problem lies in the stakeholders' ability to coordinate and exchange information without strain. To do so, they need: (i) the automatic transfer of the right information (vertical granularity), (ii) the automatic deduction of the right instruction (horizontal distribution) and (iii) the transfer of the information and the instructions at the right time to the right person, using the right vocabulary.

2.3 *The Challenges of a Continuous Communication and Coordination in Crisis Response Management*

The critical aspect of a crisis situation increases the intensity of the collaborations. Respectively, the unexpected aspect of a crisis situation increases the speed of its implementation [5]. Ensure communications (vertical granularity) and coordination (horizontal distribution), through a better collaboration among the stakeholders, is therefore needed and not innate.

2.3.1 The Two Interoperability Approaches Relevant to a Crisis Response

The interoperability is considered as [6] "the ability of a system or a product to work with other systems or products without any special effort from the user". In this context, the improvement of the stakeholders' ISs (Information Systems) interoperability will decrease the efforts needed, and reduce the time required, to maintain the two types of collaborations (horizontal distribution and vertical granularity). Three different approaches can be developed to achieve this interoperability [7]:

The **integrated approach**, which compels all the stakeholders to adapt their process to a given data format.
The **unified approach**, which proposes a common shared model to link the stakeholders information formats between them;
The **federated approach**, which does not compel the stakeholders: the interoperability is automatically done when it is needed.

The diversity of the exchanged data types and the independence managerial of each stakeholder remove the first solution. Conversely, the unified and federated approaches are both usable by the solution to support vertical granularity issues through a better interoperability.

2.3.2 The Role of Big Data in Supporting Coordination During a Crisis Response

The time required to gather the data and validate the information, in order to define a collaborative process, is sacrificed, by the stakeholders, for the response phase needs. This appears even though the overall digitalization of space is the opportunity to provide more than the necessary data. Instead, it drowns the decision makers in [8]:

An always larger data stream (**the volume**),
From multiple heterogeneous sources (**the variety**), like sensors, smartphones or social networks,
In a crisis context where the problems often appear in an unexpected and disordered way, with immediate consequences on the data flow (**the velocity**),
When the ability to create information and find new values from a large dataset (**the value**) will determine the validity of the information.

The stakeholders' ability to handle these aspects of Big Data seems to be the key to achieve coordination on the field. Besides, the efforts needed to obtain the overall picture of the situation would be eliminated by the automation of the data retrieval and the data gathering (cf. data, information and knowledge levels descriptions in Sect. 2.3). In a similar way, this aptitude to quickly manage several Big Data features, as the 4vs described earlier, will enable the continuous update of both the situation model and the collaborative process in real-time. The use of Big Data techniques will then ensure a better management of horizontal distribution issues, through a better awareness of the on-going developments.

Finally, the problematic that has to be tackled, to support both interoperability and Big Data management during the response phase, can be summarized with these three questions: (i) **How to collect aggregate and filter data to deduce a model and a visualization of the on-going crisis situation?** (ii) **How to propose a collaborative process to the decision-makers, in order to help them coordinate their teams on the field?** (iii) **How to ensure a permanent and agile monitoring of the validated collaborative process during the response phase?**

3 Supporting Interoperability During the Crisis Response

The ANR GéNéPi project proposes a MIS (Mediation Information System), based on an unified approach. It is in charge of supporting the coordination and data sharing, among the stakeholders trough their ISs. The conception of this solution is part of the MISE approach (Mediation Information System Engineering) [5], which is based on model transformations principles. First, it transforms the model of the intended collaboration into a model of the adequate collaboration process. Then the latter are transformed into workflows executable by an orchestrator. The main functions of this MIS are illustrated in Fig. 3.

As described above, the success of a crisis management lies in the crisis cell ability to manage the issues found in both horizontal distribution and vertical granularity issues. The communication flows necessitated by these two links can be divided into three categories corresponding to the three rows visible on Fig. 3 [9]: First, **the data level** corresponds to a set of results derived from observations of objects, events and characteristics of the environment. Because it is naturally raw, these data come in various shapes and sizes, from heterogeneous and multiple sources; Subsequently, **the information level** is generated thanks to relational connections used to give meaning to the data. The situation can then be interpreted and dealt with; finally, **the knowledge level** connects the information items of the second level between them, thanks to the description of additional connections. The knowledge can then be exploited to deduce decisions and actions to be carried out.

For instance, from the data: 'a fire alarm sounded in the south district', 'people have seen smoke coming from behind the town hall' and 'the place behind the town hall contains a waste container, a well and some dried plants', the related information could be 'a fire broke out in proximity of a waste container' and 'water is

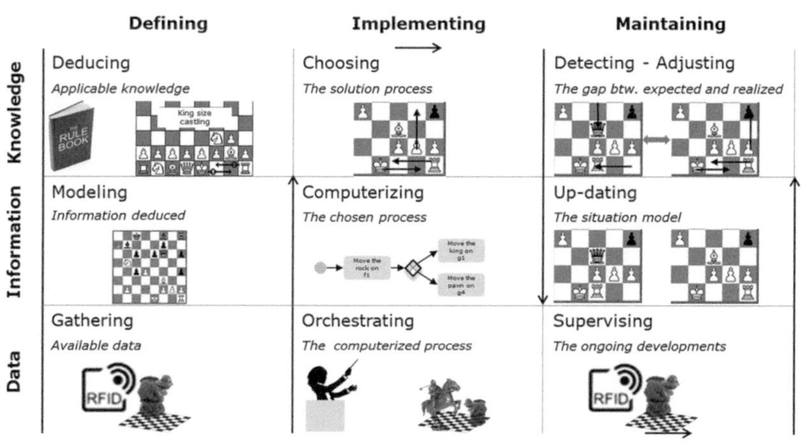

Fig. 3 Solution proposed to answer for the two issues raised about big data and interoperability, illustrated by the chess game example

available in the well nearby'. Then, the related knowledge might be 'water can extinguish fire', or, more professionally, 'do not use water to extinguish a waste container on fire'. Moreover, this also shows the importance of being able to evaluate the data sources.

The columns of Fig. 3 describe the different steps followed by the MIS: (i) the creation of a situation model thanks to the stakeholders' knowledge, and the deduction of at last one collaborative process; (ii) the transformation of the validated process into workflows run by an orchestrator; (iii) the adaptation of the running process to the on-going developments on the field. These three steps are described in 2.3.1 and the needed architecture to support this solution, along with new Big Data expectations, will be described in 2.3.2.

3.1 The Three Steps of the Proposed Solution

3.1.1 First Step: Defining at Least One Collaborative Process on the Basis of the Most Reliable Situation Model Possible

Defining the coordination necessitates a rise in the abstraction level. This means that: (i) information, in the shape of models, will have to be deduced from a data set and (ii) decisions will have to arise from exploiting the models with the knowledge previously gathered.

First, all the relevant data, describing either the crisis environmental context or the future collaborations, are gathered during the preparation phase. *At chess for instance, the information about the pieces location could be recovered.* Big data management will permit the MIS to expand its field of investigations, in terms of data sources, and therefore make it more reliable.

Then, a situation model is produced thanks to the gathered and subsequently evaluated data. *A picture of the board could be deduced from gathered data and submitted to the player. From that first model, decisions can be made far more easily than before. It is like playing chess online!* It is also the same as creating a visualization of the information deduced in the shape of a situation model. For the record, the decision-makers are fond of maps and GISs (Geographical ISs) interfaces.

Finally, thanks to the knowledge collected from the stakeholders, one collaborative process is deduced. The more relevant data are made available inside the MIS's data sets through Big Data techniques, the more business rules are made accessible by the stakeholders, the more the MIS will be able to deduce several collaborative processes, along with the best possible solutions in terms of factors like resource or speed.

3.1.2 Second Step: Implementing the Chosen Collaborative Process

This second part necessitates a drop in the abstraction level. First, the chosen collaborative process is validated by the decision-makers. The visualization of several results of collaborative processes, applied to future circumstances, will help the anticipation of the best solution. Then, the tasks of the chosen and validated collaborative process are computerized. This stage, validated by the previous PhD works of Nicolas Boissel-Dallier [10], relies on the BPMN language. Once the computerization is done, it is then possible to orchestrate the process through automatically sending the tasks to the different stakeholders. *The orchestrator is like a brain: given a particular objective, like moving a piece, he will automatically ask its arm, its hand and its fingers to move.*

3.1.3 Third Step: Maintaining the Validity of the Implemented Process Through Run-Time Adaptations

The third part necessitates a new rise in abstraction level. Imagine the player have to move one piece, without knowing what has been moved the round before or what is the new state of the board. It would be totally hazardous and entirely the same if the collaborative process was running end to end. Hence, it is necessary to learn whether the tasks' results are consistent with the tasks' expectations or whether some unanticipated events appeared on the field.

To this end, the MIS carries on collecting data from the field during the orchestration of the collaborative process. Thanks to the same process than the one described in Sect. 3.1.1, the **situation model** is updated. As before, Big data management will ensure the MIS to be synchronized with the field on-going developments, on run-time. Besides, an automation of the data retrieval during the response phase will relieve the stakeholders from sharing information.

Now, it is essential to decide whether the decision-makers must keep on performing the tasks previously defined. For this purpose, another model is designed, along with the situation model, as soon as a collaborative process is implemented. It is the **expected model**. As explained in [11], it is possible to estimate the gap between the expected model and the situation model. When the gap reaches a certain threshold, an alarm occurs inside the MIS. Then, the defining and implementing processes are run and the cycle can start again.

3.2 The Architecture Supporting the Proposed Solution

The new MIS proposed by the GéNéPi project, linked to new Big Data expectations, has two architectural requirements to be operational: (i) the architecture must be able to run an input process by directly reaching the stakeholders, whatever their protocol of communication is; (ii) the architecture must be able, at least, to manage the volume, the velocity and the variety aspects of Big Data.

SOA, Service Oriented Architecture, has been chosen to answer for the first requirement. It enables the MIS to run the input collaborative process. Moreover, an ESB, Enterprise Service Bus, will provide a high level of loose coupling among the stakeholders' ISs and it will adapt its protocol to its interlocutor's one. EDA, Event Driven Architecture, is needed in order to meet the second requirement. EDA enables the MIS to gather, filter and analyse the amount of collected data, in real time through using a CEP, Complex Event Processing, engine.

Finally, the solution makes a trade-off by using both ESB and CEP technologies to deal with both Big Data and coordination issues (cf. Fig 4.): While the orchestrator runs the chosen collaborative process, the CEP1 collects data from the field. It manages the **volume** and **velocity** of data by aggregating and filtering it, in order to produce alerts thanks to rules. The aggregation is done by following semantic and syntactic rules written in EQL language (link *a*) and the straining is done by fixing an acceptable error for each type of data (link *b*). The CEP2 uses an ontology to manage the **variety** of the collected data [12]. Once done, it transfers the information **values** to the Maintaining service of the ESB. Finally the orchestrator can adapt the running process according to the information coming from the CEP2. The chosen ESB is PETALS, developed by a French Open Source software editor [13], and both of these CEP engines are Esper, designed by EsperTech [14].

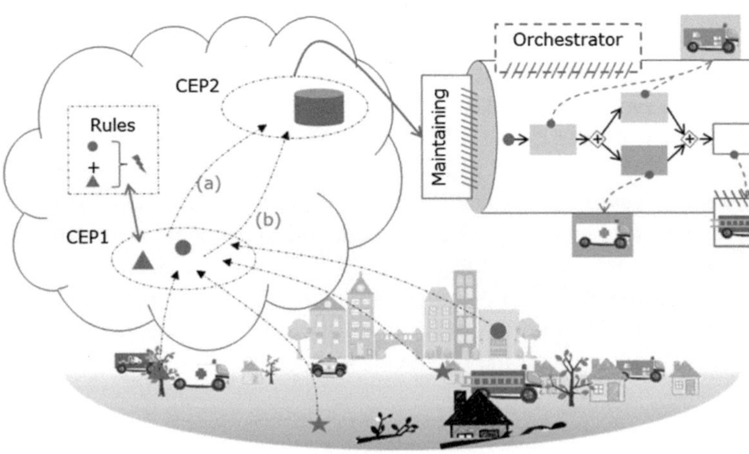

Fig. 4 Pairing EDA and SOA architectures

4 Conclusion and Future Work

A French crisis situation, illustrated here by a Loire flood case, involves several stakeholders bound by their objective to restore the situation as soon as possible. In this context, the stakeholders have to share information and coordinate in a hurry with the help of their decision-makers, while these ones respond instructions and transmit information. They need interoperability solutions to counter the lack of time and facilitate collaborations. Making their interoperability easier is the key proposed in this paper: (i) the coordination of tasks, represented by collaborative processes, is automatically suggested thanks to models; (ii) enable the system to monitor and adapt the running collaborative process to the current situation.

All the while, the two different models would be far more relevant if data were retrieve thanks to an EDA-SOA architecture dealing with the variety, the velocity and the volume of data.

To face the increase of heterogeneous sources and the request of an overall view, freely specified, by the decision-makers, the solution will have to improve this new approach of Big Data in crisis management. The main goal is now to avoid the stakeholders to be drowned by high amounts of data and information without cutting them from their vision of the crisis situation.

References

1. Drennan, L. T., Mcconnel, A., & Stark, A. (2014). *Risk and crisis management in the public sector*. Abingdon: Routledge.
2. Atlay, N., & Green, W. G. (2006). OR/MS research in disaster operations management. *European Journal of Operational Research, 175*(1), 475–493.
3. Bénaben, F., Truptil, S., Lauras, M., & Salatgé, N. (2015). Management of collaborative behavior through a service-oriented mediation system: The case of crisis management. In *IEEE International Conference* (pp. 554–561).
4. Centre Européeen de Prévention du Risque Innondation. (2014). *L'évacution massive des populations—Les territoires face à l'inondation.* http://cepri.net/Evacuation_massive_des_populations.html
5. Bénaben, F. (2012). *Conception de Système d'Information de Médiation pour la prise en charge de l'Interopérabilité dans les Collaborations d'Organisations*. Habilitation à diriger les recherches. Institut National Polytechnique de Toulouse.
6. Konstantas, D., Bourrière, J. P., Léonard, M., et al. (eds.). (2006). *Interoperability of enterprise software and applications*. Springer Science and Business Media.
7. Chen, D., Doumeingts, G., & Vernadat, F. (2008). Architectures for enterprise integration and interoperability: Past, present and future. *Computers in Industry, 59*(7), 647–659.
8. Gantz, J., & Reinsel, D. (2011). Extracting value from chaos. *IDC Iview, 1142*, 9–10.
9. Ackoff, R. L. (1989). From data to wisdom. *Journal of Applied Systems Analysis, 16*, 3–9.
10. Boissel-Dallier, N., Fréderick, B., Jean-Pierre, L., & Hervé P. (2015). Mediation information system engineering based on hybrid service composition mechanism. *Journal of Systems and Software, 108*, 39–59 (2015). doi:10.1016/j.jss.2015.05.064

11. Barthe-Delanoë, A. M., Truptil, S., Bénaben, F., & Pingaud, H. (2014). Event-driven agility of interoperability during the run-time of collaborative processes. *Decision Support Systems, 59*, 171–179.
12. Wang, T., Truptil, S., & Benaben, F. (2015). Applying a semantic and syntactic comparisons based automatic model transformation methodology to serve information sharing. *In Proceedings of the International Conference on Information and Knowledge Engineering (IKE)*, p. 3.
13. PETALS. (2015). Available from http://www.petalslink.com/produits/petals-esb
14. EsperTech. (2015). Available from http://www.espertech.com/products/esper.php

Weaving Trending, Costing and Recommendations Using Big Data Analytic: An Enterprise Capability Evaluator

Muhammad Naeem, Néjib Moalla, Yacine Ouzrout and Abdelaziz Bouras

Abstract In the life cycle of a business organization, unprecedented turmoil and turbulent timings are inevitable. This ever green truth is not only an advice but a motivation for every organization to be dynamic in its business plans. In other words, "one size fits all" strategy is short lived and seasonal for any enterprise. We in this study have proposed a dynamic platform capable of handling the challenges which are geared towards conventional stagnant business strategies in commercial world. The technical specification of the proposed platform is based on modern business intelligence tools and techniques. The study is dedicated towards exploration of investigating the Big Data capabilities to develop a platform delivering costing, sourcing and tendering capabilities for industrial parts. While doing so, we have emphasised challenging questions of interoperability in the design paradigm of inundation of structured and semi-structured data.

Keywords Big data · Enterprise collaboration · Opportunity analysis · Capability evaluation · Data asset

1 Introduction

The manufacturing paradigm is rapidly changing from whole product to component integrated production [1, 2]. The big organizations usually possess the capability of in-house manufacturing for all product components. However, Small and Medium Enterprises (SMEs) are left with the sole option of heavily relying on Request for Information (RFI), Request for Proposal (RFP), costing, tendering etc. It is

M. Naeem (✉) · N. Moalla · Y. Ouzrout
Decision and Information for Production Systems (DISP),
University Lumière Lyon 2, Lyon, France
e-mail: muhammad.naeem@univ-lyon2.fr

A. Bouras
CSE Department, Qatar University Doha, Doha, Qatar

© Springer International Publishing Switzerland 2016
K. Mertins et al. (eds.), *Enterprise Interoperability VII*,
Proceedings of the I-ESA Conferences 8, DOI 10.1007/978-3-319-30957-6_13

conclusively evident that there is an in born situation of vying environment among the corporates for the sake of enhancing their respective business values. This situation has directed the small and medium enterprises (SMEs) to be more striving and dynamic in pursuit of adapting novel business opportunities. The SMEs are facing complex situation where it becomes difficult in drawing optimized decisions for the costing, trending, reverse costing and marketing recommendations etc.

In the era of electronic age, the unstructured data is increasing more as compared to the linked data [3]. The intensity of this phenomenon in SMEs is significantly higher due to the mass production triggered by the fast means of communication resources [4, 5]. This large amount of data contains explicit and implicit patters to help out the SMEs for making essential decisions related to their business strategies. Machine learning techniques have enabled the researchers to produce real time marketing, outsourcing and marketing trends. These solutions encompass across provision of solutions to dig out useful patterns while easing out the prediction on specific scenario. However, it is an established fact that the machine learning techniques always perform better with provision of large data sets [6]. The issue of inefficiency of machine learning techniques is three dimensional. These dimensions are known as volume, velocity and variety (shortly 3v).

Big Data and its analytic have emerged as an important domain of interest for both researchers as well as industrial practitioners. It reflects the magnitude and impact of data related issues to be resolved in contemporary business enterprises. Big Data deals with three non-trivial challenges (mentioned earlier) where machine leaning techniques usually exhaust. The second challenge (variety) of dealing heterogeneous data is abundant in the business organizations. The objective of this study is to exploit the diversified features of data to address the answers related to enhance the business capability of an enterprise.

One interesting challenge in dealing heterogeneous and large amount of data is scarcity of a specific class label in application of business intelligence. There are situations when the class in interest is in acute shortage. Exemplifying this problem in customer response modeling, the number of respondents is usually much smaller than that of the number of non-respondents [7]. On the other hand, most of the machines learning algorithms behave well on roughly same fraction of classes [8]. It requires implementing a normalization technique. The normalization include under sampling, oversampling to deal with highly imbalanced classes. Some techniques have been outlined in literature review [9]. Usually the business intelligence algorithms are designed to cater the linked data while ignoring the free text in most of the cases. In fact the volume of unstructured data is growing far more rapidly in comparison to the structured data [10]. According to a report, the volume of total data being generated at the enterprises will surpass 1600 Exabyte in which less than 100 Exabyte data will be structured [10]. Overwhelming amount of the unstructured data is also one of the motivations to focus in these techniques which are more acclimatized towards handling unstructured data. The distributed natures of the task make it ensure that the Big Data with both of its dimensions Volume and Variety is handled whereas conventional strategy of dealing in processing of data with non distributed nature was arguably ineffective.

The remaining paper is structured in five more sections as follow: Sect. 2, we have briefly discussed the relevant work. Section 3 introduces the problem statement and its branches. Section 4 is related to the proposed methodology in which we have discussed each and every component of the proposed framework. Section 5 presents a case study along with the compulsory Big Data steps. We have concluded the research in the last section with some recommendations.

2 Related Work

In this study, we have examined new challenges, the untamed Big Data has posed. These challenges include.

1. Efficient retrieval of specific data out of large corpus.
2. Identification of trends and patterns.
3. Identification of Implicit relationship between various parts of evolving data.
4. Investigating out any opportunistic value culminating into an added value for a network of enterprises.
5. What are the useful aspects which only Big Data analysis can provide?
6. Scalable indexing of the data.

The researchers have shown that large enterprises have already perceived the clandestine but useful asset of the ever increasing data as we illustrated in the striking problems, challenges and issues above. Some notable examples include web giants Google, Yahoo, and Facebook. These corporates are already exploiting the data for the provision of vibrant, vigorous, dynamic and pertinent recommendations to their esteemed clients and web users. Yet the challenging research issue is formulate able and inevitable that whether Small and Medium Enterprises (SMEs) can also dig out the data in the same fashion. This study conveys the illustration of steps involved in the process of investigating the "data into assets" wherein the scope of the steps is limited to small and medium enterprises for the purpose of turning structured and semi structured data into an added value. A response to the 5th challenge/issue stays in the analogy of a serialized flow of functionality. According to this analogy, the outputs from a product recommendation system affectedly experiences the technical aspect oriented changes along increasing the size of the input space leading to Big Data. This can be highlighted by the fact given the transaction of size in million or billion, the useful and persuasive patterns can be ascertained with keen precision.

The capacity of measuring the value of the data is firmly destined to the notion of delay; herein also coined by the terms of 'latency' or 'throughput'. During the course of the processing of large volume of input data space, latency plays an ineluctable challenge. This challenge keeps an open question to the research community. Auspiciously the Big Data technical applications, infrastructure and tools have empowered the solution architects to consider this essential aspect to a great extent. It can be brought forward through keeping a balance between employing cheaper hardware to process large volume of data.

Chelmis [11] investigated the analytical treatment of Big Data technologies. Their work was targeted for collaboration in focus with highlighting some interesting research questions; these questions encompass user's communication behavioural patterns, statistical characteristics, dynamics properties and complex correlations between social assemblies and topical structures. Differentiating the value added quality of this research, the study by Chelmis [11] is limited towards the technical and functional internal processes of a SME. Its canvas is not covering the impact of Big Data for the purpose of product improvement between two or more SMEs. In commercial world, the data repositories have placed a non-trivial role and impact for the business of a SME. This aspect is backed by the fact that the data by virtue of its inherent nature acts as an intangible knowledge assets towards any corporate [12]. The nature of 'data as asset' possess an array of issues. These issues include definition of data (especially for structured or semi structured data); formulation of information assets in an enterprise; discovery of the unique patterns and characteristics associated with this input space. Adding to these issues of data assets include the definition of key concepts of input data; the adherent quality of implicit as well as explicit knowledge and information management. More important is to address the issue of the business impact of having low-quality data and information assets. In inter enterprise collaboration, negotiation plays a critical role [1]. They introduced Negotiations for Sustainable Enterprise Interoperability with Ontologies (NEGOSEIO). This framework used MENTOR [13]. The MENTOR introduced the mediator ontology concept first time which was widely adopted by the research community. Jardim et al. [1] identified the factor which are essential to be incorporated. These include information, function and behavioural factors. Their study discuss the aspects of negotiation leverage the ontological modelling.

3 Problem Statement

Enterprises in disparate, complex but essential relationship with various partners are no more capable of restricting themselves of single private information model for any particular venture. This constraint ultimately pushes them towards their interaction with different set of partners in inter-enterprise environment context. The enterprise needs to assess the value of the data which it contains. The value of the asset has numerous characteristics such as finding the degree of the value. The degree of value is a function of related stack holders. The stakeholders refer to those who are potential user of the information within the enterprise. The deceleration of value of information is marked by the nature of underlying information and the persistence of its accuracy and precision over certain range of time period. Every asset in its life cycle turn into the liability; this state of liability is referred to the state when it is no more useful to anyone at all. Keeping in view of this problem and to address the issues elaborated in the previous section; we can identify following open question.

> Can high degree of data centric and knowledge integration in enterprise collaboration can be promoted and linked with the perceived benefits of added value towards an increase in the operational efficiency of the enterprise.

The following hypothesis can be formulated based on the research question.

- If the business intelligence service detect that the data in document management system give the oscillation of pattern then enterprise collaboration provides an appropriate method to create data asset to catch assets traceability in ontological modelling.
- If enterprise collaboration yields an effective way to realize an added value then digital preservation comprised of ontologies serves the purpose of re usability.

4 Proposed Approach

This study adopts a theoretical framework to realize a platform. We first introduced an axiom of status of the SMEs in the context of this research. We propose that the enterprises (small and medium) be considered as repositories of data and knowledge. These repositories can be used in emphasizing the role of intangible resources. In quest of these aims, the goal oriented intelligent mechanism has been proposed. The target of this mechanism is improvement of the mobility mechanisms in the acquisition phase. It is a non-trivial aspect to establish a globalized standard during the pursuit of attaining the objective of inter-enterprise collaboration [14].

From our previous discussion, it is evident that the solution of a problem with heterogeneous data cannot be handled by applying a single algorithm or a single business intelligent model. The proposed solution of the problem is a logical layer as shown by Fig. 1. The nature of the problem motivates us to find out the solution in the array of correct alignment of algorithms, appropriate filtering, cleansing and development of a scalable framework which can handle large amount of variety of data. The top most layer is receiving raw data. Technically the data at this layer is a meta-base that contains free text as well as Meta attributes. Here in this case, our system is expecting data in three forms. These include unstructured data from video, images, electronic sheets and free text, the structured data related to query refinement and quality procedures. This component (centered) also contains unstructured data which is awarded of quality procedures. The left most component in this layer is typical structured data which is linked data of Supply Chain Management (SCM) system. This data is gleaned from Supplier Relationship Management (SRM), Enterprise Resource Planning (ERP), Customer Relationship Management (CRM) system and Product Lifecycle Management (PLM) system.

The machine learning algorithm can't produce useful pattern over free text and other unstructured data unless they are aggregated to a higher degree of granularity.

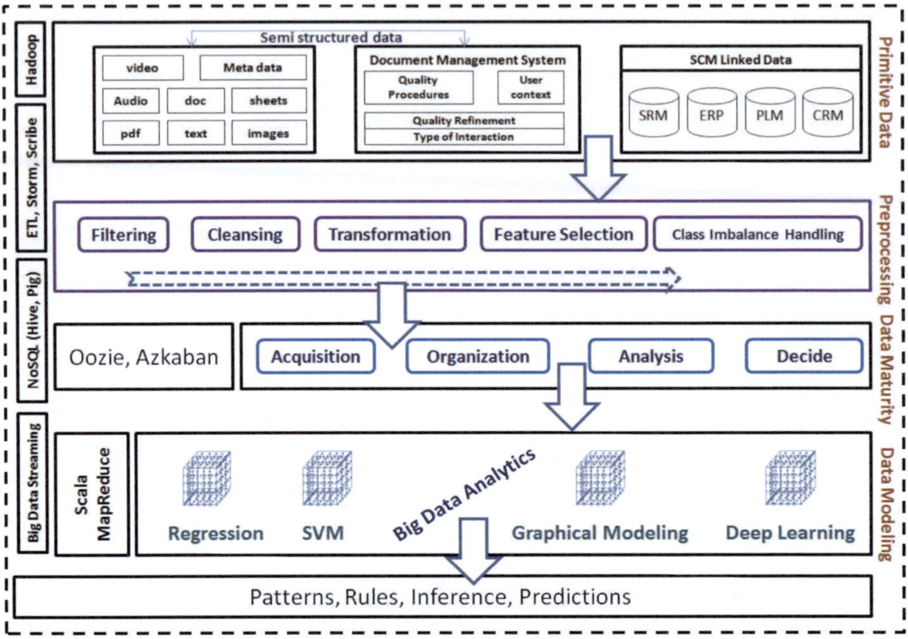

Fig. 1 Framework of proposed platform

Natural Language Processing plays a pivoted role in this situation. To summarize the free text into a meaningful information, the data needs to pass through four levels in order to tailor the unstructured data. These include lexical level, morphological level, syntactical level and semantic level. During the last two levels, variety of similarity measures have been proposed. A generic classification for all of the similarity measures have been proposed [15]. These classes include measures inspired by path length, information content, feature based and hybrid. We in this proposal suggest incorporating the measures belonging to first two classes. The technique employs the robust accurate statistical parsing graph [16]. Moreover, clustering techniques for free text has proved its significance to produce useful patterns [17]. However, we argue that the result of clustering is not directly utilizable for inference purpose when the input space and cluster realm is large enough. We fix the position of clustering in the proposed framework in the capacity of intermediate refined data. The second stream of unstructured data is images usually posted by the social media community about the product of the SMEs. We argue that the popularity of a brand can be extrapolated by means of the posting, free text and images on the social media sites. It is a fact that single image speaks hundred words and a single video is worth of hundred images. However, the perception of semantics has always been a challenging task [18]. This challenge has been taken by research community to the level that now primitive information such as figuring out the basic human features such like age group, gender, race (color) etc. can be predicted with relatively high recall [18]. A recent research solution proposed by Han et al. [18] is useful in the context as their technique can infer multiple demographic features. Once this primitive data is obtained, there is a compulsory need to perform typical data preprocessing tasks such as filtering, cleansing, transformation, feature selection and the most important handling the imbalanced classes.

To handle the imbalanced problem, smote [9] provides a reasonable solution. It employs K nearest neighbor (KNN) in its heuristics. The time complexity of KNN is Log(m) . $O(m^2)$ and space complexity is of $O(m^2)$. This baseline complexity gives problem in case of large volume of data. However, a possible solution is its integration into Hadoop echo system. Our initial analysis of reducing KNN to map reduce paradigm indicates that like simple Bayesian classifier, this problem is also effectively reducible to Big Data architecture.

The output of the second layer is a segmented data passed to Big Data platform. Here, the data is provided to the acquisition phase of the Big Data technologies. This phases include Oracle NoSQL database system and/or Oracle Online Transaction Processing System. The variety of the incoming data is unstructured, so it is demanded that NoSQL database is more suitable to this type of data. This type of DBMS system deals data line by line where every line is a complete 'value' and a corresponding 'key' is allotted. This gives the freedom from a strict schema. The output of the acquisition is subjected to the second component of the Big Data technology layer. This step is related to the organization of the data. Here the notion of the organization refers to adjust its suitability for the analysis phase. The analysis phase is the final step of this step in this research study. The analysis will identify any potential opportunity. One can observe that the granularity level of data is increasing at this stage. Logically, the information is in first three classes; however, its physical interpretation is expected to be marked by dozens of relational entities. The motivation to generate this data is twofold. The first reason is driven by application of the classification algorithms; whereas the classification systems by virtue of their supervised learning nature demands a clear position over the underlying schema. The second reason is that, the data is to be analyzed in NoSQL, which is capable of handling Big Data. The layer responsible for data modeling is showing numerous modeling architectures. The rationale is rooted in the nature of versatile data. The final layer is composed of patters, rules, inferences and prediction on Trending, Costing and Recommendations for marketing purpose.

5 Case Study

To exemplify how the proposed platform will work, we shall give an example related to the functionality of the core part of the platform. There is a scenario in which a manufacturer company produces a component by assembling. The company needs variety of components during the assembly of any finished product. There is huge variety of components. A possible collection or assembly of the component produces the situation like snowflakes where each snowflake is unique with 1018th possibility of repeating the same snowflakes. The same situation repeats in this scenario when each assembly is unique in its functional or component specifications. The whole of the information is being gleaned in a linked data repositories. The collections of the repositories consist of hundreds of relational table. There are some interesting queries which need to be answered but utilizing the information from all of the underlying linked entities. One such query is to churn out the "best prediction of the price a supplier (of a single or multiple components) is likely to quote given the final assembly to be built". Our methodology suggests that all of the information must be aggregated into a single entity. This will pose the problem of size in vertical as well as horizontal dimension. But the combination of Big Data and classification or regression will produce the most anticipated result. To get the price, we must need some prior information. Extract all meaningful information in a linked database. In this case these type of tables will be required.

1. Technical specifications
2. List of parts
3. Cost/quantity of individual parts
4. Taxation
5. Operational cost
6. Quote price

Quote pricing needs to be discretized or binned if minimum or maximum limit on price is required. Fix the empty cells (numeric to 0 and string to NA). Separate column name into a schema Map. After it, upload all of the data on HDFS. Now we shall integrate all of them using replicated left outer join taking quote price on left hand side. It will output a single file. The idea is to have a single file as it can convey all of the information. We have clearly identified the class problem which is estimated quote price. Apply the random forest over this input data given the class. Now perform the tracing of the model performance by means of ROC, Confusion Matrix or marginalization of number of trees in forest. If performance is not satisfactory, reduce the less important parameter. These steps are repeated till an optimized model is achieved. We can save this model so that we can load the same whenever required in mahout using MapReduce.

Figure 2 is showing the typical steps required to launch the Big Data in this case study. First step is related to the setup of in data house hadoop cluster in hadoop eco-system. In this case study, we are using Oracle Virtual Machine (OVM) for the experimental part. The linked data is aggregated via ETL job. There are three popular services available in the Hadoop echo-system for the creation of ETL jobs. These include Hive, Pig and MapReduce.

We restrict ourselves to MapReduce as it gives more flexibility providing a complete programming environment by high performance language Java. The consolidated data is a pre requisite for application of data analytic tools such as regression analysis, association rule mining and classification. These data analytic tools ensure the diffusion and transformation into creation of knowledge

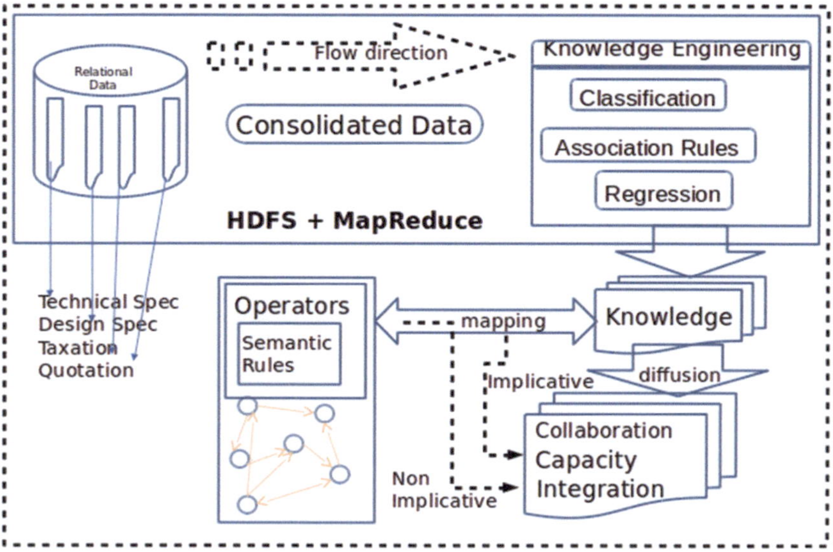

Fig. 2 Process flow of the steps in weaving trending, costing and recommendations case study

repositories. At this point, semantic engineering becomes an imperative need playing an inevitable role. We provided means of operators which can work on set of semantic rules. Intuitively, the knowledge perceived in previous step is in different shape where some knowledge is implicative and some is in non implicative (frequent item-sets vs. association rules). Moreover, some knowledge is in descriptive form and some is in inferential statistical shape achieved via clustering and regression analysis respectively. The operators ensure their diffusion into "Enterprise Capacity", "Enterprise Integration" leading to "Collaborative" perspective. The whole operation in the figure is regulated by means of Oozi work flow in Hadoop echo-system.

6 Conclusions

With the advent of modern computing tools, the service components within enterprises are generating large amount of data. The proposed platform system provides various results according to the role of end user. If the end user is a supplier in industrial sector, then he can get possibility of new chances to achieve new markets. The supplier will receive the in time and situation oriented quality feedbacks for the ranking of his offer. If the end user is a buyer then the system is likely produce the benchmark purchase prices. Moreover, the system's recommendation system will give the list of best supplier with ranks. The rank is always qualified by the specifications provided by the potential buyer. The specifications include common parameters such as dimensions of the product, technical skills and capacity, quality framework, location, etc. The proposed system offer the array of following services:

1. An improved and robust mechanism for 'Request for Information' and 'Request for Quotation'
2. Consistency analysis in vulnerable costing system of long term as well as daily life products.
3. The proposed platform is useful for technical and commercial endorsement using its recommendation system.
4. Reduction in the cost of importing for local small and medium importer companies.
5. Assistance to SMEs in optimizing investment, design to cost novel products and redesigning of existing products.

The proposed framework has illustrated that the data analytic technologies are useful for developing assets composed of opportunity analysis through exercising the Big Data technologies over ever-accumulated unstructured data.

References

1. Jardim-Goncalves, R., Coutinho, C., Cretan, A., da Silva, C. F., & Ghodous, P. (2014). Collaborative negotiation for ontology-driven enterprise businesses. *Computers in Industry, 65*(9), 1232–1241.
2. Cretan, A., Coutinho, C., Bratu, B., & Jardim-Goncalves, R. (2012). NEGOSEIO: A framework for negotiations toward sustainable enterprise interoperability. *Annual Reviews in Control, 36*(2), 291–299.
3. Philip, C. C., & Zhang, C-Y. (2014). Data-intensive applications, challenges, techniques and technologies: A survey on Big Data. *Information Sciences, 275*, 314–347.
4. Naeem, M., Bouras, A., Ouzrout, Y., & Moalla, N. (2015). Opportunity analysis for enterprise collaboration between network of SMEs. In *6th International IFIP Working Conference on Enterprise Interoperability Nîmes, France (IWEI 2015).*
5. Naeem, M., Moalla, N., Ouzrout, Y., & Bouaras, A. (2014). An ontology based digital preservation system for enterprise collaboration. In *2014 IEEE/ACS 11th International Conference on Computer Systems and Applications (AICCSA)* (pp. 691–698). New York: IEEE.
6. Hudson, S., Huang, L., Roth. M. S., & Madden, T. J. (2016). The influence of social media interactions on consumer–brand relationships: A three-country study of brand perceptions and marketing behaviors. *International Journal of Research in Marketing, 33*(1), 27–41.
7. Lee, H., Shin, H., Hwang, S., Cho, S., & MacLachlan, D. (2010). Semi-supervised response modeling. *Journal of Interactive Marketing, 24*(1), 42–54.
8. Chen, Z.-Y., Fan, Z.-P., & Sun, M. (2015). Behavior-aware user response modeling in social media: Learning from diverse heterogeneous data. *European Journal of Operational Research, 241*(2), 422–434.
9. Chawla, N. V., Bowyer, K. W., Hall, L. O., & Kegelmeyer, W. P. (2002). SMOTE: Synthetic minority over-sampling technique. *Journal of Artificial Intelligence Research, 16*, 321–357.
10. Hilbert, M., & López, P. (2011). The world's technological capacity to store, communicate, and compute information. *Science, 332*(6025), 60–65.

11. Chelmis, C. (2013). Complex modeling and analysis of workplace collaboration data. In *2013 International Conference on Collaboration Technologies and Systems (CTS)* (pp. 576–579). IEEE, New York.
12. Denicolai, S., Zucchella, A., & Strange, R. (2014). Knowledge assets and firm international performance. *International Business Review, 23*(1), 55–62.
13. Sarraipa, J., Jardim-Goncalves, R., & Steiger-Garcao, A. (2010). MENTOR: An enabler for interoperable intelligent systems. *International Journal of General Systems, 39*(5), 557–573.
14. Lin, H. K., & Harding, J. A. (2007). A manufacturing system engineering ontology model on the semantic web for inter-enterprise collaboration. *Computers in Industry, 58*(5), 428–437.
15. Meng, L., Huang, R., & Gu, J. (2013). A review of semantic similarity measures in wordnet. *International Journal of Hybrid Information Technology, 6*(1), 1–12.
16. Rei, M. (2013). Minimally supervised dependency based methods for natural language processing (UCAM-CL-TR-840).
17. Larsen, K. R., Monarchi, D. E., Hovorka, D. S., & Bailey, C. N. (2008). Analyzing unstructured text data: Using latent categorization to identify intellectual communities in information systems. *Decision Support Systems, 45*(4), 884–896.
18. Han, H., Otto, C., Liu, X., & Jain, A. (2015). Demographic estimation from face images: Human vs. machine performance. *IEEE Transactions on Pattern Analysis and Machine Intelligence, 37*(6), 1148–1161.

The Industry Cockpit Approach: A Framework for Flexible Real-Time Production Monitoring

N. Oertwig and P. Gering

Abstract Products are becoming more and more individualized. Additionally, customer demands are increasingly more influential on the production process. The individualization is no longer restricted to the final products, but increasingly influences the processes of the entire company. Thus, an additional requirement for a manufacturing process also indirectly requires individual procedures in planning, billing, documentation and controlling. Current cockpit solutions, programmed with great effort, are inadequate since they can only be applied for a limited process variety and thus the necessary flexibility cannot be provided. However, real-time monitoring of business processes faces several challenges. Observing process executions and at the same time focusing on other tasks cannot be tackled with state-of-the-art, visualization-based process monitoring systems. In this paper, a model-based Industry Cockpit is presented which can represent data accurately and flexibly in a contextual manner for the whole company, including its products, processes, roles and machine systems. With this solution, every employee receives real-time information at the right place and at the right time.

Keywords Cockpit · Dashboard · Enterprise model · Decision support · Monitoring · Real-time

1 Introduction

The industrial sector is changing rapidly. Due to the increasing transparency of global markets and the resulting intensity of competition, companies are faced with new frameworks for resource efficiency, increasing flexibility and optimizing decision making. This increase of the inherent flexibility and the implementation of economic mutability are among the greatest challenges that manufacturing enterprises face in the turbulent environment of today [1]. It is essential for most

N. Oertwig · P. Gering (✉)
Fraunhofer IPK, Pascalstraße 8-9, 10587 Berlin, Germany
e-mail: nicole.oertwig@ipk.fraunhofer.de; patrick.gering@ipk.fraunhofer.de

© Springer International Publishing Switzerland 2016
K. Mertins et al. (eds.), *Enterprise Interoperability VII*,
Proceedings of the I-ESA Conferences 8, DOI 10.1007/978-3-319-30957-6_14

companies to be able to monitor the executions of their business processes in real time in order to adapt to sudden changes or to react quickly to undesirable situations [2, 3]. This includes recognizing the need for adaptation to the future at all levels of a company, e.g., to develop the optimal configuration of a factory in a short time and to start the implementation. To fulfill these challenges, an excellent knowledge of the current state of the factory is required. This needs to include production-related indirect areas, quickly and easily comprehensible as well as applicable planning tools to prepare and hedge changes in operating procedures [4, 5].

The substantial increase in data as well as data flows that need to be processed complicate the task of evaluation, aggregation and decision making tremendously. The challenge goes beyond the vertical and horizontal integration of the systems involved. Since management models, planning processes and data are not static, they constantly have to be adapted to the changing requirements and must be made available to the recipient in a focused manner. Enterprise modelling can make a major contribution towards resolving this complex task. It represents the companies' structures, processes, systems, organization and the human resources including their interdependencies and interactions [3]. Therefore, with the help of enterprise modelling in conjunction with the existing IT-systems and the data provided by the control and the command level, a constant, integrated, multi-perspective illustration of data and processes in a dynamic visualization environment is possible. Even Information Technology is located in a complex and dynamic environment. Analyses have shown that IT development reflects the growing globalization of the company [6]. The contribution of this paper to this development is an extendable framework for a model-based performance monitoring device called Industry Cockpit. The focus of the present work is to develop a method which allows a dynamization of enterprise modelling, so that the typical phases of the process identification, description, implementation, optimization, execution and analysis can be managed and that they allow an appropriate contextual and role-based real-time monitoring for all actors involved (from worker to sales manager up to the managing director) just via configuration mechanisms. In conjunction with adaptive machinery, the entire process network is ready for the integration of individual wishes.

The presented framework aims to (i) reduce the effort of setting up dashboard views, (ii) facilitate a better information transfer, (iii) create a dynamic performance measurement environment and (iv) produce better usability due to a model-based configuration of dashboards. These objectives are addressed by a modular, model-based approach where every module can be reused in multiple scenarios. If a process a module is changed, the cockpit adapts automatically. Moreover, the possibility to have a single-point of configuration via a model-based approach increases the usability and flexibility in setting up a dynamic real-time cockpit.

After this short introduction to the subject, Sect. 2 describes the challenge, objective, approach and resulting requirements of the Industry Cockpit. In Sect. 3, the solution concept is presented in detail followed by a prototypical use case implementation in Sect. 4. Finally, Sect. 5 concludes the paper.

2 The Industry Cockpit

2.1 Challenge

Short-term customer orders in connection with short delivery times and an explosive increase in the number of customer-specific product variants are everyday challenges for companies [7]. As shown in Fig. 1, existing rigidly programmed management cockpit solutions are no longer able to handle these challenges in an adequate way. Situation-specific adaptations always have to be implemented individually for the information demand of every role involved by the responsible developer. The time and coordination efforts are economically unreasonable and impossible to realize in the short reaction time the customer claims for. However, to implement a flexible and customized production, the key factor is to have a complete overview about the production processes and mechanisms at any time in order to plan and visualize processes and relevant data. This can only be achieved if a dynamic, holistic and context-sensitive monitoring environment, which has not to be programmed anymore, connected with production planning and process models, is available. Every user has to be able to configure his individual information demand ad hoc, based on a common backbone of process and information. Therefore, process management and operational systems have to be combined to track customer related process changes and to filter and process information in a model-based manner.

2.2 Objective

Companies can only meet short-term customer demands if they are organized in a flexible process network. In order to reliably control and monitor such highly dynamic processes, the model-based Industry Cockpit has been developed. Configuration instead of programming leads to shorter reaction times for individual

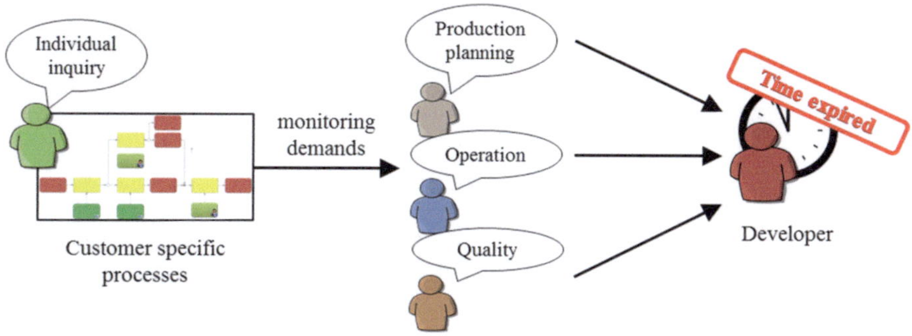

Fig. 1 Challenge of the Industry Cockpit

customer requirements. Moreover, the idea of plug and monitor allows standardized products instead of individual programmed dashboards. Through the process models and by merging heterogeneous data in real-time, an entire corporate structure view is given. In addition, a role-based dashboard presents and supports every user with relevant information. Any participating employee is able to become a "process controller" and has the possibility of ad hoc creation of customized evaluations on processes.

2.3 Approach

The Industry Cockpit depicts the structures of an intelligent and networked production and links, on the basis of enterprise models, people, products, equipment, information systems and connected organizations holistically together. Therefore, the visualization components are already included in model elements' "backpacks" as illustrated in Fig. 2. To fill this "backpack", every process module is connected with the module-relevant evaluations which refer to operational data within distributed system. The combination of several process modules is used to build up the entire enterprise model. By combining single modules, the maximum set of evaluations which can be represented in the cockpit view has been defined as well.

It depends on the user role within the organization (e.g., detailed process data for the operator or just quality relevant evaluations for the quality manager) which of these evaluations are shown subsequently. This is also defined via a view concept derived from a framework for contextual enterprise modelling [8].

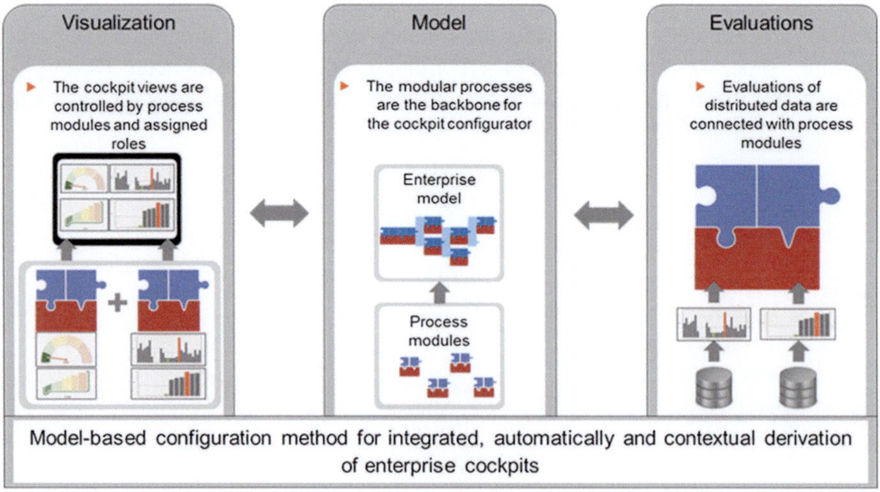

Fig. 2 Approach for model-based configurations of enterprise cockpits

In order to enable decision making, it always delivers an exact overview of the overall situation of the operation and all processes, as well as concerning the state of the production facilities to the users of all levels. If changes are being made in the processes under one's responsibility, the cockpit automatically changes the dashboard view, enabling a transparent communication and presentation of the tasks ahead. With this system, companies are not only much more flexible—it will additionally set new standards in terms of cooperation. The collaboration between different disciplines is strongly supported—every employee remains in his discipline, doing what he or she does best. At the same time, the impact of his or her decisions is going to be visible. Real-time data from different systems (e.g., Manufacturing Execution System (MES) or Customer Relationship Management (CRM)) can be mapped continuously. The corporate data is necessary for decisions to be recognized across hierarchy, regardless of location, and thus across enterprises and supply chains. In order to ensure the ability to make decisions, an assistant system supports the user in the individual configuration of necessary overviews. This is another advantage benefitting cooperation, since the cooperation between areas, nowadays often being separated within the company, is supported on a model basis [9].

2.4 Requirements

Business processes are the connection between products, organizations, roles and supporting systems in and in between organizations [10]. They represent the chaining or networking of all business activities. With clients' products being individual, the business processes have to be individual too. Therefore, planning and controlling instruments in particular must be realized dynamically at the process level. This raises the question of how such a system has to work in detail. The customers increasingly not only determine appearance and properties of the final product, they indirectly influence administrative operations too. In addition to the impact of customer requirements on the administrative process level, there are also demands for the technical processes that are not included in the business process models. Therefore, a specific customer-based manufacturing, which is traceable on an individual product basis, is required.

It needs to capture new process steps in order to realize individual customer requirements in the shortest time possible and pass them onto operational planning and control immediately in order to ensure their monitoring. A model-based monitoring makes a description of the occurring entities (e.g., people, resources, products and information) within a business in the form of model elements necessary in order to provide a reference to the corresponding data in the planning and control system. Moreover, networked information flow instead of a flood of information at the data level is advantageous. All data must be processed quickly and efficiently. In many cases, real-time information is needed. The data must be available everywhere, for example on the machine as well as on mobile devices on a trip. Likewise, a networked information flow instead of information overload at the organizational level is requested. Therefore and due to the processed data, decisions need to be made at all levels. All organizational units have to be integrated, because their gearing has a direct impact on product quality and production flexibility.

In order to meet such requirements, without risking their own profitability, companies need a flexible process network, which can be adapted quickly when needed. Targeted decisions, based on substantiated information, have to be made safe and implemented quickly [11]. It is therefore necessary to consult former values and current data from the real factory and forecast values.

3 Solution Concept for the Industry Cockpit

In the following section, the implementation of a systematic cockpit management using the Industry Cockpit approach is described. The solution covers the whole process of enterprise modelling, defining and configuration of evaluations, integration of operational systems and visualizing components in the dashboard application. This will be the prerequisite for a seamless synchronization between the enterprise model and its individual monitoring. Therefore, a major aspect is the modular framework of the method.

This solution concept can be explained within five steps, the first one being an initial step that only needs to be executed once:

1. Set initial enterprise model

The focus of the configuration method lies on the enterprise modelling. It forms the basis of the cockpit configurator and requires the following: information about the processes within the organization as well as interrelation between the single processes in terms of business logic and data of the different processes in the enterprise model. In a first step, an initial enterprise model has to be defined by the organization. As a basic model, it includes all necessary processes and their interaction within the preliminary defined system boundaries (see Fig. 3).

The representation of an enterprise that refers to aspects of structure, activities, process, information, resources, people and behaviour is defined as an enterprise model [13]. To create such a model, the Integrated Enterprise Modelling (IEM) [14] is used, because of its object orientation, and facilitates the integration of steering and control aspects in connection with the business logic. In addition, the IEM is also conform to the international standard EN/ISO 19440 (constructs for enterprise modelling) [15] and is fully supported by the enterprise modelling tool MO^2GO [16–18].

2. Set up process modules in the module library

In a second step, the initial process model will be divided into defined process modules (see Fig. 3). Such a process module consists of a defined process input and

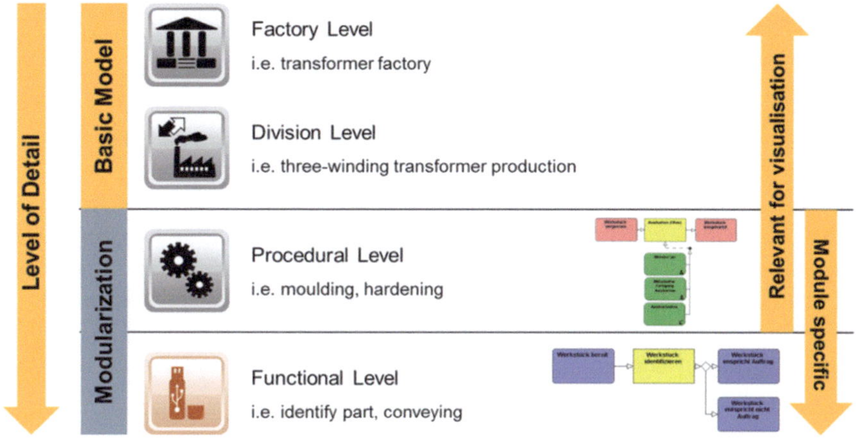

Fig. 3 Holistic enterprise model [12]

output, a defined process content, relevant resource for fulfilling the process and information objects. Flexile process networks are built up by connecting process modules of the type information module (e.g., administrative processes), product module (e.g., value adding processes) and resource modules (e.g., set up processes). Order specific process modules can be created, instantiated and stored in a module library, so that they can be used within a special process management client to generate a customized process model from predefined process modules.

3. Define evaluations for modules

Distributed data of the company is linked through evaluations to process modules. Thereby the information and data within the company and the representation in the cockpit are defined (from sales to production up to billing). On the basis of those process modules, the information that should be visualized and its presentation are uniquely defined by an evaluation generator, so that they can be reused. In addition to the impact of customer requirements on the administrative process level, there is also a high demand for the visualization of technical processes. Those processes, usually programmed and configured by MES suppliers, are modelled on the functional level of our holistic approach (see Fig. 3). We don't aim to replace existing MES solutions rather than making those technical processes accessible to a wider range of users. Therefore, the process management client generates the MES configuration based on the technical processes.

4. Integrate/change modules in enterprise models

Process modules of the product flow can be connected at this point with so-called "work plan containers". These include individual specific steps for the manufacturing of a component and can be assembled throughout the overall process logic routing. Furthermore, autonomous handling of sub-processes is provided. The recourse of existing processes and experiences and the fast flexible configuration of overall processes is a significant aspect. Situation-specific enhancement of processes and the restriction of the changes to sub-processes support the handling of modules.

5. Present results in cockpit

The cockpit views are controlled by the process modules and offer individual monitoring to support decisions. An individual access management offers each employee a unique interface with which he or she can fulfil the tasks entrusted to them by the network. Further properties like simple and intuitive handling as well as reactions, alarms and measures support the monitoring in real time. The cockpit accesses diverse data sources and information that are specifically tailored to the role and situation. A dynamic switching (e.g., view for active job) and the dynamic fade in and out of sensitive information requires clear presentation of results via different technologies e.g., browser, mobile phones or tablet.

4 Use Case Implementation

A prototypical implementation has been realized for the following scenario: A big bicycle manufacturer increasingly receives requests from its clients if existing components can be integrated into a new bicycle. Due to the increasing environmental awareness, the reuse of bicycle frames in particular is part of the clients' demands. In order to make this wish come true, an adaptation of the production processes is necessary. There already is a well-developed process management and an MES for the efficient control of the production in place for the company. Figure 4 illustrates the main components of the use case implementation schematically.

If the customer expresses the wish to reuse the existing frame for a new bicycle for the first time, a process module "Prepare customer frame" will first be created and connected with relevant evaluations coming from the operational data in the MES. This module will replace the module "manufacture frame" in the process flow. In the next step, a new work plan wallet will be created in the MES based on the existing model.

The corresponding detailed production steps (e.g., measure frame; examine weld seams' quality) for the customer-related process module are assigned within the MES so that with the process sequence of production, coming from the enterprise model via a configuration file, the overall work plan is defined. Beside the derivation of the routing, the connected process modules with the assigned roles lead to the contextual dashboard views. The monitoring components (e.g., completion date, quality characteristics or measurement results) for the process module "prepare customized frame" as well as their displayed position in the cockpit are defined through the cockpit configurator.

By activating the referenced order in the production process (e.g., via barcode-scan), the cockpit illustration for this order will change automatically. The relevant information for this order and the customer specific process step "prepare customer frame" are occurring until the task has been finished.

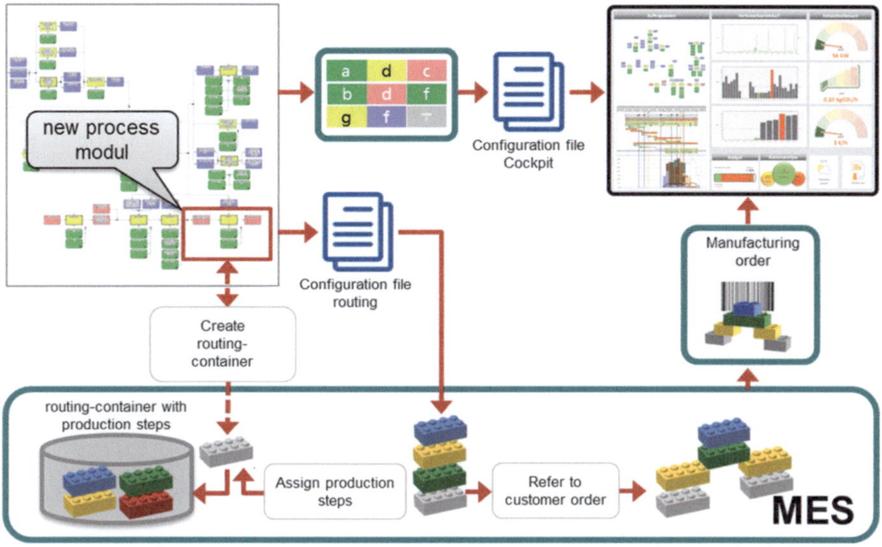

Fig. 4 Use Case implementation

The reusability of the one-time created process modules will lead to considerably shortened response times for individual customer requirements. This is specifically the case since the necessary control mechanisms will no longer be programmed but only configured. This also means that the ad hoc creation of customized evaluations offers optimal support to users in their analysis and decision making processes within manufacturing industry.

5 Conclusion

A key factor for successful enterprise management is the availability of evaluation tools for performance monitoring and tracking from a qualitative and quantitative point of view. An important point of leadership is the ability to collect, evaluate and understand the correct data in order to drive change effectively. Sharing this data with the right people in the right time is equally important. Through the close link of the Industry Cockpit and the digital enterprise model, changes in the real factory are transmitted via a change in the digital enterprise model, which in return directly changes the visualization in the cockpit. It means that the developed system will differ significantly from the cockpit solutions available on the market.

Since an enterprise consists of several different units and elements which are related to each other on several levels, it is necessary to consider all influences and possible side effects within the planning process [18]. Therefore, interfaces to all common systems support and form the base of a cockpit. The economic objective of a company to promptly and optimally adapt to a constantly changing environment requires tools to support the parties that accompany the continuous process of change. The Industry Cockpit fulfills those requirements by showing the need for action and encourages model maintenance so that a current database and planning results are provided and represented in an intelligent and legible form for all participating actors.

Acknowledgments This research and development project is funded by the German Federal Ministry of Education and Research (BMBF) within the Framework Concept "Research for Tomorrow's Production" (funding number 02PJ4044) and managed by the Project Management Agency Karlsruhe (PTKA). The author is responsible for the contents of this publication.

References

1. Drucker, P. F. (1999). *Management challenges for the 21st century*. New York: Routledge.
2. Kung, P., Hagen, C., Rodel, M., & Seifert, S. (2005). Business process monitoring & measurement in a large bank: Challenges and selected approaches. In: *16th International Workshop DEXA, Copenhagen* (pp. 995–961).
3. Harmon, P., & Wolf, C. (2014). The state of business process management 2014. http://www.bptrends.com/bpt/wp-content/uploads/BPTrends-State-of-BPM-Survey-Report.pdf. Accessed May 5, 2015.
4. Eckerson, W. W. (2011). *Performance dashboard—Measuring, monitoring and managing your business* (2nd ed.). Hoboken, New York: Wiley.
5. Kapp, R. (2006). Echtzeitfähiges Fabrik-Cockpit für den produzierenden Mittelstand. *Industrie-Management, 22*(2), 49–52.
6. Neumann, R., Zborg, F., & Dumke, R. (2009). *Cockpit-basierte management-systeme*. Magdeburg: Otto-von-Guericke-Universität.
7. Kapp, R., Weimer, T., & Westkämper, E. (2006). *Fabrik-Cockpit: Eine echtzeitfähige Planungsumgebung*. Stuttgart: Fraunhofer IPA.
8. Wintrich, N., & Meißner, M. (2015). Framework for a contextual representation for enterprise model application (C.R.E.M.A). In: *3rd International Conference on Model-Driven Engineering and Software Development, Angers*.
9. Laqua, I. (2012). Production Intelligence—Mit dem Produktionscockpit die Fertigung im Blick. *Zeitschrift für Wirtschaftlichen Fabrikbetrieb, 107*, 416–420.
10. Business Process Modeling. *Bindeglied zwischen IT und operativem Management?*. http://www.it-daily.net/it-technologie/system-service-management/62-it-management-service-management/1631-business-process-modeling-bindeglied-zwischen-it-und-operativem-management
11. Oertwig, N., & Knothe, T. (2015). Industrie Cockpit. *Futur, 1*, 6–7.
12. MetamoFAB. http://www.metamofab.de
13. Fox, M. S., & Gruninger, M. (1998). Enterprise modeling. *AI Magazine, 19*, 109–121.
14. Spur, G., Mertins, K., & Jochem, R. (1996). *Integrated enterprise modelling* (1 ed., pp. 60–83). Berlin.
15. International Organization for Standardization. (2008). Industrial automation systems—Enterprise integration: Constructs for enterprise modeling. ISO 19440.
16. Mertins, K., Jochem, R., & Jäkel, F. -W. (1997). A tool for object-oriented modelling and analysis of business processes. In *Computers in Industry*.
17. Jäkel, F. -W. (2014). MO^{2}GO. In *Enterprise Modeling, Analysis and Optimization*.
18. Fraunhofer Institute for Production Systems. http://www.prozessassistent.de

Part V
Services for the Enterprise Interoperability

iFloW: An Integrated Logistics Software System for Inbound Supply Chain Traceability

Nuno Santos, Diogo Barbosa, Paulo Maia, Fábio A. Fernandes, Márcio P. Rebelo, Pedro V. Silva, M. Sameiro Carvalho, João M. Fernandes and Ricardo J. Machado

Abstract Visibility plays an important role in supply chain management. Such visibility is not only important for better planning, but especially for real-time execution related with the traceability of goods. In inbound supply chain management, logistics planners need to trace raw materials from their requests in order to properly plan a plant's production. The iFloW (Inbound Logistics Tracking System) integrates logistics providers IT applications and Global Positioning System (GPS) technology to track and trace incoming freights. The Estimated Time of Arrival (ETA) is updated in real-time allowing an improved materials planning process. This paper presents the iFloW project and describes how these issues are addressed and validated in a real pilot project.

Keywords Logistics software system · Inbound logistics · Track and trace · Supply chain visibility · Traceability · GPS

N. Santos (✉) · D. Barbosa · P. Maia · F.A. Fernandes · M.P. Rebelo
M. Sameiro Carvalho · J.M. Fernandes · R.J. Machado
School of Engineering, ALGORITMI Research Center, University of Minho, Braga, Portugal
e-mail: nuno.a.santos@algoritmi.uminho.pt

P.V. Silva
Bosch Car Multimedia Portugal S.A., Braga, Portugal

M. Sameiro Carvalho
Department of Production and Systems, University of Minho, Guimarães, Portugal

J.M. Fernandes
Department of Informatics, University of Minho, Braga, Portugal

R.J. Machado
Department of Information Systems, University of Minho, Guimarães, Portugal

© Springer International Publishing Switzerland 2016
K. Mertins et al. (eds.), *Enterprise Interoperability VII*,
Proceedings of the I-ESA Conferences 8, DOI 10.1007/978-3-319-30957-6_15

1 Introduction

As supply chains are more global and interconnected, they are becoming more complex, costly and vulnerable and to deal effectively with risk and meet business objectives they need to be a lot smarter [1]. According to the supply chain leaders, there are five top challenges for supply chain management: visibility, cost containment, customer intimacy, risk and globalization. The supply chain visibility is highlighted as the most important challenge—70 % of those leaders have reported that this challenge impacts their supply chain to a significant or very significant extent and the external integration and visibility as very important or critically important [2].

The supply chain visibility enables to track goods from the manufacturer to the final destination, either in an inbound or outbound logistics perspective. Tracking goods allows one to follow routes of their transportation, so that a visualization software are able to show roadmaps of those routes. The optimization of routes is another common way of using the data from tracking any trade item [3], as well as for planning and management improvement [4].

Tracking routes of shipments in real-time allows one to estimate more accurately the time of arrival of goods in a production network with several locations. The estimation is performed by predicting arrival times at plants based on real-time data from shipments actual position and speed [5]. Thus, production planners get a better information to perform synchronization of material delivery at the plant with the production scheduling. As a result, these tasks may be performed in less time and, if needed, planners have more time to reschedule production. Storage the data of routes also provides additional data for future planning [6].

Issues that may occur within the inbound logistics processes of a manufacturing plant are the time logistics planners spend in freight tracking tasks throughout their activities, and complaints to suppliers caused by transportation process with negative effects in production (like production stoppage and special freight requests—i.e., non-planned freight that occurs, usually, when the freight is in risk of compromising the fulfilment of the supplier delivery plan). Therefore, there is a need to reduce e-mail and phone calls exchange, and any other form of communication. To overcome these issues, some integration and interoperability between the plant's and forwarders' systems is required, originating the Inbound Logistics Tracking System project (hereafter called iFloW project), performed at the plant located in Braga, Portugal of Bosch Car Multimedia (Bosch BrgP).

The main goal of the iFloW project is to develop an integrated tracking platform, which tracks in real-time all the raw materials in transit, from their starting point to the arrival at the Bosch BrgP's facilities, by their part number. This platform currently under development controls and monitors the raw material flow from suppliers to Bosch BrgP's warehouse. Additionally, the iFloW system alerts users in case of any deviation to the Estimated Time of Arrival (ETA), quantities or part number. This project aimed executing a pilot scenario, where the software system was deployed to be able to address a specific scenario (one specific forwarder and in one of their routes). This way, the project assessed: its usage in a tracking task, where it is required improvements, and ultimately its usefulness to the inbound logistics process.

This paper is structured as follows: Sect. 2 addresses related work on solutions for inbound logistics and traceability; Sect. 3 describes the iFloW project, from the problem statement to the specification of the solution; Sect. 4 presents the results obtained in the pilot scenario and Sect. 5 the conclusions.

2 Related Work

Information Technology (IT) has a strong impact on logistics, where in the last years many software solutions have been developed [7]. According to the report from Gartner Group [8], some of these solutions are delivered as (cloud computing-based) services, namely Business-Process-as-a-Service (BPaaS). Many forwarders deliver track and trace services of their freights through BPaaS in complex Supply Chain Management (SCM) environments, like UPS [9], Kuehne + Nagel [10], TNT [11], Panalpina [12], DHL [13], FedEx [14] or Lusocargo [15]. The report from Gartner Group also refers BPaaS SCM solutions as the fourth highest BPaaS-based services market share [8].

A more efficient management of logistics systems is required to allow organizations to be able to obtain the needed materials, at the right time and place [7]. Within logistics systems, Kandel, Klumpp, and Keusgen categorize tracking solutions as discrete and continuous tracking [6]. Barcoding and RFID technologies are categorized as discrete tracking solutions. Alternatively, continuous tracking solutions, like the Global System for Mobile Communications (GSM) and the Global Positioning System (GPS) technologies, allow tracking the position at any time. Yin, Wang, and Zhang propose a topology based in Electronic Business using eXtensible Markup Language (ebXML) for integration of logistics information and e-tag, Radio Frequency Identification (RFID), GPS and General Packet Radio Services (GPRS) technology [16].

The integration of these tracking technologies in a common logistics platform is referred as Integrated logistics information management system (ILIMS) [17]. Platforms like ILIMS allow different parties transmitting, capturing, sharing and gathering the required logistics data via internet. These systems facilitate the processes of both logistics service providers as well as supply chain participants. An approach like a logistics information hub is used for standardizing the definition of third party logistics (3PL) service providers and integrating RFID technology [18, 19], GPS technology [6], and others.

The integration of logistics information obtained with tracking technologies is addressed by many authors. Doukidis et al. propose an integrated web-based RFID-Electronic Product Code (EPC) compliant logistics information system, aiming to discover and share RFID/EPC data [17]. The usage of RFID and GPS technologies, alongside with information models and web-services, is proposed in [20]. In this work, RFID events are integrated with geographical information and associated with the cargo, in order to track and trace that cargo. Cargo transports can also be tracked and traced using logistics solutions for GPS-based track and trace services [20, 21]. Platforms, like Track-Trace.com [22], are able to provide users with track and trace functionalities that are integrated with the forwarders' information systems.

Inbound logistics additionally face product traceability issues. An integrated view on RFID and barcodes is presented in [23, 24], where interoperability for product traceability are addressed. Chen [25] presents issues the handling of large amounts of data on the initial stages of the traceability process using a dedicated RFID middleware. Also the research work in [26] addressed interoperability for customizing products in multi-company operations environments, where also track and trace of shipments and of composite products is addressed.

These proposed solutions focus in defining and developing platforms for receiving information from devices and then allow users to manage the received data as better suits their business. However, the data only flows from suppliers to plants, and afterwards there is not any change to the planning work (which still is performed using phone or e-mail communication, for instance). Additionally, neither of these platforms allow to track and trace products by the part number used within the plant. In comparison, iFloW receives freight information from GPS systems and Electronic Product Code Information Services (EPCIS) events, and integrates with forwarder's systems tracking information for updating ETA values in different scenarios (i.e., iFloW may deal with information of material transported by land, sea or air, and that may be pass in European, African and Asian ports). In addition, allows Bosch BrgP logistic planners to perform tracking searches by part numbers and allows forwarders and Bosch BrgP logistic planners to negotiate special freights requests by directly using iFloW system. The integration of the iFloW system with the forwarder's systems allows logistic planners to be able to track and trace products by using iFloW as a single point of information, where information is available to all users in a standardized format, instead of using the existing BPaaS platforms that only allow to track and trace products from their specific brand. Track-Trace.com platform [22] also integrates the existing BPaaS platforms but only allows to visualize the tracking data, while both Bosch BrgP and the forwarders are able to negotiate freights using iFloW.

3 The iFloW Project

3.1 Problem Statement

Bosch BrgP has a large number of suppliers, which are dispersed all over the world. These suppliers send raw materials to Bosch BrgP's warehouses as a response to requests from the system that controls and manages the stock levels in the plant. However, at a certain point in time, Bosch BrgP does not have access to the exact location of these raw materials. The information related to the inbound flow of logistics is not consolidated and systemized. Bosch BrgP has access to a lot of information from suppliers and 3PL providers, but this information is dispersed over many alternatives, like emails, reports, and physical paper-based documents. Additionally, if any deviation occurs in the transportation process, logistics planners face many difficulties for accessing such information. This problem affects several stakeholders, from operational employees to the top management. This process faces many other concerns, like the fact that people from different areas are involved, information may not be available to users at the same time and in different formats. The planner must control information on all levels, however many times repeated information deviate the process. Also, the dependency from many different partners is a risk that must be managed.

Logistic planners spend at least 10 % of their time in track and trace activities. In the inbound flow of raw materials, the transit time ranges are between four hours for local milkruns and eight days for air transportation, with the exception of the sea transportation that takes five to six weeks. Currently, the tracking of the raw materials is mainly performed by e-mail and phone, in order to communicate with suppliers, carriers and forwarders. The logistic planners have access to a wide set of software applications (e.g., spreadsheets, forwarder's tracking websites, ERP system) to support their decisions.

The plant receives, on a daily basis, a large number of freight shipments from different suppliers dispersed all over the world, and there are many issues that may arise during each process. For instance, ETA of a particular part can be affected by various type of problems associated with the transportation or transshipment processes. Also, the lack of visibility of incoming freights results in larger efforts in the materials planning team that needs to manage dispersed and not normalized information (emails, phone calls, web sites) from their 3PL providers to get the best time of arrival estimation for their materials. Furthermore, delays on the parts′ time of arrival can have a huge impact on the production scheduling, increasing the overall production costs (either by rescheduling production plans whenever a shipment is not on time at the plant or using special air freights to speed up parts delivery at the plant).

Therefore, a new solution is required to integrate freight information from all stakeholders involved in the inbound logistics process providing Bosch's planners with more accurate data and smarter decisions.

3.2 The System Architecture

The software system is a tracking tool that provides its users a global view of the materials in transit, thus fulfilling the following goals:

1. To improve inbound logistics visibility in terms of raw material traceability;
2. To systemize the ordering process and consequently improve its efficiency;

3. To reduce costs related with the occurrence of special freights requests and production downtime occurrences due to failures in the transportation process;
4. To improve the accuracy of inbound logistics indicators, such as time spent by each carrier in a given activity and the degree of participation in the various activities.

Figure 1 illustrates the architecture of the iFloW system, where one can see the integration of the iFloW main server with other systems. It uses information that provide from many technologies, like GPS, Personal Digital Assistant (PDA) devices or SAP-OER (Object Event Repository). The 'middleware server' relates to the enterprise interoperability challenges, aiming to standardize the way information between Bosch BrgP and providers is exchanged. Each supplier or forwarder has their own system, protocol or data structure to integrate. These challenges were overcome with joint development and continuous meetings between iFloW project members and freight providers, by performing mappings and alignment between data formats, used in Bosch BrgP logistics, suppliers and forwarders. The middleware was also developed in order to assure the interaction with GPS systems and EPCIS events aligned with Bosch security policies. EPCIS is a GS1 standard to support information sharing about a (physical) movement between trading partners. The use of this standard also allows visualizing the status of products throughout the supply chain. Finally, the 'iFloW server' executes all business logic, where Bosch BrgP's users, as well as the remaining stakeholders, use these features as a single point of information via a web-based user-interface (web app).

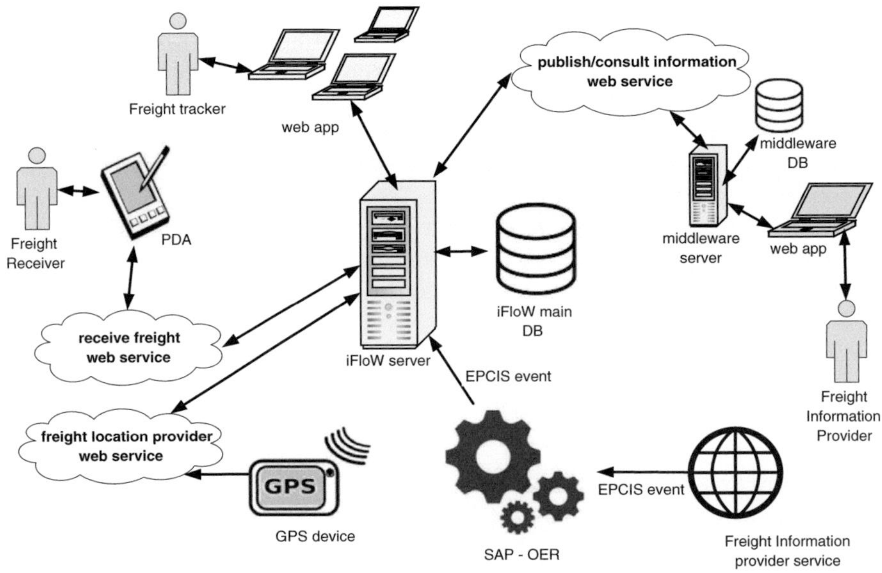

Fig. 1 System architecture

The 'freight location provider web service' is the connection to the GPS system. A GPS device is placed in each vehicle and then activated. The location of the freight is provided to the iFloW system by means of a web service that was developed to that specific purpose. The SAP-OER addresses centralized control and visibility requirements, in this case for product tracking and authentication. The information is exchanged by a EPCIS event. The freight information provider service is the system that sends information about freights in transit. Finally, a PDA device inputs the information of the freights that arrive to the Bosch BrgP facilities and sends to the iFloW system ('receive freight web service').

4 Current Results

A pilot study (small scale preliminary study) was performed in order to evaluate the solution's feasibility. The pilot was developed in order to assess the software system's usage in a tracking task, the need for improvements within the initial design, and ultimately its potential, prior to the development of the full-scale iFloW system. The pilot project characterizes routes performed in the Asian scope, mainly transported by sea and transported by *Forwarder A* (see example in Fig. 2). The route includes land transport between the Asian suppliers and an Asian port, sea transport between the Asian(s) port(s) and a European port and land transport between the European port and Bosch BrgP. The pilot solution allows the real-time tracking of raw material in transit from a departure point to Bosch BrgP facilities, giving additional information to perform the tracking of goods effectively and efficiently. This pilot is able to provide the required insights since: (1) 42 % of suppliers are located in Asia; (2) 95 % of incoming orders are transported by sea and performed by *Forwarder A*; (3) after the first Asian port, the transport is Bosch's "responsibility"; (4) transit time is 5 to 6 weeks; (5) is related to 79 % of special freights requests that arise from transportation process; (6) in Bosch BrgP is one of the route with more time spent in freight tracking tasks; (7) high risk to production downtime due to extended transit time; (8) *Forwarder A* has Application Programming Interfaces (APIs) that facilitate detailed information to be obtained; and (9) to get detailed information with the GPS technology.

The pilot phase of iFloW project is at this date concluded. The software system provides the location and detailed information of a given part or freight over the route and the respective part details and freight details (Fig. 3). From Fig. 3, we would like to highlight the following aspects: the status (e.g., moving or stopped in an intermediate point—used to intercept a freight if necessary), the ETA to the next intermediate point, the trace of the part along the entire route and respective check points (the user can switch between map view and table view) and very important to the production planning—the ETA to Bosch BrgP facilities.

Fig. 2 Example of a route used in pilot

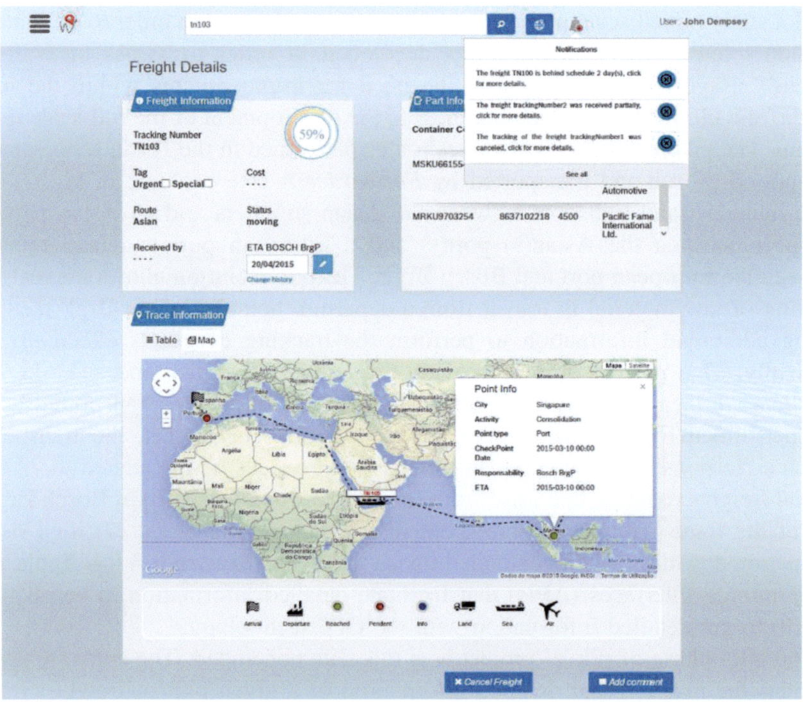

Fig. 3 Freight Details screenshot

Additionally, as can be observed in the top right menu of Fig. 3, the iFloW system alerts ETA changes (e.g., the new ETA and the number of days in that the freight is delayed), or related with freight's cancelation or changes in freight quantities. These notifications add relevant value in freight tracking and production planning tasks. During the pilot phase, the iFloW system was able to successfully:

- Obtain freight information and real-time location from European port and six Asian ports from *Forwarder A*'s system (this task replaces, for instance, e-mail exchanges).
- Search freight location and show its general information, trace freight (shows the events and progress), edit freight information, issue alerts (for ETA changes and for updates in freight quantities).
- Export freight information to Legacy System (including, for instance, Excel files).
- Other features, like: produce statistics information; confirm that the freight has arrived; make freight information available to the Freight Information and Location Provider; consult freight information to prepare, for instance, the delivery plan for a set of freights; and edit and validate the delivery plan for a set of freights for a given day.

These functionalities allowed the iFloW system to optimize in 2 % the flow of transport arrivals, due to the freight negotiation performed in the platform. Freight negotiation relates to cases like, for instance, decide to store materials in one of the intermediate ports when weekly production planning is updated.

Although this phase has ended, it is difficult at this point to measure the project's impact within Bosch BrgP's inbound logistics processes. The pilot allowed perceiving that the iFloW system has indeed potential to improve inbound logistics processes and supply chain management, by providing planners with an improved context for performing the planning. Additionally, allowed depicting risks and obstacles that the prior full-scale development faces, and required encryption efforts in messages' exchange. The pilot faced problems on dependency and communication with third-parties, since the core functionalities rely on inputs from suppliers and forwarders, mitigated with the use of GPS technology from the European port to the freight's arrival at the plant in Braga. Stakeholders were permanently involved so there was common understanding on the approach to use.

5 Conclusions

This paper presents the iFloW project, an integrated logistics software system for inbound supply chain traceability. iFloW is a real-time tracking software system of freights in transit from suppliers to Bosch BrgP facilities, that receives freight information from GPS systems and it integrates with forwarder's systems for retrieving tracking information for update ETA's value and notifications of deviations, as well as enables visibility of the global inbound supply chain traceability process. Visibility and traceability were challenging since the project is executed in a heterogeneous environment, where each supplier or forwarder have their own system, protocol or data structure to integrate. The project required joint efforts between project team, suppliers and forwarders in order to define mappings between data formats, as well as secured interactions with GPS and EPCIS events.

The proposed functionalities were sufficient to successfully enable the tool to perform within the proposed pilot project. Thus, remaining effort will be devoted to implement additional functionalities. The pilot allowed to validate the approach and to depict the risks that the full-scale development faces, namely the dependency form third-parties and security. Since the pilot phase has just been recently concluded, it is not possible to measure its impact within Bosch BrgP facilities. It is expected that its use results in the reduction of track and trace activities, the number of complaints caused by transportation process and of the number of special freight requests due to freight monitoring failures, because Bosch BrgP planners, suppliers and forwarders use the iFloW system as a single point of communication. After the conclusion of the project, we will measure and address these impacts in the plant in future works.

Acknowledgments This research is sponsored by the Portugal Incentive System for Research and Technological Development PEst-UID/CEC/00319/2013 and by project in co-promotion n° 36265/2013 (Project HMIExcel—2013–2015).

References

1. Butner, K. (2010). The smarter supply chain of the future. *Strategy & Leadership, 38*(1), 22–31.
2. IBM. (2010). The smarter supply chain of the future.
3. GS1. (2007). The GS1 Traceability Standard: What you need to know. Retrieved from http://www.gs1.org/docs/traceability/GS1_tracebility_what_you_need_to_know.pdf
4. Martins, J. C., & Machado, R. J. (Eds.) (2012). Ontologies for product and process traceability at manufacturing organizations. In: *Proceedings of the 8th International Conference in A Software Requirements Approach. Quality of Information and Communications Technology (QUATIC)*, IEEE.
5. Brewer, A., Sloan, N., & Landers, T. L. (1999). Intelligent tracking in manufacturing. *Journal of Intelligent Manufacturing, 10*(3–4), 245–250.
6. Kandel, C., Klumpp, M., & Keusgen, T. (Eds.) (2011). GPS based track and trace for transparent and sustainable global supply chains. In: *Proceedings Of The 17th International Conference In Concurrent Enterprising (Ice)*. IEEE.
7. Ghiani, G., Laporte, G., & Musmanno, R. (2004). *Introduction to logistics systems planning and control*. New York: Wiley.
8. Gartner Group. (2012). *Hype cycle for supply chain management*. Stamford: Gartner Group.
9. UPS: Tracking System. Available from http://www.ups.com/WebTracking/track?loc=en_us. Accessed October, 2015.
10. KN Portal. Available from http://www.kn-portal.com/. Accessed October, 2015.
11. TNT Express Tracker. Available from http://www.tnt.com/webtracker/tracker.do?navigation=1&respLang=en&respCountry=US&genericSiteIdent. Accessed October, 2015.
12. Panalpina. Available from http://www.panalpina.com/. Accessed October, 2015.
13. DHL. Available from http://www.dhl.com/en/express/tracking.html. Accessed October, 2015.
14. FedEx Tracking. Available from https://www.fedex.com/Tracking. Accessed October, 2015.
15. Lusocargo. Available from https://www.lusocargo.pt/track.php#!/home. Accessed October, 2015.

16. Yin, M., Wang, X., & Zhang, R. (Eds.) (2008) Standardization and integration of information system in international container logistics wireless communications. In: *Proceedings of the 4th International Conference on Networking and Mobile Computing WiCOM'08* IEEE.
17. Doukidis, G. I., Chow, H. K., Choy, K., Lee, W., & Chan, F. T. (2007). Integration of web-based and RFID technology in visualizing logistics operations-a case study. *Supply Chain Management: An International Journal, 12*(3), 221–234.
18. Choy, K. & Ng, S. W. (Eds.) (2006). Improving supply chain traceability with the integration of logistics information system and RFID technology. *Materials Science Forum*. Switzerland: Trans Tech Publ.
19. Choy, K., So, S. C., Liu, J. J., & Lau, H. (2007). Improving logistics visibility in a supply chain: an integrated approach with radio frequency identification technology. *International Journal of Integrated Supply Management, 3*(2), 135–155.
20. He, W., Tan, E. L., Lee, E., & Li, T. (Eds.) (2009). A solution for integrated track and trace in supply chain based on RFID & GPS. In: *Proceedings of the IEEE Conference on Emerging Technologies & Factory Automation (ETFA 2009)* IEEE.
21. Carlino, M., Confessore, G., & Liotta, G. (Eds.) (2009). Potentials of satellite-based traceability services in logistics. In: *Proceedings of the European Navigation Conference–Global Navigation Satellite Systems.*
22. Track-trace.com. Available from http://www.track-trace.com/. Accessed April, 2015.
23. Terzi, S., Cassina, J., & Panetto, H. (2006). Development of a metamodel to foster interoperability along the product lifecycle traceability. In: D. Konstantas, J. P. Bourrières, M. Léonard & N. Boudjlida (Eds.), *Interoperability of enterprise software and applications*. Germany: Springer.
24. Terzi, S., Panetto, H., Morel, G., & Garetti, M. (2007). A holonic metamodel for product traceability in product lifecycle management. *International Journal of Product Lifecycle Management, 2*(3), 253–289.
25. Chen, X. (2009). RFID middleware design research. Applied computing, computer science, and advanced communication (pp. 50–56). Springer.
26. Holmström, J. & Främling, K. (Eds.) (2005). Design patterns for loosely coupled track, trace, configuration, and check operations in multi-company environments. In: *Proceedings of EUROMA' 2005 conference.*

Qualitative Evaluation of Manufacturing Software Units Interoperability Using ISO 25000 Quality Model

Henri Basson, Mourad Bouneffa, Michiko Matsuda, Adeel Ahmad, Dukki Chung and Eiji Arai

Abstract In the context of development and evolution of heterogeneous industrial systems, the quality of interoperability among manufacturing software units is a major concern. Qualitative characteristics of the interoperability such as reliability, functionality, performance, and security are necessary for the assurance of high quality of the developed manufacturing applications. The implemented interoperability among various software units of manufacturing applications often have various quality criteria. Therefore, an adequate qualitative modeling is required to satisfy the qualitative evaluation of the interoperability. This paper presents the use of quality model developed in respect of quality standard ISO 25000 series as a means to permit the comprehensive evaluation of interoperability of the applications developed regarding the ISO 16100 standard series.

Keywords Manufacturing software units · Quality of interoperability · ISO 25000 series · Quality model · ISO 16100 series

H. Basson (✉) · M. Bouneffa · A. Ahmad
LISIC - Laboratoire d'Informatique Signal et Image de la Côte d'Opale EA 4491, Univ. Littoral Côte d'Opale, Calais F- 62228, France
e-mail: basson@lisic.univ-littoral.fr

M. Matsuda
Kanagawa Institute of Technology, 1030 Shimo-Ogino, Atsugi-shi 243-0292, Kanagawa, Japan

D. Chung
Rockwell Automation, 1 Allen-Bradley Drive Mayfield Hts, Cleveland, Ohio 44124, USA

E. Arai
Graduate School of Engineering, Osaka University, 2-1 Yamadaoka, Suita 565-0871, Osaka, Japan

© Springer International Publishing Switzerland 2016

K. Mertins et al. (eds.), *Enterprise Interoperability VII*, Proceedings of the I-ESA Conferences 8, DOI 10.1007/978-3-319-30957-6_16

1 Introduction

The functional and qualitative characteristics of interoperability cover a wide range of possible capabilities according to the considered types of system, their description level, and their organizations. In the manufacturing domain, along with the evolution of software applications in various activities domains, the interoperability has become a major critical aspect influencing the quality of working systems. Various definitions have been proposed for the interoperability, as illustration, Kosanke cited 22 different definitions extracted from the literature on the topic [1]. For its specified domain, an important work has been realized by the US DoD to propose a schema called LISI (Levels of Information Systems Interoperability) [2] where the interoperability was defined as "*The ability of systems, units or forces to provide services and accept services from other systems*". The IEEE definition designates the interoperability as the "*ability of two or more systems or components to exchange information and to use the exchanged information*".

The present work considers mainly the manufacturing applications for which the constituting manufacturing software units may require interoperability facilities among units for the application development and use. More precisely, the interoperability mechanisms such as data exchange, messages communication, services calls, etc. may need to focus on the quality of interoperability and its functioning between concerned manufacturing application units.

The interoperability multiplicity of levels inevitably encourages the standardization activities [3]. Effectively, the governing bodies have always encouraged and supported the standardization to insure an integration of multiple heterogeneous systems to operate as a coherent whole.

In this paper, we focus on the standards that are dedicated to interoperability of manufacturing software units and the standards that can be used for the qualitative evaluation. The major goal is to use adequate recognized standards to examine the proposed practices for system integration through interoperability and to evaluate its quality using ISO recommended practices. Primarily, our work considers the ISO 16100 series standard [4] to propose an approach and reference model to assure better qualitative level of interoperability solutions to reach better integration of manufacturing systems. The present work considers the association of these applications with the qualitative characteristics of the ISO/IEC 25000 standard [5] providing one of the most exhaustive model for software quality characteristics. The major goal remains to enhance the competency of application designer using the ISO 16100 series approach while considering the quality of interoperability of working application software units.

The rest of the paper is organized as follows: the Sect. 2 is dedicated to describe the ISO 16100 series. The corresponding approach advocates the use of standardized manufacturing software units to build integrated manufacturing applications. The Sect. 3 presents the ISO/IEC 25000 series software quality standards. In Sect. 4, we briefly show, how to associate, the quality characteristics as designed in reference to ISO/IEC 25000, with the manufacturing software units developed according to the ISO 16100. The Sect. 5, describes the prototype scheme of the proposed approach where the implementation is based on the linked data and ontologies. The conclusion is given in Sect. 6.

2 The Interoperability of Manufacturing Applications: Approach as per ISO 16100 Standard

The ISO 16100 standard series defines an approach to insure the interoperability of software applications that can be used in manufacturing domain. The central element of this approach consists of the implementation of a repository or distributed catalogue of software units called MSUs (Manufacturing Software Units). The MSUs are used by the vendors and the application developers of the manufacturing systems to construct their applications. These MSUs are indexed by profiles so that they could be located according to a search criteria. These criteria describe the functionalities assured by the MSUs and other quality criteria such as reliability, performance, etc. Obviously, the MSUs are distributed on multiple vendors that can be geographically apart. It is therefore necessary to adopt a standardized vocabulary to elaborate the indexation profiles of these MSUs. In this regard, the ISO 16100 standard adopts the following approach:

- Define a data dictionary, called MDD (Model Data Dictionary), which can be considered as an ontology regrouping the most significant elements of a given domain. These elements represent the processes, activities, and the resources that can be used in a given domain.
- Elaborate the taxonomy of the regrouped capability elements in a hierarchical structure of capability classes. A capability class encapsulates activities or functions i.e., "elaborate a fabrication sequence" to be performed. It must also take into account the assurance of certain qualitative criteria such as response time, optimal use of resources, etc. In its definition the capability class is linked to the MDD elements.

The profile of a software unit in this regard is an instantiation of one or more capabilities classes, which might be serialized in XML format, etc. A more detailed description of ISO 16100 can be found in [6].

The Table 1 shows the structure of a profile in terms of capabilities. This structure represents a template that a profile may instantiate with the associated values of each element. Different elements involved in a profile describe also the MSU identification, used language, and its vendor, etc. The various profile indications on MSU structure and some quality criteria are related to capability classes that are implemented by the MSU along with its associated constraints. We can also remark that the quality characteristics in regard to the interoperability (intrinsic to the MSU) can only be invoked across a list of objects or entities of the MDD. We can assume that

Table 1 Structure of a "software unit profile"

- Common Part
    - Template ID
    - Capability Class Name and Reference CCS
    - Software Unit ID
        - Vendor Name
        - Version Number & History
        - Computing Facilities Required
            - Processor
            - OperatingSystem&Options
            - Language
            - RuntimeMemory
            - DiskSpace
            - MultiUserSupport
            - RemoteAccess
            - AddOns&PlugIns
        - Measured Performance of the Unit
            - ElapsedTime
            - NumberOfTransactionsPerUnitTime
        - Reliability Data of the Unit
            - UsageHistory
            - ....
        - Support Policy
        - Price Data
    - Capability Class Reference Dictionary Name
    - Number Of Profile Attributes
    - ......
    - ....
- Specific Part for Capability Class
    - Reference MDM Name
    - MDD Description Format
    - MDD Description
        - Set Of MDD Objects
        - List Of MDD Objects
        - Time Ordered MDD Objects
        - Event Ordered MDD Objects
    - **Interoperability MDD Objects**
    - List Of Capability Class Attributes
    - List Of Capability Class Methods
    - List Of Capability Class Resources
    - ....

such objects may represent a software artifact like a data bus, a database connection, or a web service, etc. For instance, it is difficult to establish the associated quality of interoperability not only within a software unit, but also in the constituted application. In the following section, we present our approach that emphasizes the use of quality standard ISO 25000 to elaborate the guidelines for the evaluation of interoperability of MSUs developed according to the ISO 16100 series.

3 Adaptation of ISO 25000 Quality Model for Interoperability Validation of MSUs

The interoperability validation process requires to examine the correspondence between the specified qualitative requirements of interoperability between MSUs and its relevant implementation.

A major aim of our work is to provide guidelines and to illustrate the series of ISO 25000 quality standards addressing all aspects related to: Quality requirements specification, Quality model, Product Quality, Quality evaluation, and Quality measurement. For each quality aspect, a set of ISO standards has been elaborated where a set of standards constitutes a "Division" aiming at a specific aspect related to system or software quality [5].

The ISO 25000 series of standards has been supported with the SQuaRE (Software product Quality Requirements and Evaluation) framework. The objective of SQuaRE is to help the practice of main concepts and models provided by ISO 25000. SQuaRE is composed of five divisions intended to cover the wide range of aspects related to System and Software Quality Specification, as well as its modeling, assessment, and measurement (Fig. 1).

The ISO norms of SQuaRE make distinction between three major types of quality reflecting the perception of the quality by the developers and users. The major quality types are the quality in use, the external quality, and the internal quality. SQuaRE proposes to examine the conformity of the each quality type between expected specified quality of system or software and the effective quality of implemented system or software, in order to validate the developed interoperability of manufacturing application. At each hierarchical quality level a quality model is built in terms of quality characteristics, sub-characteristics, and properties.

For this goal, ISO/IEC 25010 proposes a generic quality model along with the corresponding characteristics and sub-characteristics for the concerned major quality levels i.e., internal quality, external quality, and quality in use. These quality models are to be instantiated and adapted in respect to the interoperability of the MSUs of the manufacturing application under consideration. The generic quality model, to be instantiated, proposes six characteristics and 27 sub-characteristics. These indicate the links of dependency between each individual characteristic and the corresponding sub-characteristics. The six major quality characteristics are adapted to interoperability, as shown in Table 2.

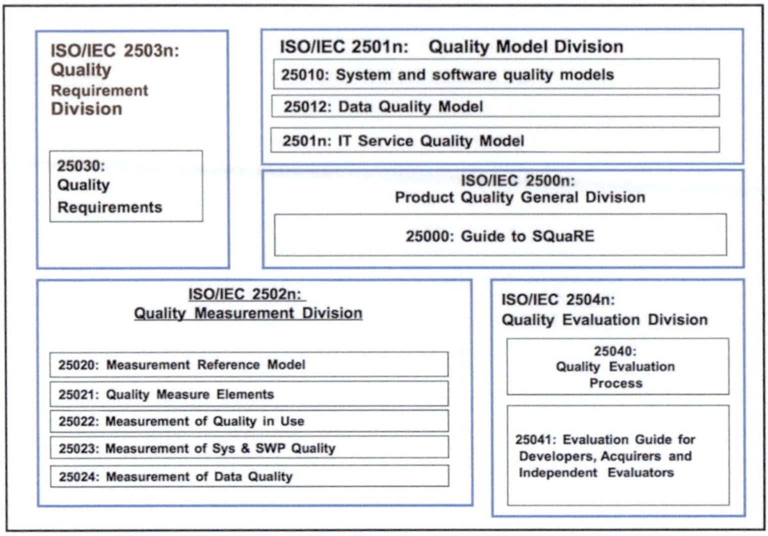

Fig. 1 Global architecture of ISO 25000 SQuaRE

Table 2 Characteristics and sub-characteristics of SQuaRE schema adopted for the quality of interoperability between MSUs

Quality characteristics of MSU interoperability	Description	Quality sub-characteristics of MSUs interoperability
Functionality	The capability of the MSUs to provide functions and interoperability which meet stated and implied needs when the MSUs are used under specified conditions	Suitability Accuracy Security
Reliability	The capability of the MSUs to perform functions and interoperability which meet stated and implied needs when the MSUs are used under specified conditions for a specified period of time	Availability Fault tolerance Recoverability
Usability	The capability of the MSUs interoperability to be understood, used, and appreciated by the developer	Understandability Learnability Testability
Efficiency	The capability of the MSUs interoperability to provide the required performance relative to the amount of used resources under stated conditions	Time behavior Resource utilization
Maintainability	The capability of the MSUs interoperability to be modified	Reusability Analysability Changeability Testability
Portability	The capability of the MSUs interoperability to maintain its behavior when hosted on a different hosting environment	Adaptability Installability Replaceability

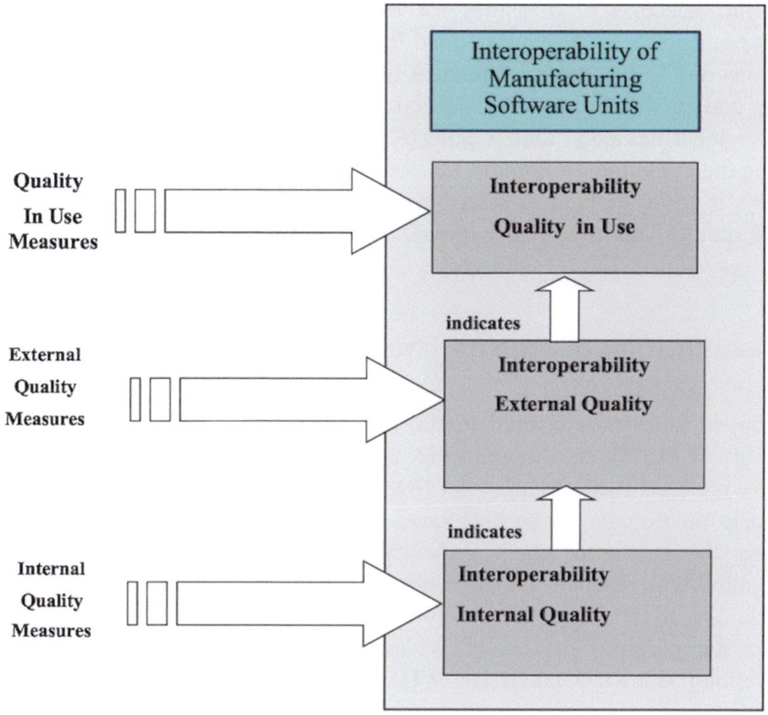

Fig. 2 Evaluation of interoperability quality

SQuaRE proposes the breaking-down of the characteristics into sub-characteristics, which is adapted to the interoperability quality, as stated below:

The quality engineer of the current application defines each individual sub-characteristic. It specifies its contextual meaning and the way to assess the extent to which each characteristic is met. Moreover, in a given application, sub-characteristics may have different priorities. The priority of each individual sub-characteristic depends on the hierarchical scale established between sub-characteristics according to the specific functional and qualitative priorities of the current manufacturing application.

The quality engineer adapts a set of metrics for each individual sub-characteristic. The adapted set of metrics is addressed to quantitatively assess the extent to which a sub-characteristic is met in reference to the expected values decided at the phase of qualitative requirements specifications.

The qualitative evaluation of interoperability is based on the three quality tree-like hierarchical structures instantiated by the quality engineer for the quality in use, external quality, and internal quality. Figure 2 shows the procession of the interoperability validation as indicated by SQuaRE schema.

For the selection of measures for the characteristics of interoperability, the Quality Measurement Division defined in ISO/IEC 25020 covers the mathematical definitions and guidelines for practical measurements of internal quality, external quality, and quality in use. In addition, it includes the definitions for the basic primitive measurements, and is able to formulate the all possible measures. By applying the guidelines of Quality Measurement Division, a set of basic and derived measures is to be defined to select metrics for evaluation of the internal quality, external quality, and the quality in use.

4 Classification of Quality Evaluation Metrics

We propose to adapt a layered model [7] for the qualitative evaluation of interoperability of MSUs. As shown in Fig. 3, it is comprised of a core and two layers. The core is called Basic Metrics Set (BMS). BMS is composed of elements where each basic metric can not be calculated using other defined metrics. Each metric examines the extent to which a quality characteristic (or sub-characteristic) of interoperability is reached. It can be observed that the basic metrics are insufficient for evaluating diverse interoperability qualities. Two layers of metrics can be used to serve the purpose: the set of Widely-Used Metrics (WUM) and the set of User-Defined Metrics (UDM). The WUM is comprised of the proposed metrics in

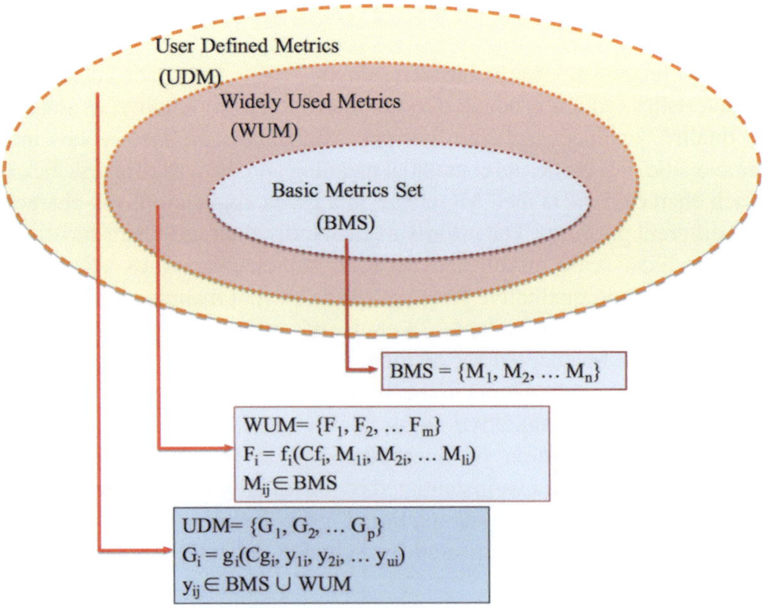

Fig. 3 Classification of interoperability quality metrics

Table 3 Some basic metrics (BMS)

Metric	Semantic	Formula
FUN_INDE	The number of input data elements of MSU_i (the ith MSU of the current application)	FUN_INDE (MSU_i)
FUN_OUTDE	The number of output data elements of MSU_i	FUN_OUTDE (MSU_i)
MSU_PORT_IN	The number of ports of message reception of MSU_i	FUN_PORT_IN (MSU_i)
MSU_PORT_OUT	The number of ports of message sending of the MSU_i	MSU_PORT_OUT (MSU_i)
NB_MinR_MSU	Lowest number of messages to be received by an MSU_i	NB_MinR_MSU (MSU_i)
MSU_XMLR	Number of XML files read by an MSU_i	MSU_XMLR (MSU_i)
MSU_XMLW	Number of XML files written by an MSU_i	MSU_XMLW (MSU_i)

the literature and widely used in the industry. An element of WUM is a function parameter of either one or more MSUs for interoperability evaluation, or one or more metrics belonging to the set BMS.

The Table 3 summarizes some basic metrics frequently used to measure software systems interoperability.

5 Prototyping Issues

In order to validate the proposed approach of evaluating the qualitative characteristics of MSU interoperability, a prototype tool is under development to validate the proposed approach based on ISO 16100 and further extended in reference to ISO 25000 model. The general architecture of the framework and its components is described in Fig. 4. The implementation strategy considers that the vendors of MSUs publish their profile on Internet in form of linked data. It allows the construction of a large collection of MSUs, which can be accessed via the semantic web tools, such as the SPARQL queries [8]. The adapted implementation approach considers mainly following constituents:

A knowledge base which is composed of two major parts:

1. A database of linked data which is a distributed collection of RDF (Resource Data Framework) [9] files, representing the template classes of capability, their instances, which in fact are the profiles of MSU and the inter-linking qualitative characteristics.
2. A set of three ontologies, specified by the OWL (Web Ontology Language) [10] and which explicitly describes the semantics of each capability class, the MDD, and the modeling of the interoperability quality. Indeed, the profile templates are

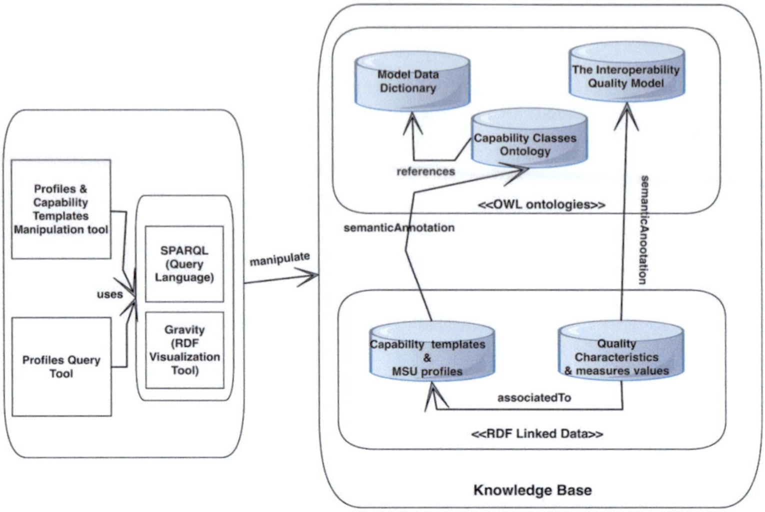

Fig. 4 Global architecture of the framework

the RDF files, where the elements or tags have the semantics defined by the ontologies of capability classes. These are referred to the entities defined by the MDD. Similarly, the semantics of the qualitative characteristics are defined by ontologies of quality.

Profiles and Capability Templates Manipulation tool which allows the manipulation (creation, modification, and deletion) of the profiles and templates, along with Profiles Query tool to search the MSU profiles. These two tools use the SPARQL language to query the RDF files and OWL. We also use Gravity[1] that allows a graphical exploitation of the linked data, in graph-oriented forms.

The MSUs vendor may use the profile and capability templates manipulation tool to find the best template instance, in order to create a profile for its MSU. In case if the template exists already, the vendor may proceed to instantiate it with the provision of respective specific data.

6 Conclusion and Perspectives

For an adequate modeling of the qualitative evaluation of the interoperability among the Software Units of a manufacturing application, we adapted and used a quality model developed in respect of quality standard ISO 25000 series. The used model is exhaustive, therefore, it requires a good experience in software quality

[1]http://semweb.salzburgresearch.at/apps/rdf-gravity/.

evaluation. The proposed use of metrics examine the extent of each interoperability characteristic or sub characteristic. Metrics are structured into layers for better traceability of adapted quality measurements. The work aims at permitting the developers to search the MSUs based not only on the functional capabilities characteristics but also on qualitative estimation of MSU interoperability characteristics. The proposed work is directly related the activities of ISO-TC 184/SC 5/WG4 aiming at the development of ISO 16300 series with the perspective of integrating the work, on Data Quality in ISO 8000, elaborated by ISO-TC 184/SC4.

References

1. Kosanke, K. (2007). ISO standards for interoperability: A comparison. In D. Konstantas, J.-P. Bourrières, M. Léonard & N. Boudjlida (Eds.), *Interoperability of enterprise software and applications* (pp. 55–64). London: Springer.
2. Kasunic, M., & Anderson, W. (2004). *Assuring systems interoperability: Challenges and opportunities*. Pittsburgh: Carnegie Mellon University.
3. Folmer, E., & Verhoosel, J. (2011). *State of the art on semantic IS standardization, interoperability & quality*. UT, CTIT, TNO en NOiV. ISBN 9789090260303.
4. Matsuda, M., & Wang, Q. (2010). Software interoperability tools: Standardized capability-profiling methodology ISO16100. In *Enterprise architecture, integration and interoperability* (pp. 140–151). Heidelberg: Springer.
5. ISO/IEC 25012 (2008). Software engineering-software product quality requirements and evaluation (SQuaRE)—Data quality model.
6. ISO/IEC 1600-N(1..6). (2011). Industrial automation systems and integration—manufacturing software capability profiling for interoperability.
7. Ahmad, A., Basson, H., Deruelle, L., & Bouneffa, M. (2010, October 3–5). Towards an integrated quality-oriented modeling approach for software evolution control, In *Proceedings of the 2nd International Conference on Software Technology and Engineering (ICSTE)*, (Vol. 2, pp. V2–320-V2-324). San Juan, Puerto Rico, USA.
8. Harris, S., & Seaborne, A. (2013). The SPARQL 1.1 query language, W3C recommendation. http://www.w3.org/TR/sparql11-query
9. Cyganiak, R., Wood, D., & Lanthaler, M. (2014). RDF 1.1 concepts and abstract syntax, W3C recommendation. http://www.w3.org/TR/rdf11-concepts/
10. McGuinness, D. L., & Van Harmelen, F. (2004). OWL web ontology language overview, W3C recommendation. http://www.w3.org/TR/owl-features

Process Modelling Approach for the Liquid-Sensing Enterprise

C. Marques-Lucena, J. Ferreira, M. Sesana, K. Fisher and C. Agostinho

Abstract A mutual interaction between the real, the digital and the virtual worlds in the Future Internet networked society provides a fertile playground for IoT-related research and innovation. This conceptual view enables enterprises to separate concerns and be able to define better strategies on how to act, react, plan and optimize their activities. Nowadays, business process and specific technologies can already support each stage of process implementation. However, none is focused on this conceptual division provided by the IoT advances. Representing and implementing the interactions between the three worlds is achieved by means of interoperable models and tools where business objectives can be aligned with software applications and services at the single enterprise or network level. This paper presents a model-driven approach and a tool to support the modelling and instantiation of the IoT-enabled liquid and sensing processes in an inter-world osmosis style.

Keywords Processes modelling · Osmosis processes · Model-driven architecture · Model transformations · MDSEA

C. Marques-Lucena
Faculdade de Ciencias e Tecnologia, UNL, Campus da Caparica,
2829-516 Caparica, Portugal
e-mail: clucena@campus.fct.unl.pt

J. Ferreira (✉) · C. Agostinho
UNINOVA-GRIS, Group for the Research in Interoperability of Systems,
UNINOVA, Campus da Caparica, 2829-516 Caparica, Portugal
e-mail: japf@uninova.pt

C. Agostinho
e-mail: ca@uninova.pt

M. Sesana
TXT e-Solutions, 20126 Milan, Italy
e-mail: michele.sesana@txtgroup.com

K. Fisher
DFKI GmbH, 66123 Saarbrucken, Germany
e-mail: Klaus.fisher@dfki.de

© Springer International Publishing Switzerland 2016
K. Mertins et al. (eds.), *Enterprise Interoperability VII*,
Proceedings of the I-ESA Conferences 8, DOI 10.1007/978-3-319-30957-6_17

1 Introduction

The progressive adoption of Business Process Management (BPM) paradigm by enterprises puts the spotlight on the business process lifecycle and on tools and technologies to support each stage of process modelling [1]. A business process can be defined as a set of activities performed in coordination in an organizational environment to reach a business objective [2]. Therefore, to better align the implementation and support of a process lifecycle, a separation of the business points of view from the technical and physical means to realize it is required [3], hence promoting the coordination and cooperation between teams.

Research in Enterprise Interoperability suggests that organizations can seamlessly interoperate with others at all stages of development, as long as they keep their business objectives aligned, software applications communicating, and the knowledge and understanding of the domain harmonized [4]. However, business process modelling is frequently over-comprehensive and hard to accomplish in a collaborative way, due to the extended domains and activities/services each enterprise is providing. With the advances and integration of IoT (intelligent) devices providing services for the network, that gap is becoming more obvious. The community is in need for an approach that puts in practice the separation of concerns to shorten the domain of analysis of business objectives, bringing together enterprises with different *know how*, to collaborate in processes modelling for the Future Internet dynamic market.

Part of the problem can be tackled following a model driven approach, which instead of writing the code directly, enables software products to be firstly modelled with a high level of abstraction in a platform independent way. It provides many advantages like the improvement of the portability, interoperability and reusability through the architectural separation of concerns [5]. In fact, Ducq et al. [3] adapted this concept to manufacturing services design and development, with the Model Driven Service Engineering Architecture (MDSEA). It follows the Model Driven Architecture (MDA) and Model Driven Interoperability (MDI) principles [6], supporting the modelling stage and guiding the transformation from the business requirements (Business Service Model, BSM) into detailed specification of components that need to be implemented (Technology Specific Models, TSM). The MDSEA approach implies that the different models, obtained via model transformation from the upper-level ones, should use dedicated service modelling languages that represent the system with different concerns, i.e., ICT, Human and Physical levels [7]. Indeed, MDSEA provides the building blocks for enterprise system development in the scope of an ecosystem of collaborating enterprises, providing: (1) The capability to transform a business specific model into a functional one so it can be perfected by a system architect detailing the necessary resources; (2) The capability to transform a functional model into a technology specific one envisaging the generation of concrete services.

Considering MDSEA is an efficient strategy for collative service design in a manufacturing environment, it is still missing an integrated approach to develop services with concerns related with the IoT-enabled real, digital and virtual worlds of the Liquid-Sensing Enterprise (LSE) [8]. This new form enterprise as special needs, emerging when objects, equipment, and technological infrastructures exhibit advanced networking and processing capabilities, actively cooperating to form a sort of 'nervous system' within the enterprise, which can be capitalized to define better strategies on how to act, react, plan and optimize their activities.

1.1 Osmosis Processes Concept

The OSMOSE Project (www.osmose-project.eu) is developing a reference architecture for modelling and managing sensing liquid enterprises [9], this is achieved by interconnecting three worlds: the Real World (RW), the Digital World (DW), and the Virtual World (VW). The osmosis concept is a process of passing the molecules from a less concentrated solution (individual perception of each world) to a more concentrated one [10]. The project follows this concept, where each enterprise has a special type of business processes used to moderate the information exchanged between the different worlds, named Osmosis Processes. Although having a different meaning inside the project they can me modelled using the same strategies of regular ones. The six Osmosis processes considered are [11]: (1) **Digitalization** (RW-DW)—Model and representation of real world data in a computer-tractable form; (2) **Actuation** (DW-RW)—Plan and implement highly distributed decision-making; (3) **Enrichment** (VW-DW)—Extends the computational and experiential capabilities of the Digital World annotations and projections coming from simulations and what-if hypothetical scenarios; (4) **Simulation** (DW-VW)—Instantiate and run hypothetical future scenarios fed by Digital World data; (5) **Virtualization** (RW-VW)—Provides data for simulation of hypothetical simulations from the real world and runs the simulation; and (6) **Augmentation** (VW-RW)—Annotates Real World objects with Virtual World information.

The work presented in this paper follows the need to have an integrated approach and tool to model the exchange of information between worlds (intrinsic of the osmosis processes) and the conditions (or events) that trigger the need of this exchange of information. In Sect. 2, the general approach towards the osmosis LSE processes modelling is presented. Sect. 3 consists in the detailed explanation of the OSMOSE modelling environment, and it is followed by a use case scenario in Sect. 4. Finally, some conclusions and future remarks are provided.

2 General Approach Model and Instantiate the OSMOSIS LSE Processes

The Osmosis LSE processes is a special type of enterprise process, embedded in the organization's business process, they extend the business processes running in the different worlds to link them with each other by crossing the membrane between the different worlds. They can be associated to certain services, which in turn can trigger events as well as act upon organizational resources.

Following the work of Ducq et al. (in [3, 12]) on model driven service engineering, and as explored by Agostinho et al. (in [7]) one can consider that the design of services is inherent to the manufacturing Liquid-Sensing Enterprise (LSE). Hence, the LSE design could benefit from the methodology behind MDSEA in order to accelerate the transition of the traditional enterprise to the "internet-friendly" and context-aware organization envisaged in OSMOSE. Thus, even though the initial objectives where different (processes modelling instead of services) the MDSEA can be adapted to support the osmosis process development, envisaging modelling at three abstraction levels that are integrated among themselves following semi-automatic top-down transformation mechanisms (Fig. 1):

- **Osmosis Business Models (OBM)**, specifying the business service model (BSM), and identifying the innovation requirements and behaviour for the novel service (using business language). This level is extended with specific annotation possibilities (OBM Annotated) that provide the possibility to separate the concerns among the IoT-worlds. This enables to identify osmosis processes. For these reasons OBM was divided in two parts, the BSM (Business Service Modelling) to design the Business model, and the OBM Annotated;
- **OSMOSE Technology Independent Models (OTIM)**, complementing the upper level model with detailed technology independent functionally. This modelling level is indented for product engineers or software architects to model the osmosis processes, their full behaviour and constraints. In transformation from OBM are oriented and pre-prepared for the osmosis process instantiation;

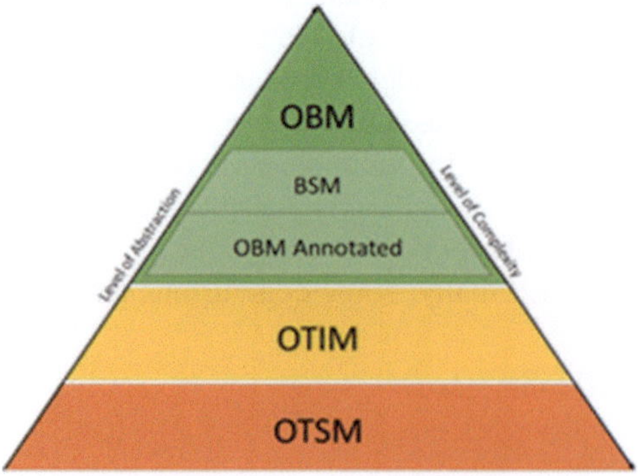

Fig. 1 OSMOSE process manager modelling architecture

- **OSMOSE Technology Specific Models (OTSM)**, which is the last level and consists in the instantiation of the identified behaviours and constrains with architectural components and in the process execution.

This three-layer paradigm facilitates the innovation potential involving multi-disciplinary teams, and bringing together skills that go from business and marketing, to design and implementation, progressively deriving and reusing knowledge down-to the implementation level.

A guidance on how the OSMOSE paradigm is adopted in the processes modelling and execution task is described step by step following the waterfall approach illustrated in Fig. 2. The first two steps are part of the OBM layer. The first one consists in the definition of the application goal, and the second one is focused on the identification of basic (business) participants involved (e.g., organizations and resources). The categories of participants that can be identified at this level are represented in the centre of the left part of Fig. 2.

The next two steps are related to the OTIM layer. The step 3 (Enrichment of business level participants and developments) starts with the identification of the involved worlds and separation of concerns to identify the osmosis processes. Then, the definition of the right events is essential. Some of them can be already in the business model and should be recognized. Others have to be created from scratch exposing new items or functionalities to the system. The participants identified in the OBM layer can also be enriched with technical information (e.g., relations between services and activities, since services where only identified at this point). In the OTSM steps, the required elements of the architectural components are instantiated to accomplish the OTIM level specifications and the processes are executed.

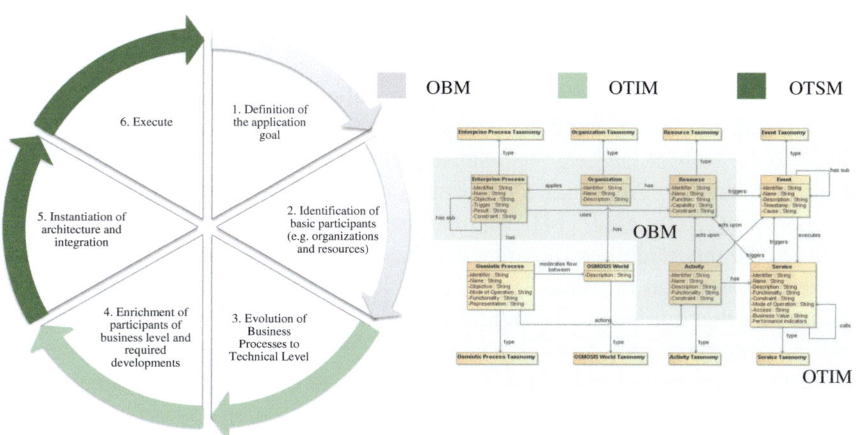

Fig. 2 Waterfall approach to instantiate OSMOSE processes modelling and execution

3 LSE Modelling Environment

As the general approach is instantiating a model driven framework, extending MDSEA, it is a logical choice that the model-driven process modelling framework is a continuation of MSEE's for service modelling. Hence, the LSE toolbox for process modelling and development is taking up the MSEE service lifecycle toolbox [13]. Several modelling languages have been chosen at each level to enable the osmosis processes modelling. At the OBM level, following the MSEE's MDSEA, the main language used is Extended Actigram [14]. The advantage to use this language at this level is its simplicity and easiness in the use for the companies' business members. At the TIM level, the modelling languages must allow a more detailed representation of each concept and process. Hence, also following MDSEA, BPMN 2.0 is the appropriate choice. At the TSM level, the jBPM environment for processes execution, was elected, accompanied by the Esper complex event processor for the events management, and RabbitMQ for messages profiles (E-mail and SMS) creation and management.

3.1 OBM-OTIM Process Transformation

In order to reach the OTIM layer, an ATL[1] engine is used to automatically execute the predefined transformation rules between the OBM and OTIM models. ATL is one of the most used transformation languages, having a large user base and being well documented [15]. It is composed by a set of rules that define how the source model elements are linked and how the target model elements are instantiated. Subsequently, BPMN 2.0 is used (via the BPMN2.0 Visual Editor for Eclipse[2]) to model technical details of osmosis processes. Thus, every time a user finalizes the design of their Business model, the tools offers to the user the possibility of transforming the business model into the technical model and enrich it with more technical information.

To enable this feature, it was necessary to define the mappings between both models used in the transformation. These mappings include a mapping from EA* language to the BPMN language and subsequent extension to handle the modelling of osmosis behaviour. An initial transformation from EA* to BPMN can be found on the work of Bazoun et al. [14]. Taking this work as a basis, some changes were made to handle the modelling of osmosis behaviours that are summarized in Fig. 3.

As can be observed in Table 1, the EA* process transformation to BPMN 2.0 triggers a rule that automatically creates all the necessary elements to describe the worlds interactions (pools, processes and lanes). The resources transformation was also changed. In the current result, both the EA* resource participants and

[1]https://eclipse.org/atl/.

[2]https://www.eclipse.org/bpmn2-modeler/.

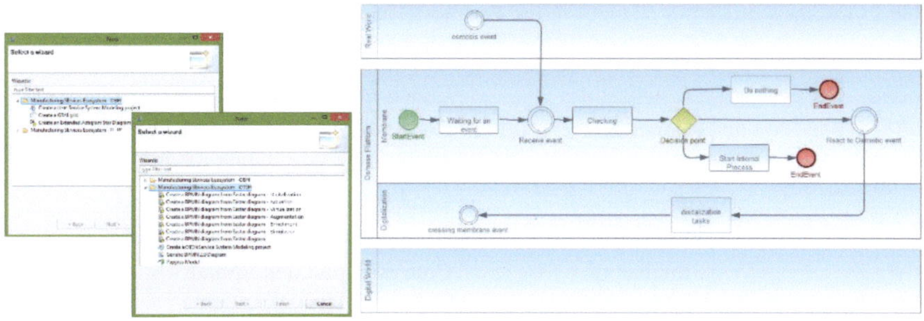

(a) (b)

Fig. 3 Osmosis process selection in tool interface

responsibles are transformed to elements belonging to the osmosis process staring world. The way that the current transformation affects the LSE modelling environment is explained in more detail in the following sections.

Table 1 Summary of changes made in EA* to BPMN 2.0 transformation

EA* (OBM)			BPMN 2.0 (OTIM)	
Process			Pool	Starting world pool
				Ending world pool
				Osmose platform pool
			Process	Middleware process
				Starting world process
				Ending world process
			Lane	Membrane Lane
				Osmosis process lane (e.g., digitalization)
				Ending world lane
Resources	Resource	Responsible for	Lane	Lane in starting world
		Participant in	Resource	Resource (added to the list of resources of a task)
	Human	Responsible for	Lane	Lane in starting world
		Participant in	Resource	Resource (added to the list of resources of a task)
	IT	Responsible for	Lane	Lane in starting world
		Participant in	Resource	Resource (added to the list of resources of a task)

Table 2 Osmosis processes changes in the EA* to BPMN 2.0 transformation

EA* (OBM)	Condition	BPMN 2.0 (OTIM)
Process	Digitalization	Creates real and digital world pools
	Actuation	Creates digital and real world pools
	Virtualization	Creates real and virtual world pools
	Augmentation	Creates virtual and real world pools
	Enrichment	Creates virtual and digital world pools
	Simulation	Creates digital and virtual world pools

3.1.1 Osmosis Process Modelling as Collaboration Diagrams

One change in the transformation is related to the fact that the transformation of an OBM EA* will **always** result in a OTIM BPMN 2.0 Collaborative Diagram. The advantage against using a regular BPMN, where the elements are represented within a pool or participant [16], is that a collaborative process allows to not only represent the activities of each osmosis word or process, but also the interactions between them. Thus, it can be considered, in a global point of view, that not only a perspective of a particular world is represented, but also the osmosis events that might cause the transition between worlds and also the messages exchanged between participants.

As represented in Table 2, when the transformation between EA* and BPMN is executed, independently of the osmosis process selected, the output BPMN diagram will have the *Osmosis Platform,* destination world and staring world pools and the *Membrane* (with its logic included) and osmosis process lanes, like represented in Table 2.

3.1.2 Osmosis Processes and Worlds

Another change in relation to MSEE's developments is related to the tool interface. As illustrated in 3a, at the moment a user decides to evolve the business model to its technical representation, he/she need to know *a prior* which is the Osmosis Process to represent. Thus, accordingly with the osmosis process selection, the conditions provided in 3 are triggered (e.g., the selection of a digitalization process will result in the creation of the RW and DW pools, as represented in 3b).

4 OSMOSE Use Case Example

The use case provided is related to the monitoring of a camshaft manufacturing tool which as a short life-time left. When it reaches a certain number of years or performance levels, a recalibration/replacement is needed and the camshaft production should be interrupted. This monitoring is made by a RW smart toolbox which has the capability to evaluate if the manufacturing tool is near expiration.

4.1 OSMOSE Business Model

Following the first two steps of the waterfall approach presented Sect. 2, the modelling starts by the identification of the application goal (Enterprise Process) and in the identification of the business participants:

- Enterprise Process—Monitoring of the manufacturing tool.
- Organizations—Manufacturer Organization.
- Resources—Monitoring System; Tool; Maintenance and Production Manager; Scheduler System; Technician;
- Basic Activities—Monitoring Lifetime of Manufacturing tool, Block Tool Usage, Notify Responsible, Schedule Maintenance Procedure for Tool, Select Appropriate Technician, Notify Technician, Execute Maintenance Procedure.

To describe the relations between the identified participants and the business strategy, an EA* model should be provided next (see Fig. 4). This model starts with the monitoring of the tool activity, and then, depending on the identified issues, other activities should be carried on: block the tool usage and notify the responsible entity if it is a severe problem; or just notify the responsible entity if the issue does not justify the tool shutdown. After that, a maintenance procedure should be scheduled, the appropriate maintenance technician selected and notified, so the actual maintenance can be executed and the camshaft produced.

4.2 OSMOSE Technological Independent Model

The steps 3 and 4 of the waterfall approach are executed by the technical team, which has the knowledge about the OSMOSE concepts and technical modelling skills.

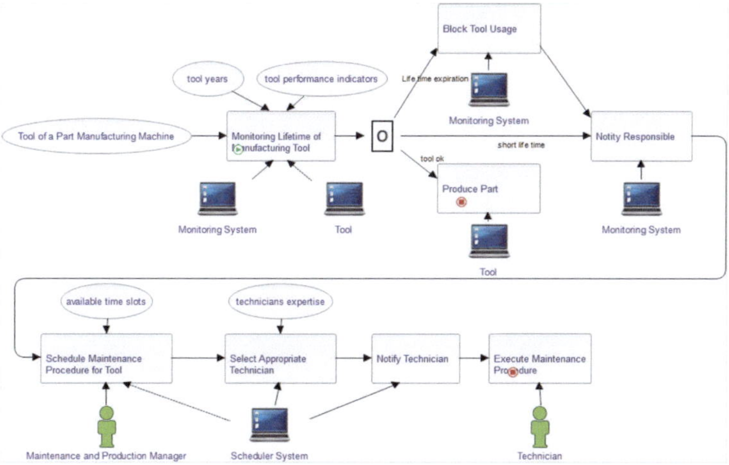

Fig. 4 EA* (OBM) model of Monitoring of the manufacturing tool

Thus, after the evolution of the business process to the technical level through the automatic transformation of EA* to BPMN 2.0 (step 3), the technical team is in charge of enrich the model with information about participants not identified at the business level (step 4). Although, for each OBM one or more OTIM models can be derived, for the sake of simplicity, in the example provided it is only considered the flow between the "Monitoring of Lifetime of Manufacturing Tool" as a RW activity and the need to "Notify Technician", a DW activity. After the identification of the flow as a digitalization osmosis process, new participants are identified:

- Osmosis Process—(Digitalization) Maintenance Schedule update;
- Worlds—Real and Digital Worlds;
- Events—Info available, Maintenance Required (osmosis event), Maintenance info available;
- Activities—Process tool information, Get tool location, Store data.

As illustrated in Fig. 5, the output of the transformation is composed by the OBM participants (only the ones in the identified flow) plus the elements

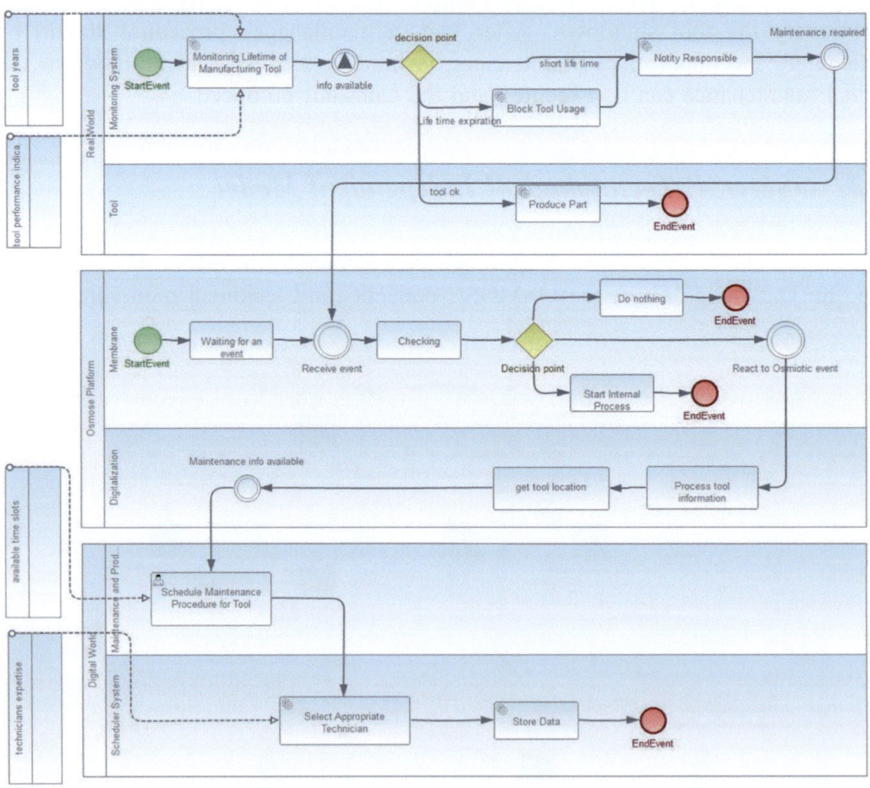

Fig. 5 OTIM model of the example

characteristic of the osmosis behaviour explained in Sect. 3. Afterwards, the diagram is complement with further technical detail using the newly (OTIM) participants identified.

4.3 OSMOSE Technological Specific Model

After the conceptual modelling phase, it is the implementation model, which is the design of the BPMN to be launched to execution. Between the OTIM and OTSM and because both are using BPMN 2.0, a simple export/import of the BPMN file is done. Hence, it is not really a transformation, however, if needed some predefined rules could be implemented together with the export functionality.

Then, in the model is made the configuration of the BPMN with the respective calls of the services and events, which are going to be used in the run-time mode. Finally, the model is complemented with more detail and ready to be tested and then ready to be launch in execution mode, making the osmiotic process runnable.

5 Conclusions

In this paper, the authors presented the *Osmosis Processes* concept and its associated modelling challenges for the liquid-sensing enterprise. In order to address them, a three layer paradigm based on the MSDEA approach is used to facilitate the coordination and cooperation potential between multi-disciplinary teams. This three-layer paradigm is instantiated by a 6 steps waterfall approach that allows the modeller to separate concerns between the real, digital and virtual worlds. It starts by the definition of the process application goal, to the identification of activities in each of the worlds, its modelling using a business and technical language (successively), and finally the osmosis process execution.

The 6 steps of the waterfall approach are supported by a modelling tool adapted from MSEE's project results to the purpose. The changes made to the tool follow the need to model the osmosis processes concept, namely the interactions between worlds, and the middleware membrane decision logic. As described in Sect. 3, the evolution ranges from the EA* to BPMN 2.0 mappings and transformations, to the interface itself (see Table 1). These changes enable enterprise's business members to use a simple and easy to modelling language like EA* to model the business strategies, and then, the technical team, with their knowledge about the osmosis worlds concept and technical modelling skills, to enrich the business model with the osmosis behaviours and constrains that need to be instantiated with technical components in the third layer paradigm level.

As future work, the authors intend to enrich LSE environment tool with the osmosis events pallet, so the osmosis processes modelling can be facilitated. The authors also want to improve the transformation by developing an application that allows the selection of the activities that will belong to each osmosis world's pools. These feature would enable users to access contextual information from the different worlds, providing to them control over the full process design, and worlds' components dynamics modelling and integration.

Acknowledgments The research leading to these results have received funding from the European Union 7th Framework Programme under grant agreement: OSMOSE nr 610905 (www.osmose-project.eu) and MSEE nr 284869 (www.msee-ip.eu).

References

1. Delgado, A., Ruiz, F., de Guzmán, I., & Piattini, M. (2010). MINERVA: Model driven and service oriented framework for the continuous business process improvement and related tools. In A. Dan, F. Gittler, & F. Toumani (Eds.) *Service-Oriented Computing. ICSOC/ServiceWave 2009 Workshops* (Vol. 6275, pp. 456–466). Berlin, Heidelberg: Springer.
2. Weske, M. (2007). *Business process management: Concepts, languages architectures*. Berlin: Springer.
3. Ducq, Y., Chen, D., & Alix, T. (2012). Principles of servitization and definition of an architecture for model driven service system engineering. In M. van Sinderen, P. Johnson, X. Xu, & G. Doumeingts (Eds.), *Enterprise interoperability* (Vol. 122, pp. 117–128). Berlin, Heidelberg: Springer.
4. Berre, A.-J., Elvesæter, B., Figay, N., Guglielmina, C., Johnsen, S. G., Karlsen, D., et al. (2007). The ATHENA interoperability framework. In R. Gonçalves, J. Müller, K. Mertins, & M. Zelm (Eds.), *Enterprise interoperability II* (pp. 569–580). London: Springer.
5. Grangel, R., Bigand, M., & Bourey, J. P. (2008). A UML profile as support for transformation of business process models at enterprise level. In *1st International Workshop on Model Driven Interoperability Sustainable Information Systems, MDISIS 2008—Held Conjunction with CAiSE 2008 Conference* (Vol. 340, pp. 73–87).
6. Lemrabet, Y., Bigand, M., Clin, D., Benkeltoum, N., & Bourey, J.-P. (2010). Model driven interoperability in practice: Preliminary evidences and issues from an industrial project. In *Proceedings of the First International Workshop on Model-Driven Interoperability* (pp. 3–9).
7. Agostinho, C., Bazoun, H., Zacharewicz, G., Ducq, Y., & Boye, H. (2014). Information models and transformation principles applied to servitization of manufacturing and service systems design. In *2014 2nd International Conference on Model-Driven Engineering and Software Development (MODELSWARD)* (pp. 657–665).
8. Agostinho, C., & Jardim-Goncalves, R. (2015). Sustaining interoperability of networked liquid-sensing enterprises: A complex systems perspective. *Annual Review in Control, 39*, 128–143.
9. Lampathaki, F., Athens, H. P., Kokkinakos, P., Tucci, C., Alvertis, I., Psarras, J., et al. (2015). FutureEnterprise a roadmap for sensing enterprise. Available from http://ceur-ws.org/Vol-1414/paper9.pdf.
10. Gusmeroli, S., Agostinho, C., Lucena, C., Sesana, M., Felic, A., & Fischer, K. (2015). {OSMOSE}: A Paradigm for the Liquid-Sensing Enterprise. In *Proceedings of the Workshops of the {IWEI} 2015 Conference Co-located with the 6th International {IFIP} Working Conference on Enterprise Interoperability {IWEI} 2015, N{î}mes, France, May 27, 2015.*
11. Marques-Lucena, C., Agostinho, C., Marcelino-Jesus, E., Sarraipa, J., & Jardim-Goncalves, R. (2015). Collaborative management of requirements using semantic wiki modules. In *2015 20th International Conference on Control Systems and Computer Science (CSCS)* (pp. 665–672).

12. Ducq, Y., Agostinho, C., Chen, D., Zacharewicz, G., & Jardim-Goncalves, R. (2014). Generic methodology for service engineering based on service modelling and model transformation. In *Manufacturing Service Ecosystem: Achievments of the European 7th Framework Programme FoF-ICT Project*
13. MSEE (2013). Specifications and Design of Service Lifecycle Management ToolBox. Available from http://cordis.europa.eu/docs/projects/cnect/0/284860/080/deliverables/001-MSEED153SpecificationsandDesignofServiceLifecycleManagementPlatformM18.pdf.
14. Bazoun, H., Zacharewicz, G., Ducq, Y., & Boye, H. (2013). Transformation of Extended Actigram Star to BPMN2.0 and simulation model in the frame of model driven service engineering architecture. In *Proceedings of the Symposium on Theory of Modeling & Simulation—DEVS Integrative M&S Symposium* (pp. 20:1–20:8).
15. Bézivin, J., Dupé, G., Jouault, F., Pitette, G., & Rougui, J. E. (2003) First experiments with the ATL model transformation language: Transforming XSLT into XQuery. In *2nd OOPSLA Workshop on Generative Techniques in the context of Model Driven Architecture* (Vol. 37).
16. White, S. A. (2004). Introduction to BPMN. In *BPTrends* (pp. 1–11).

Part VI
Ontologies and Concepts for Enterprise Interoperability

A MetaMeta Level Formal Manufacturing Ontology for Meta Level Production Methods

Z. Usman and R.I.M. Young

Abstract In a manufacturing environment, engineers and scientist often need to access and reason about general types of manufacturing methods without complex specifications and numbers. They may also want to disseminate the general process plans without sensitive details. This general knowledge is typically contained in the structure of knowledge bases which is composed of classes and can be referred to as Meta Level Knowledge. Typical manufacturing knowledge bases are designed for populated knowledge only and do not support capturing and reasoning over Meta level knowledge. Modelling to capture Meta level knowledge has been investigated in broader software community but the same is not true in manufacturing domain particularly in process planning. Moreover, an exploration to capture and reason about such knowledge in formal manufacturing ontologies has not been conducted. In this paper, a new formal ontology is proposed to capture meta level production methods. The ontology makes use of "powertypes" and "clabjects" to treat classes as objects. Meta level production methods can be captured using the proposed ontology. Manufacturability of features within a part family can be found at the meta level. Various tests are conducted to examine the ability to access, infer and reason about the meta level production methods to show the effectiveness of proposed ontological model.

Keywords Clabjects · Meta level manufacturing knowledge · Production method ontology

---

Z. Usman (✉)
Institute for Advanced Manufacturing and Engineering, Coventry University, Coventry, UK
e-mail: ac1095@coventry.ac.uk

R.I.M. Young
School of Mechanical and Manufacturing Engineering, Loughborough University, Loughborough, UK

© Springer International Publishing Switzerland 2016

K. Mertins et al. (eds.), *Enterprise Interoperability VII*,
Proceedings of the I-ESA Conferences 8, DOI 10.1007/978-3-319-30957-6_18

1 Introduction

Effective preservation and provision of manufacturing knowledge is imperative to the success of any manufacturing organisations [1, 2]. Knowledge bases (KBs) should facilitate the availability of knowledge at the required levels of abstraction. Manufacturing KBs are typically designed to capture and provide the populated knowledge only which contains complex details that are not required in many cases. For example, detailed knowledge is not required during early stages of product development. Production engineers may want to query the manufacturability of certain new features within a previous part family without going into the specific details of machines, fixtures, cutting tools, dimensions and tolerances. Detailed specification and information is also not wanted when engineers wish to share general process planning information without any senstitive details. The same is true when an overview of process plan is required.

In such cases, abstracted versions of process plans that represent general types of activities and resources e.g., the kind of operations such as milling and drilling, their sequence and the kind of machine tools are required. This requirement has also been established in the literature [3, 4]. Such general knowledge is contained within the structure of KBs. Thus, the structure of a KB is also knowledge albeit at an abstract level. Such knowledge can be referred to as "Meta level Knowledge (MLK)" [5]. Knowledge composed of the final instances that cannot further instantiate can be called "Individual Level Knowledge (ILK)" [5].

Whereas, all manufacturing KBs and ontologies facilitate access to and reasoning over ILK, the same is not true for MLK. This issue is more pronounced in formal (computer-interpretable) ontologies. Although researchers have demonstrated capturing manufacturing semantics and supporting knowledge sharing through formal manufacturing ontologies [6, 7], the work has been limited to ILK. Researchers have highlighted the importance of formally capturing and sharing process knowledge and have made significant contributions [8, 9]. The work resulted in the form an ISO standard i.e., Process Specification Language [10]. However, models to access and reason over MLK process plans remain to be explored.

The approach used to model MLK is referred to as multi-level modelling [11]. In multi-level modelling classes are modelled to have both class and object facets and can be referred to as "clabjects" [3]. This enables the system to reason over classes i.e., clabjects. In broader software community, multi-level modelling has been gaining attention. Conceptualization for representing abstractions of objects through clabjects and powertypes have been discussed in the literature [12–14]. Several researchers have published work on new features for multi-level modelling [11, 15], formalising multi-level knowledge [16] and their applications [17, 18].

The use of multi-level modelling has recently been reported in manufacturing domain as well. Aschauer [19] proposed multi-levels modelling for manufacturing automation by presenting Unified Modelling Language (UML) models. Altendorfer [20] analysed the current modelling approaches and presented a clabject based modelling approach in a production scenario. Both [19 and 20] presented semi-formal models with little attention to MLK. A comparison of multi-level manufacturing ontologies was presented by [21]. However, in [21], manufacturing was only used as an example to compare different multi-level modelling approaches. Moreover, this work only compared light weight ontological models without any formal semantics. This highlights the potential to explore the formal capture and sharing of manufacturing MLK.

In this paper a new formal ontology to capture and reason over manufacturing MLK manufacturing is presented. It is proposed that formal production method ontology based on clabjects and powertypes can facilitate capturing and reasoning over manufacturing MLK. In order to do this, the semantics of manufacturing concepts (i.e., classes) have been captured at the MetaMeta level. The developed formal ontological model has been tested for capturing, accessing and reasoning over MLK production methods.

The rest of the paper is organised as follows. The following section explains the need to access and reason over MLK production knowledge. This is followed by UML representation of the proposal ontology. The next section reports the formal logic modelling of the production method ontology. This is followed by experimental validation. At the end, a set of conclusion and future research direction is detailed.

2 Need to Access and Reason Over Manufacturing MLK

Figure 1 shows examples of ILK and MLK process plans on the right and in the middle respectively. ILK process plans contain detailed specifications of operations, their order, date and time, specifications of machines, fixtures and tools. Such details make things more complex and are not required in many cases e.g., early product development, disseminating generic non-sensitive process plans, for over view of production methods etc. In such cases, abstracted or MLK process plans (mid-section Fig. 1) that represent general types of activities and resources are required. MLK process plans only represent the type of things and not the actual objects. For example, the MLK process plans in Fig. 1 are composed of types of machine, fixtures and tools without any specific details model numbers and specifications.

In order to reason about the MLK process plans, they should be instantiated from an underlying structure. This means that classes that can instantiate the MLK are required. Such classes will, therefore, be defined at an even higher level of abstraction. Given that normal classes are present at Meta Level [22], the concepts from which the MLK is instantiated can be considered to be at the MetaMeta level. Figure 1 shows an example of a MetaMeta level structure on the extreme left which instantiates the MLK and provides a reference for accessing and reasoning over the MLK. This is a simplistic representation and proper modelling and formalisation is required to define MetaMeta level structure. This requires concepts and

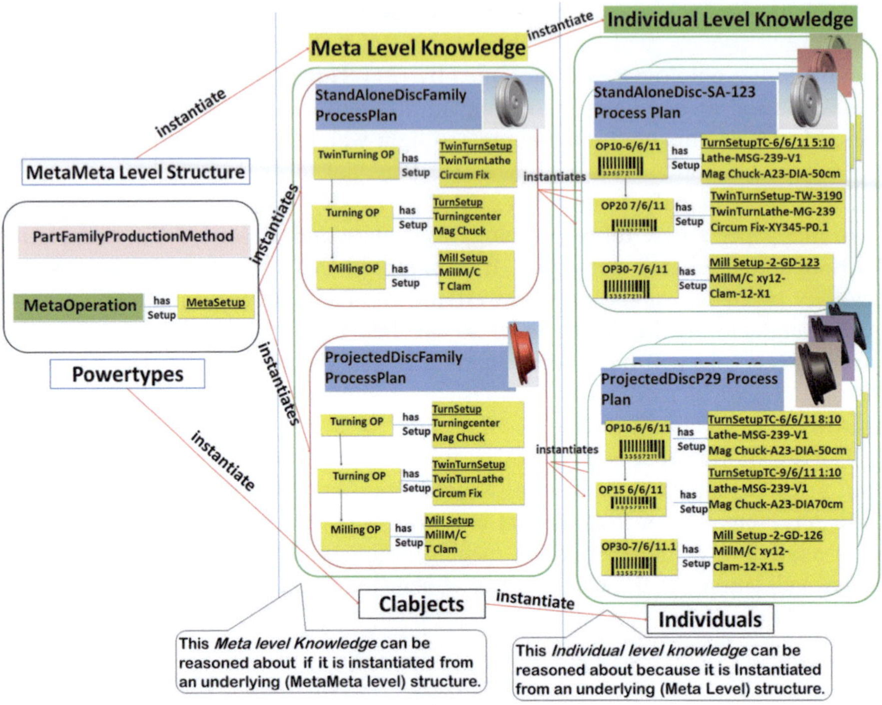

Fig. 1 Manufacturing methods for part at individual and meta levels

relations which will be modelled in formal logic to define the production method ontology that facilitates accessing and reasoning over MLK production methods. This highlights the need to define the MetaMeta level structure and its classes and relations.

3 Developing MetaMeta Level Production Method Ontology

The work done on capturing multi-level modelling [3, 12–14] provides help in defining the *MetaMeta level* structure. It is understood that the MLK can be modelled using clabjects. However, this requires classes that can instantiate clabjects i.e., 'powertypes' [3, 12]. Concepts that subsume clabjects i.e., super-classes are also required. This raises the following research questions.

1. What are the required concepts i.e., clabjects, powertypes, super classes and relations that can support the capture and reasoning over manufacturing MLK?
2. How can the manufacturing MLK be formally represented, captured and reasoned about using the identified concepts and relation?

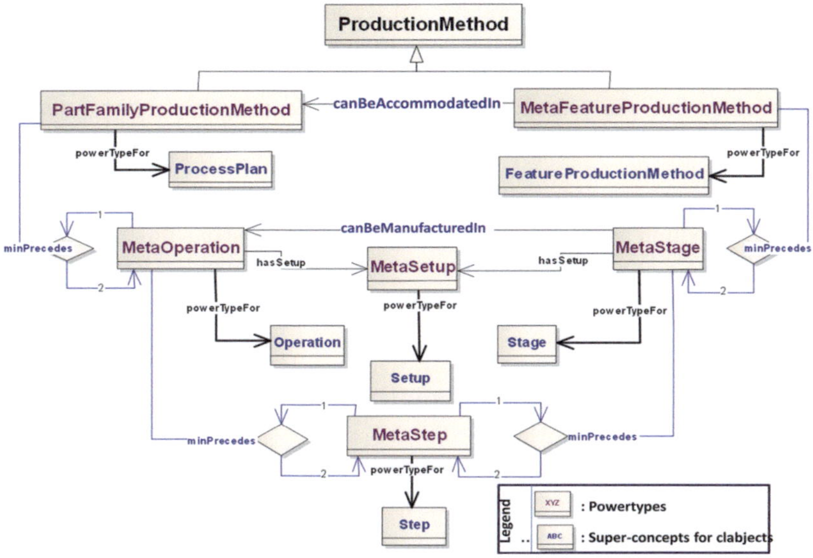

Fig. 2 Ontology for capturing and reasoning over MLK production methods

This first step towards that is to identify and define the required meta level concepts and relation. After detailed research and an industrial investigation a model for MLK was prepared as shown in Fig. 2. In this model *'ProductionMethod'* is defined as "a sequence of events which describe the procedure of production". *ProductionMethod* for parts is termed as '*ProcessPlan*' and that for Features as '*FeatureProductionMethod*'.

ProcessPlan is defined as "a sequence of *operations*" *and Operation* as "an event in a *ProcessPlan* that has a unique *Setup*" based on the understanding from [25, 26]. *FeatureProductionMethod* is defined as a "Sequence of stages" based on [26] and *Stage* as "an event in the *FeatureProductionMethod* with a unique setup". Each *Operation* and *Stage* consists of a sequence of *Steps* where a *Step* is "an event in an that is performed with a specific *CuttingTool*".

The relation *minPrecedes* is adapted from PSL [10] to capture the sequencing of *Operations, Stages* and *Steps*. The relation linking *Operation* and *Setup* is '*hasSetup*'. The relation *Setup* itself links *FeatureProductionMethod* to *ProcessPlan*. A *Setup* provides the basis to reason about the manufacturability of *FeatureProductionMethods* in *ProcessPlans*.

The concepts described so far only enable capture and reasoning over ILK. In order facilitate capturing and reasoning over MLK production methods, the defined concepts are converted to clabjects. This requires powertypes which are identified as *MetaOperation, MetaStage, MetaSetup* and *MetaStep*. These are used to define the MetaMeta level model for MLK production methods as shown in Fig. 2.

Figure 2 shows the *ProductionMethod* for part families on the left and the *ProductionMethod*(s) for features on the right. The use of powertypes (purple text in Fig. 2), provides the structure for MLK production methods. In Fig. 2 the relations *minPrecedes* and *hasSetup* have been defined over powertypes to ensure their applicability at MLK and ILK.

In order to find the manufacturing of a *MetaFeatureProductionMethod* in a *PartFamilyProductionMethod*, the manufacturability of its *Stages* should be investigated within the *Operations* of the *PartFamilyProductionMethod*. It is known that both *Operation* and *Stage* have unique *Setups*. Therefore, for a *Stage* to be manufactured within an *Operation* its *Setup* should be same or similar to the *Setup* of an *Operation*. This logic is captured through the relation *canBeManufacturedIn* as shown in Fig. 2.

4 Common Logic Based Formalization of the Model

Figure 2 presents a MetaMeta level UML model without any attention to semantics. In order to enable reasoning over MLK, the reasoning rules are captured in formal logic. For this purpose, Common logic which is an international standard for logic based languages [24] is used in the form of Knowledge Frame Language (KFL) [25]. Common logic is used because it provides higher expressive power and supports better inference and reasoning ability as compared to the languages like XML, RDF/RDFS and OWL. It is not possible to detail the formalisation of all the concepts and rules, therefore, a selected set of concepts, relation and their axioms will be detailed. The KFL declaration for a powertype and its clabject is as follows.

```
Powertype                             Clabject
:Prop  PartFamilyProductionMethod     :Prop  StandAloneDiscProcessPlan
:Inst  MetaProperty                   :sup   Type                    :Inst
PartFamilyProductionMethod   :sup   ProcessPlan
:metaPropFor  ProcessPlan
```

MetaProperty represent a powertype in KFL. The clabjects i.e., *StandAloneDisc ProcessPlan*'s declaration shows that clabjects are instances of *PartFamilyProductionMethod* and sub-concepts of *ProcessPlan.*

Most important part of formalisation was to define the axioms to find whether a MLK *FeatureProductionMethod* can be accommodated within MLK *PartFamily ProductionMethod(s)*. From the understanding gained from the definitions of production methods for feature and parts, the required logic can be stated as:

> "A *FeatureProductionMethod* can be accommodated within a *PartFamilyProductionMethod* if all the *Stages* of a *FeatureProductionMethod* can be manufactured with correct sequencing in the *Operations* of a *PartFamilyProductionMethod*."

Manufacturability of *Stages* in *Operation* is captured through the relation *canBeManufacturedIn* which leads to defining the most important relation i.e., *canBeAccommodatedIn*. This relation combines all the logic in the ontology to find the manufacturability of features within a part family.

For *canBeAccommodatedIn* to hold true, the system has to map the *Stage* sequence of a *FeatureProductionMethod* against the *Operation* sequence of the *ProcessPlan*. For example, it may be possible to accommodate all *Stage*s within the *Operation*s, however, the correct ordering of *Stages* should also be ensured. This logic is captured by defining a relation '*canBeAccommodatedIn_false*' which captures the false sequence. For a *FeatureProductionMethod* to be accommodated in a *PartFamilyProductionMethod*, *canBeAcommodatedIn_false* should hold false. This relation is formalised in KFL as follows.

```
(<= (canBeAccommodatedIn_false ?mfpm ?pfpm)
    (and (MetaStage ?st1) (MetaStage ?st2)(minPrecedes ?st1 ?st2 ?fpm)
              (MetaOperation ?op1)(MetaOperation ?op2)
              (minPrecedes ?op1 ?op2 ?pfpm)
              (canBeManufacturedIn ?st1 ?fpm ?op2 ?pfpm)
              (canBeManufacturedIn ?st2 ?fpm ?op1 ?pfpm)))
```

The above axiom states that a clabjects 'mfpm' of *MetaFeatureProduction Method* can be accommodated in a clabject 'pfpm' of *PartFamilyProductionMethod* even if a clabject '?st2' of *MetaStage* that occurs later than the clabject '?st1' can be manufactured in clabject '?op1' of *MetaOperation* earlier than ?st1. However, this is not possible in reality, hence the directive '_false'. If the above axiom holds true, *FeatureProductionMethod* cannot be accommodated in *PartFamilyProduction Method* even if all the stages are manufacture-able.

If the above relation is false, the system proceeds to final inference rule which states that a *FeatureProductionMethod canBeAccommodatedIn* a *PartFamilyProduction Method* if; (1) all the stages of *FeatureProductionMethod canBeManufacturedIn* the *Operation*s of *PartFamilyProductionMethod* and (2) *canBeAccommodatedIn_false* holds false.

```
(<= (canBeAccommodatedIn ?fpm ?pfpm)
     (and     (MetaStage ?st)  (MetaOperation ?op)
              (canBeManufacturedIn ?st ?fpm ?op ?pfpm)
              (not (canBeAccommodatedIn_false ?fpm ?pfpm)))))
   :rem "The FeatureProductionMethod ?fpm can be accommodated in the
PartFamilyProductionMethod ?pfpm"
```

The relations *canBeManufacturedIn* and *canBeAccommodatedIn* enable querying the manufacturability of *FeatureProductionMethod* within a *PartFamily ProductionMethods*.

5 Experimental Verification

Objectives of the experimentation were (1) to verify the capture of MLK feature and part family production methods (2) to verify the reasoning capability over the MLK production methods.

The experimentation environment is Integrated Ontology Development Environment (IODE). The manufacturing KB is developed as an eXtensible Knowledge Server (XKS). The ontologies are loaded as KFL files in IODE.

Because the MLK is composed of clabjects, therefore, it is asserted as part of the ontology. Different MLK *ProductionMethod*s for *Feature(s)* and *PartFamily*(s), were captured using KFL as part of the ontology. The proposed modelling approach should make it possible to query MLK like other facts. The KB is loaded with several clabject based production methods for features and part families. The two production methods being queried are *StandAloneDiscFamilyProductionMethod* and *RimProductionMethod*. This required ability of proposed ontology is tested through the following queries;

1. Acquiring the MLK ProductionMethod for StandAloneDiscFamily ProductionMethod
2. Querying the manufacturability of an MLK feature production method i.e., RimProductionMethod within the asserted MLK PartFamilyProduction Methods.

Due to space limitations, the results from 1st query are not which provided the exact sequence of *Operation*s clabject involved in the *StandAloneDiscFamily ProductionMethod.*

The second query was made to find the manufacturability of a clabject of *MetaFeatureProductionMethod* within clabjects of *PartFamilyProductionMethod. In order to do that, s*everal Meta level production methods were captured in the KB. The query was based on the key relation *canBeAccommodateIn.* Figure 3 depicts the query (section-1) and its results (section-2). As shown in Fig. 3, the *RimProductionMethod* can be accommodated in two clabjects of *PartFamily-ProductionMethod* i.e., *StandAloneDiscFamilyProductionMethod* and *Projected-DiscFamilyProductionMethod.*

The complex logic working at the back end of this query has already been explained. The system can answer the manufacturability of a *MetaFeature-ProductionMethod's* clabject in a *PartFamilyProductionmethod's* clabject, the manufacturability of the *Stages* of *MetaFeatureProductionMethod's* clabject in the *operation*s of the *PartfamilyProductionProductionMethod's* clabject. This is done automatically within the system based on the defined logic.

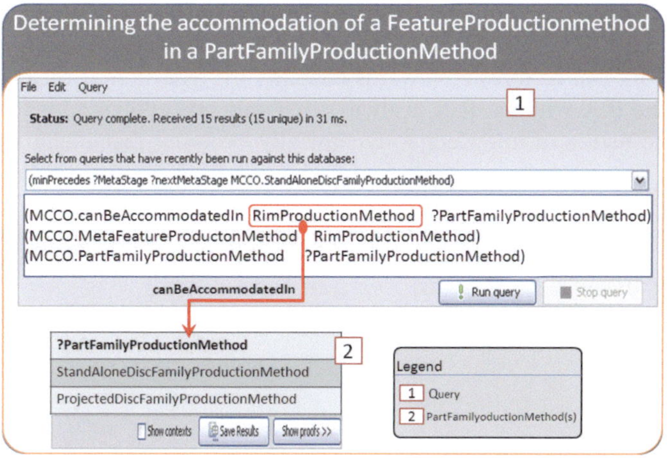

Fig. 3 Determining the accommodation of a *FeatureProductionMethod* in a *PartFamily ProductionMethod*

Before the system can answer the manufacturability of a *MetaFeature-ProductionMethod's* clabject in a *PartFamilyProductionmethod's* clabject, the manufacturability of the *Stages* in the *Operation*s is to be found. This is done within the system based on the semantics of the relation *canBeManufacturedIn*. This verifies the ability of the proposed MetaMeta level ontology to support capturing and reasoning over MLK production methods.

6 Conclusions

This paper explicitly explained the requirements to capture and reason about Meta level manufacturing methods. A formal production method ontology has been defined to capture and reason about production methods for features and part families at the *MLK*. The proposed ontology is a step towards the understanding of the use of formal ontologies to support capturing and reasoning about production methods at multiple levels of knowledge.

A comprehensive set of core manufacturing concepts and relation (Fig. 2) have been identified and formalised in KFL to capture production methods for features and part families at multiple levels of knowledge abstraction and particularly for *MLK*.

Relation *minPrecedes* has been tailored from PSL [10] to be applicable at the MLK. Key relation *canBeManufacturedIn and canBeAccommodatedIn* have been identified and their complex logic is successfully defined and captured in formal logic. KB has been equipped with ability to reason about the manufacturability of production methods for features within the production methods for part families *at MLK*.

Future work can be directed towards enabling the assertion of MLK in a similar manner to the ILK without. ILK is always dynamic and evolving, to accommodate this dynamic nature of knowledge, dynamically evolving manufacturing ontology should also be explored.

Acknowledgments This work has been funded by the EPSRC's Innoavtive Manufacturinag and Construction Research Centre (IMCRC) through the Interoperable Manufacturing Knowledge Systems (IMKS) project.

References

1. Hansen, M. T. (2002). Knowledge networks: Explaining effective knowledge sharing in multiunit companies. *Organization Science, 13*(3), 232–248.
2. Mesmer-Magnus, J. R., & DeChurch, L. A. (2009). Information sharing and team performance: A meta-analysis. *Journal of Applied Psychology, 94*(2), 535–546.
3. Atkinson, C., & Kühne, T. (2000). Meta-level independent modelling. In *International Workshop on Model Engineering at 14th European Conference on Object-Oriented Programming*, Sophia Antipolis and Cannes, France.
4. Schlenoff, C., Gruninger, M., & Ciocoiu, M. (1999, December). The essence of the process specification language. *Transactions of the Society for Computer Simulation, 16*(4), 204–216.
5. Usman, Z. (2012). *A manufacturing core concepts ontology to support knowledge sharing* (Ph. D. Thesis, Mechanical and Manufacturing Engineering, Loughborough University).
6. Usman, Z., Young, B., Chungoora, N., Palmer, C., Gunendran, G., Case, K., et al. (2013, July). Towards a reference ontology for manufacturing. *International Journal of Production Research.*
7. Imran, M., & Young, B. (2013). The application of common logic based formal ontologies to assembly knowledge sharing. *Journal of Intelligent Manufacturing.*
8. Ghallab, M., et al. (1998). PDDL: The planning domain definition language v.2. Technical Report CVC TR-98-003, Yale Center for Computer Vision and Control.
9. Tate, A. (1998, March). Roots of SPAR-shared planning and activity representation. *The Knowledge Engineering Review, 13*(1), 121–128 (Special Issue on Putting Ontologies to Use, Cambridge University Press).
10. ISO 18629-1:2004 Industrial automation systems and integration—Process specification language—Part 1: Overview and basic principles.
11. Atkinson, C., & Gerbig, R. (2014). Level-agnostic designation of model elements. In *Proceedings of ECMFA 2014* (Vol. 8569, pp. 18–34). LNCS. Berlin: Springer.
12. Henderson-Sellers, B., & Gonzalez-Perez, C. (2005). Connecting powertypes and stereotypes. *The Journal of Object Technology, 4*(7), 83–96.
13. Henderson-Sellers, B. (2011). Bridging metamodels and ontologies in software engineering. *Journal of Systems and Software, 84*(2), 301–313.
14. Palmer, C., Gunendran, A. G., Young, B., Chungoora, N., Usman, Z., Case, K., et al. (2012). Exploiting UML as a preliminary design tool for common logic-based ontologies. *International Journal of Computer Integrated Manufacturing.*
15. Neumayr, B., Jeusfeld, M. A., Schrefl, M., & Schätz, C. (2014). Dual deep instantiation and its concept base implementation. In *Proceedings of CAiSE 2014* (Vol. 8484, pp. 503–517). LNCS. Berlin: Springer.

16. Rossini, R., de-Lara, J., Guerra, E., Rutle, A., & Wolter, U. (2014). A formalisation of deep metamodelling. *Formal Aspects of Computing*, 1–41 (in press).
17. de-Lara, J., Guerra, E., Cobos, R., & Llorena, J. M. (2014). Extending deep meta-modelling for practical model-driven engineering. *The Computer Journal*, *57*(1), 36–58.
18. Jordan, A., Grossmann, G., Mayer, W., Selway, M., & Stumptner, M. (2012). On the application of software modelling principles on ISO 15926. In *Proceedings of the Modelling of the Physical World (MOTPW) Workshop at MODELS 2012*. ACM.
19. Aschauer, T., Dauenhauer, G., & Pree, W. (2009, August). Multi-level modeling for industrial automation systems. In *Proceedings of the 35th Euromicro Conference on Software Engineering and Advanced Applications, SEAA '09* (pp. 490–496, 27–29).
20. Altendorfer, S., & Zsifkovits, H. (2013). A model-driven engineering approach for production systems illustrated on an automotive test case. In *Journal Proceedings in Manufacturing Systems* (pp. 159–164) (Rainer Hampp Verlag, München, Mering).
21. Neumayr, B., & Schrefl, M. (2009). *Comparison criteria for ontological multi-level modeling*. Forschungsbericht Research Report 8.03, Linz, Austria.
22. Gonzalez-Perez, C., & Henderson-Sellers, Brian. (2006). A powertype-based metamodeling framework. *Software & Systems Modeling, 5*(1), 72–90.
23. ISO/IEC-24707. (2007). International Standard, First edition 2007-10-01 Information technology-Common Logic (CL): A framework for a family of logic based languages.
24. IODE, Highleet. (2010). Ontology library reference. In *Highfleet's IODE/XKS Documentation. Baltimore*.
25. ISO-10303-49. (1998). Industrial automation systems and integration—Product data representation and exchange—Part 49: Integrated generic resources: Process structure and properties. International Organization of Standards.
26. Gunendran, A. G., & Young, R. I. M. (2010). Methods for the capture of manufacture best practice in product lifecycle management. *International Journal of Production Research, 48* (20), 5885–5904.

Towards an Interoperable Decision Support Platform for Eco-labeling Process

Da Xu, Hedi Karray and Bernard Archimède

Abstract Along with the rising concern of environmental performance, eco-labelling is becoming more popular. However, the complex process of eco-labelling demotivated manufacturers and service providers to be certificated. In this paper, we propose a decision support system aiming at further improvement and acceleration of the eco-labeling process in order to democratize a broader application and certification of eco-labels. This decision support system will be based upon a comprehensive knowledge base composed of various domain ontologies covering the whole life cycle of a product or service. Through continuous enrichment on the knowledge base in modular ontologies and by defing standard RDF and OWL format interfaces, the decision support system will stimulate domain knowledge sharing and have the interoperability to be applied into other practice.

Keywords Eco-labeling · Knowledge sharing · Interoperability · Modular ontology · Decision support system

1 Introduction

Since the late 1980s, there has been a growing demand from customers for products that do less harm to the environment. On the other hand, the public willingness to use buying power as a tool to protect the environment provides manufacturers with an opportunity to develop new products [1]. From a global point of view, promote of environment-friendly produce-consume-recycle progress will contribute not only

D. Xu (✉) · H. Karray · B. Archimède
Laboratoire Génie de Production ENIT, Tarbes, France
e-mail: dxu@enit.fr

H. Karray
e-mail: mkarray@enit.fr

B. Archimède
e-mail: Bernard.Archimede@enit.fr

© Springer International Publishing Switzerland 2016
K. Mertins et al. (eds.), *Enterprise Interoperability VII*,
Proceedings of the I-ESA Conferences 8, DOI 10.1007/978-3-319-30957-6_19

to the life quality but also the economy itself. But how does a consumer, faced with numbers of products in the market, judge and make a good choice to reduce environmental impacts? How should we assess the validity of a statement about a product or service's environmental impacts? The need of evaluating a product's environmental performance has led to the establishment of eco-labels certificating a product or service that meets certain environmental criteria.

For an eco-label applicant, usually a manufacturer or a service provider, it is easy to provide the required information in whatever formats. However, the difficulties encountered in the evaluating processes are representative in decision-making processes. To efficiently assess certain product or service, we need indeed to manipulate different types of voluminous data; take into account different criteria and conduct a multi-criteria analysis; consider different phases of product or service life cycle. Usually, a bunch of human experts coming from various domains will work together and the evaluating process will take a long time, and errors and conflicts may exist. In addition, the evaluation result is actually a good resource that could have been made better use of.

In this paper, we are interested in developing a decision support system in the scope of certificating or labeling process. The heterogeneous data and knowledge crosscovered in such process will be represented in ontology. The objective of this project is to build a decision support system that improve and accelerate the evaluation process of eco-labeling to help the domain experts make wiser decisions as well as reduce the costs of the process in order to finally democratize eco-labeling achieving a more eco-logical economic. Our approach is based on interconnected knowledge base composed of the identified domain knowledge by means of ontologies, which will provide a structured description of the domain concepts, relationships and rules covering the whole lifecycle of certain product or service categories. Taking advantages of ontology's semantic and interoperable nature, we can establish reference ontology that could be reused in other systems by extracting and refining modules from our system's knowledge base. In this approach, we will also develop distributed reasoning and inference mechanisms capable of traceable argumentation generation.

2 State of Art

2.1 Eco-Label and EU Eco-Label

According to Global Eco-labelling Network (GEN), "Eco-labelling" is a voluntary method of environmental performance certification and labelling that is practiced around the world. An "eco-label" is a label which identifies overall, proven environmental preference of a product or service within a specific product/service category. They usually concern the whole life cycle of the product and are issued by third party [2]. The International Organization for Standardization (ISO) has identified three broad types of voluntary labels, with eco-labelling fitting under the Type I designation [3].

Since the first eco-label with the name Blue Angel[1] awarded in Germany in 1978, many eco-labels covering various environmental aspects has been developed. To better manage and recognize eco-labels coming from different markets and countries, a Global Eco-labelling Network[2] (GEN) was even established in 1994 as a worldwide non-profit interest group whose goal is to foster co-operation, information exchange and harmonization among members. Driven by the guidance of government and society organizations, the number of products or services that are certificated by eco-labels are also increasing rapidly.

Eco-labelling has numbers of benefits from various points of view. First, eco-labeling is a good way to inform consumers of the environmental impacts of selected products. In the practice of some existent eco-labeling, the fitness of use and human health aspects are also included as well as the environmental performance. All these information will help a consumer make decision out of different willingness. Then, eco-labeling is generally cheaper than regulatory controls. By empowering customers and manufacturers to make environmentally supportive decisions, the need for regulation is kept to a minimum. This is beneficial to both government and industry [4]. Eco-labeling will also stimulate market development and encourage continuous improvement on product and service.

After a brief review of generic eco-label, we focus on the EU eco-label which relates to most of our research work. Created in 1992, the EU Eco-label is the only official European ecological label authorized for use in every member country of the European Union [5]. Until 2011, there are over 1300 enterprises that have been issued EU Eco-label licenses. By September of 2014, there are already over 43,000 products or services being labelled [6]. However, compared to the enormous Europe market, the awarded eco-labels are still too few. We consider that qualified enterprises should be encouraged to obtain eco-labels to become more competitive.

The Commission mandates the EUEB (European Union Eco-Labelling Board) is responsible to develop and regularly review eco-label criteria. EUEB will set up an advisory body including representatives coming from different stakeholders. Feasibility study will be carried out to draft the environmental criteria. At last, representatives from every member state will be summoned to vote to approve the criteria or the guideline [7]. The guideline developed by the advisory body, together with the possible amendment or annex will be the baselines for our knowledge base.

Throughout the product categories, the multi-criteria or guideline referred by EU Eco-label is usually stricter than the domain regulation. Such differences between EU Eco-label and standards consolidate its effect as a stimulation to the market and somehow a driving force as regards to the producer.

[1]https://www.blauer-engel.de/en.

[2]http://www.globalecolabelling.net/.

2.2 *Knowledge-Based Decision Support and Modular Ontology*

The general decision supporting counts as a practical sub-branch of Artificial Intelligence. While it is still hard to define a clear border for decision supporting as various application of different levels can be found in various domains. Generally speaking, three fundamental components of a decision support system architecture are: database, model and user interface [8]. For each mentioned component, we can find good support from both theoretic and practical standpoint, as the traditional computer science and software engineering have been mature enough. In our research scope, we care about the database component, since the other two components are much related to specific business process, which depend on the field we will apply the decision supporting to.

Traditional database based on relationship model is becoming clumsy, especially in data exchange and inference aspects. With the rapid development of Semantic Web[3] and Linked Data,[4] the interoperability, reusability and modularity of knowledge are becoming more and more important. As a result, ontology and ontology engineering has attract much attentions and efforts. In Computer Science, we refer to an ontology as a special kind of information object or computational artifact [9]. Studer et al. [10] gave definition stating that: "An ontology is a formal, explicit specification of a shared conceptualization." For the notion of a conceptualization according to Genesereth and Nilsson [11], who claim: "A body of formally represented knowledge is based on a conceptualization: the objects, concepts, and other entities that are assumed to exist in some area of interest and the relationships that hold among them. A conceptualization is an abstract, simplified view of the world that we wish to represent for some purpose."

In the last decades, so many ontologies and knowledge repositories have been developed, however, much problems are encountered when knowledge engineers as well as general users want to understand and employ the ontologies into their own software development. One reason of such difficulties is the semantic confusion among domains. Another reason, according to the author, is that there is still lack of a comprehensive and widely accepted standard system for ontology construction, e.g., ontology languages are developed based on logics having different expressiveness, which somehow block the compatibility for data exchange and reasoning. One of the thorniest problems is how we build ontology and utilize it to furthest maintain a well reusability. Due to the initial nature of being shared, certain formation of knowledge shall be meaningless if it could not be exploited and reused.

An interesting approach that deals with ontology reusability is the implementation of modularity, which reminds us of the similar term in software engineering. Modularization materializes the long-established complexity management

[3]http://www.w3.org/standards/semanticweb/.

[4]http://www.w3.org/standards/semanticweb/data.

technique known as divide and conquer. It is found in various areas of computer science. In the application of ontology, there is also a definite need from applications to gather knowledge from several, not just one, ontological sources. It is known that, when knowledge is distributed, the idea to collect all knowledge into a single repository (i.e., the integration approach) is very difficult to implement, because of semantic heterogeneity calling for human processing [12].

In our research, we will put much focus on how to connect existing modular ontologies, rather than the partitioning and extraction of modules. To achieve ontology modularity in a distributed scenario, E-Connection is proposed as a set of "connected" ontologies. An E-Connected ontology contains information about classes, properties and their instances, but also about a new kind of properties, called Link Properties, which establish the connection between the ontologies [13]. Another interesting approach is the use of Distributed Description Logics (DDL) framework [14] and the distributed reasoner DRAGO (Distributed Reasoning Architecture for a Galaxy of Ontologies) [15] as formal and practical tools for composing modular ontologies. Also, we have Package-Based Description Logics as another formalism that can support contextual reuse of knowledge from multiple ontology modules [16]. While these methods and formalisms more or less set up a logics syntax barricade that should limit a large scale reasoning and modification between heterogeneous and distributed modular ontologies, e.g., the underlying logics formalism of E-Connection is OWL-DL (i.e., SHOIN); while, logics formalism for DDL is SHIQ; when it comes to Package-Based Description, it turns into SHOIQ. From a practical perspective, these methods have not been applied in such a considerable scope that we could have successful application cases for a good study.

3 An Interoperable Decision Support Approach

Figure 1 presents a simplified outline of our platform for eco-labeling decision support. There are three roles as participant involved in our decision making process. First, at the top left of this schema, we have a human icon representing the applicant manufacturer or service provider who will initiate the eco-label application. On the right side another human icon represents the experts who take the results of the system and make the final decision whether the product or service is qualified or not. In the existent evaluating process, the member country's authorized Competent Body, human icon located at the left bottom, should first give advice and guideline to the applicant, in the meanwhile the applicant should prepare the required documents. In our schema, we trim off such routine operations because the concrete communication and administrative affairs have few concerns to the decision making.

Now we will follow a complete process to illustrate how we make use of the knowledge base to facilitate a product/service evaluation with traceable argumentation. Firstly, the applicant will provide a detailed description of their product or

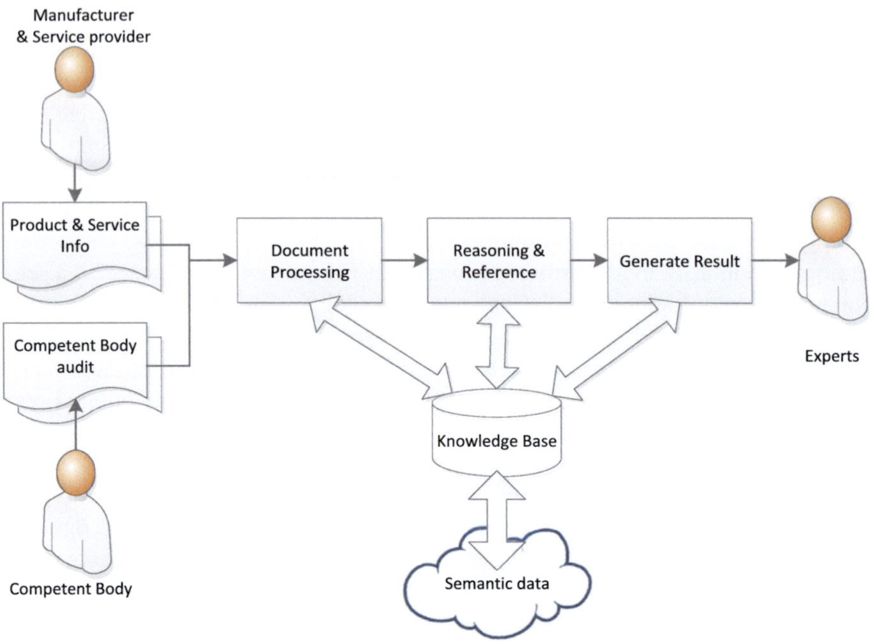

Fig. 1 Outline of our platform for eco-labeling decision support

service. The competent body's audit result will also accompany with the description. In order to let the machine to understand the description, we propose a parser component to retrieve concerned information from the description to form a machine readable structured document. According to the understanding of the system, which is supported old cases and experience, the structured document will be transferred into a user ontology and then the system will select necessary domain ontology from the knowledge base. Towards these domain ontologies, a modularization and refinement formation proceeds to gather the very necessary knowledge parts to build a merged or integrated criterion ontology. In the next step, the inference component takes both user ontology and criterion ontology to test and verify if the user ontology that contains all the key description of the product comply with the corresponding guidelines. At last an argument tracer component will parse and translate the conflicts between user ontology and criterion ontology so as to generate the final report for human experts review. Then the task of our decision support system is finished and the following procedure will be the experts judge the results of the system and feedback to the applicant.

All the document processing, reasoning or generating process will be supported by a comprehensive knowledge base. The knowledge base is connected to other effective data source locally or remotely. All the data and knowledge will be serialized in multiple unified formats such as RDF or Turtle. To achieve a better interoperability performance, the knowledge base is equipped with public semantic data source accessing interface, which allows the ontology and data stored locally being accessed by other endpoint. In the opposite direction, our knowledge base is designed to be able to browse other knowledge base or ontology repository so as to acquire extra information. With such a open information sharing mechanism, we guarantee that part of our knowledge base shall be shared. This will be the cornerstone of interoperability when our decision support system is about to cooperate with other systems or is to be integrated into other systems.

4 The Knowledge Base in Modular Ontology

Single ontology may be not yet powerful enough to set up a complete conceptualization about the real world of Eco-labeling. Besides, as for knowledge's reuse possibility, modularize ontology into pieces is a reasonal choice and also a challenge. In our research, we choose only one product group at first implementation. Let's say "laundry detergent". If our methodology and system works well on single product group, then we should include more product groups.

For example, in EU Eco-labeling's conceptualization of the world of laundry detergent (Fig. 2), we have a main conceptualization modular named "detergent" which hold the general concepts and properties of this domain. While, for other

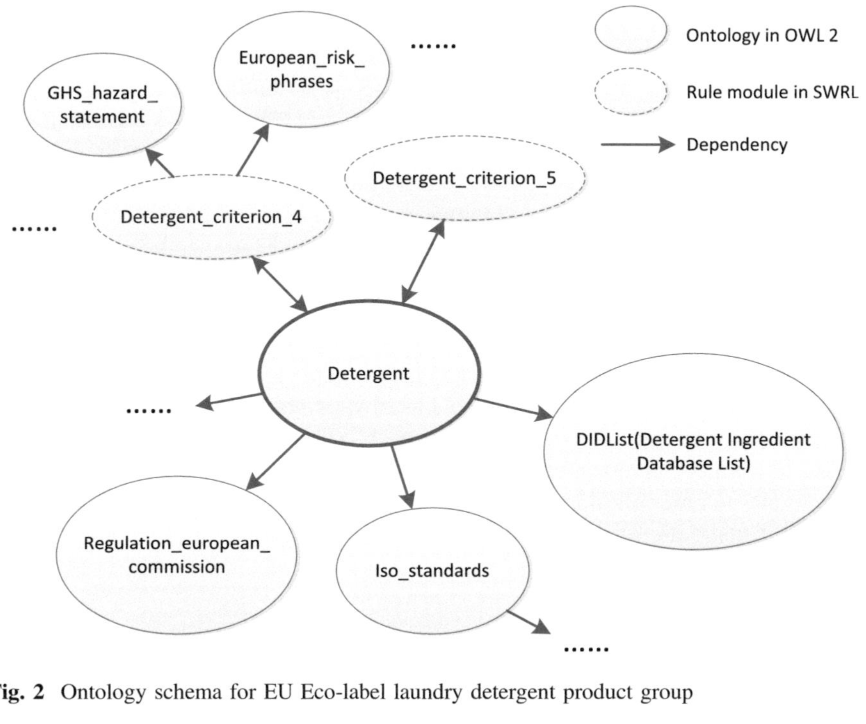

Fig. 2 Ontology schema for EU Eco-label laundry detergent product group

more specific concepts, it reaches to other modulars via several dependencies (as with OWL 2, we can implement dependency by using “import” syntax, which means current ontology will use external concepts or relationships from the imported ontology. In software engineering, reducing dependencies between modules is one of the basic design rules). It’s not difficult to see that several advantages exist in this modular design. As more coherent concepts and relationships are gathered together to form modules, it’s easier to manage knowledge and data in large scale. Complex conceptualization can be achieved by composing multiple small modules. Also, it is easy to configure and replace modules rather than to alter small parts directly in a large structure. Take the same example in Fig. 2. We have a general conceptualization of laundry detergent product which is stored in ontology module “detergent”. We also have some general rules stored in the same “detergent” ontology. This principle ontology will “import” or make use of information of european commission regulation, ISO standards, and DIDList (Detergent Ingredient Database List), which could be possiblely used and imported by other ontology of other product group. Particularly, once DIDList ontology module is properly defined and developed in laundry detergent product group, it doesn’t have to be redeveloped in hand washing detergent as it has identical reference to DIDList, thus we arrive at reusing some ontology modules such as DIDlist.etc.

Modularization implies separation and conceptualization, if we follow this path of thinking we can see that it will be practical to extract rules from ontology modular. In other words, it’s better to keep objective constraints and world description separated. Take the detergent ontology example shown in Fig. 2, ontology represented in ellipse with solid border are concept-centred, which means the main function of these ontology is to describe the concrete world. These ontology contains concepts and relationships that are edited to describe or record the fact about real world. There are indeed rules contained in such ontology modules, while they also serve mainly to describe rather than to judge or calculate. However, as for a product group guideline or criteria, quite a part of information are involved with human objectives. For the same detergent example, the concentration of different chemical ingredients has to respect to certain limit. In such cases, we can hardly say that such objective elaboration or goal-oriented specification are plain description of the world. Moreover, such rules or objective information may change time after time. This actually happens because, the product guideline keeps being updated as new EU Commission has always been generating new amendments or revise. In our approach, we have each criteria item to be an independent module (not totally independent, as we can see that these rule modules also have dependencies to other ontology modules), thus we can easily replace them with new rules and manage them in a configurable way (Fig. 3).

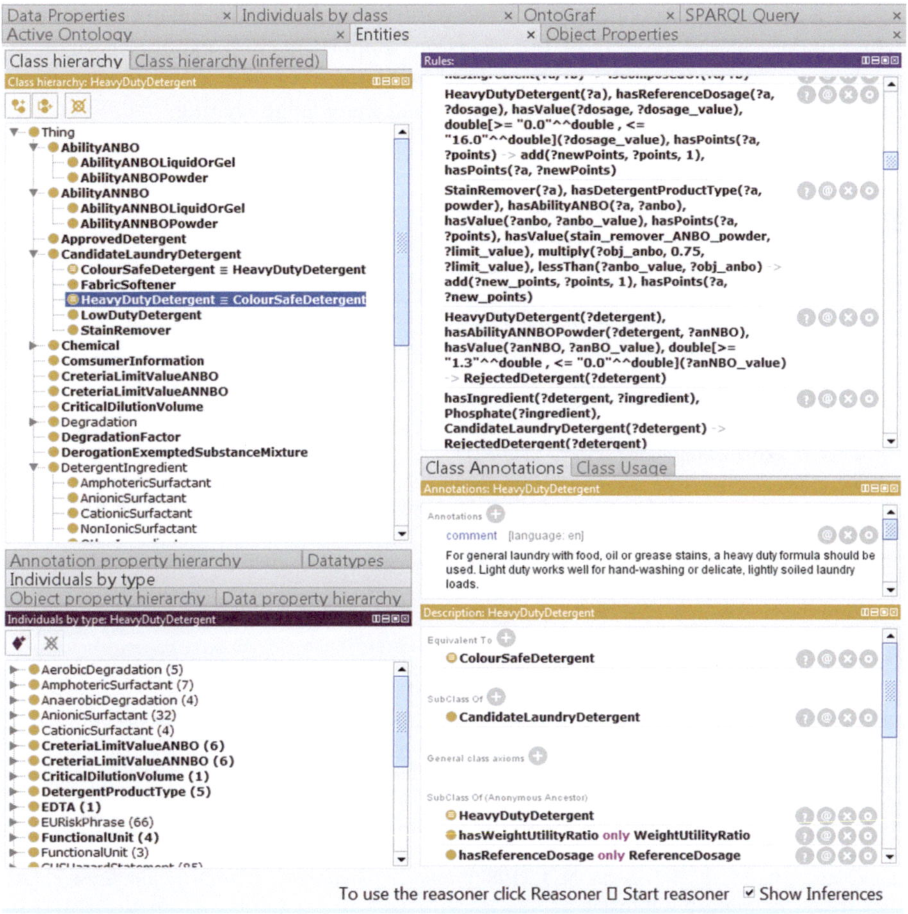

Fig. 3 Composition of detergent ontology developed in Protege editor

5 Conclusion

In this paper we've seen what is eco-labelling. How an eco-labeled product or service shall contribute to the environment and economics. More specifically, a current status of EU Eco-label and its future trend are also presented. To popularize eco-labeled products and services in order to achieve a more competent and ecological economic, a better eco-labelling process is needed. We propose a decision support system that should improve and accelerate the evaluation process for eco-labeling to help the domain experts make wiser decisions as well as reduce the costs of the process. Our approach is based on a knowledge base composed of the identified domain knowledge by means of ontologies, which provides a structured description of the domain concepts, relationships and rules covering the whole lifecycle of certain product or service categories. The modularized knowledge base, which is key to the success of our decision support process, exposes part of its modules as reference ontologies that could be browsed or reused by other systems in order to achieve an data interoperability and knowledge sharing.

References

1. Eco-labeling Markets. International Institute for Sustainable Development. http://www.iisd.org/business/markets/eco_labeling.aspx
2. Lefébure, A., & Rosales Muñoz, R. (2011). Communicating to consumers in Sweden with eco-labels: Is the message getting through? http://www.diva-portal.org/smash/get/diva2:430238/FULLTEXT02.pdf
3. GEN: Global Ecolabelling Network: What is Ecolabelling? http://www.globalecolabelling.net/what_is_ecolabelling/index.htm
4. Ihamäki, V., van Ooij, E., & van der Panne, S. (2014). Green public procurement in the European Union and the use of eco-labels. http://ec.europa.eu/environment/gpp/pdf/toolkit/module1_factsheet_ecolabels.pdf
5. The EU Eco-label scheme—What is it? http://www.ecolabels.fr/en/the-eu-eco-label-scheme-what-is-it
6. Facts and Figures—Ecolabel—European Commission. http://ec.europa.eu/environment/ecolabel/facts-and-figures.html
7. How are guidelines drawn up?—Ecolabel. http://www.ecolabels.fr/en/the-eu-eco-label-scheme-what-is-it/how-are-guidelines-drawn-up
8. Druzdzel, M. J., & Flynn, R. R. (2003). Encyclopedia of library and information science. In M. Drake (Ed.), *Decision support systems* (Vol. 1). CRC Press.
9. Guarino, N., Oberle, D., & Staab, S. (2009). What is an ontology? In S. Staab & R. Studer (Eds.), *Handbook on ontologies* (2nd ed., pp. 1–17). Heidelberg: Springer.
10. Studer, R., Benjamins, R., & Fensel, D. (1998). Knowledge engineering: Principles and methods. *Data & Knowledge Engineering, 25*(1–2), 161–198.
11. Genesereth, M. R., & Nilsson, N. J. (1987). *Logical foundations of artificial intelligence*. Los Altos, CA: Morgan Kaufmann.
12. Parent, C., & Spaccapietra, S. (2009). An overview of modularity. In *Modular ontologies* (pp. 5–23). Berlin Heidelberg: Springer.
13. Grau, B. C., Parsia, B., & Sirin, E. (2009). Ontology integration using ε-connections. In *Modular ontologies* (pp. 293–320). Berlin Heidelberg: Springer.
14. Serafini, L., & Tamilin, A. (2009). Composing modular ontologies with distributed description logics. In *Modular ontologies* (pp. 321–347). Berlin Heidelberg: Springer.
15. Serafini, L., & Tamilin, A. (2005). Drago: Distributed reasoning architecture for the semantic web. In *The semantic web: Research and applications* (pp. 361–376). Berlin Heidelberg: Springer.
16. Bao, J., Voutsadakis, G., Slutzki, G., & Honavar, V. (2009). Package-based description logics. In *Modular ontologies* (pp. 349–371). Berlin Heidelberg: Springer.

Framework of Design Principles and Standards for Enterprise Interoperability Service Utilities

Irene Matzakou, Ourania Markaki, Kostas Ergazakis and Dimitris Askounis

Abstract The Interoperability Service Utility (ISU) concept constitutes an innovative approach for the provision of collaborative web services promoting the establishment of Enterprise Interoperability (EI) across different and diverse enterprise information systems. The purpose of this paper is to investigate the formulation of a framework on how to effectively design, develop and implement an ISU that would successfully serve certain purpose and context. A thorough study on current ISU implementations has been conducted in order to investigate and identify existing standards, common patterns and good practices. This information has been systematically classified into a set of basic standards and design principles that could serve as roadmap to support every interested party that wishes to implement an ISU. The basic conclusion is that the ISU is open to several interpretations and may be instantiated through diverse implementations of different purpose and scale, depending on the particular requirements it has to fulfill.

Keywords Interoperability service utility · ISU · Enterprise interoperability · Design principles · Standards · Information systems · Smes

1 Research Context and Scope of the Paper

1.1 The Interoperability Service Utility (ISU) Concept

Interoperability constitutes an issue of particular interest and importance in both the entrepreneurial and political agenda. More specifically, the EU research agenda on interoperability is embodied through two strategic documents; the Enterprise Interoperability Research Roadmap—EIRR [1], as well as the Digital Agenda for Europe [2]. The latter, drives the evolution towards the creation of a new society

I. Matzakou (✉) · O. Markaki · K. Ergazakis · D. Askounis
School of Electrical and Computer Engineering, Decision Support Systems Laboratory, National Technical University of Athens, 9, Iroon Polytechniou Street, 15780 Athens, Greece
e-mail: imatzakou@epu.ntua.gr

© Springer International Publishing Switzerland 2016
K. Mertins et al. (eds.), *Enterprise Interoperability VII*,
Proceedings of the I-ESA Conferences 8, DOI 10.1007/978-3-319-30957-6_20

model where interoperability is recognized as a commoditized functionality, that can be invoked and used on demand, in a “plug n’ play” way, constituting in parallel, one of the most important pillars for the development of a single digital market for European citizens and enterprises.

Thus, the European Commission introduced the term Interoperability Service Utility (ISU) in order to denote a basic software infrastructure that collects, composes and provides a common set of web services that enable and support the smooth and efficient information exchange between diverse knowledge sources, software applications and web services. That way, it delivers basic interoperability capabilities among enterprises, independent of particular IT solution deployment, offered as utility-like capabilities [1–3].

1.2 Research Challenge

Currently, the need to move away from monolithic, centralized architectures and software paradigms that focus only on the enterprise operation alone and not on cross-enterprise and inter-organizational collaboration needs still remains eminent. Therefore, new architectural models are to be defined and applied, in order to achieve loosely coupled enterprises, able to formulate virtual alliances and respond effectively and swiftly to competitive challenges. Various research efforts have been made so far towards EI enactment through the ISU notion, especially on the architectural, framework and technological aspects of the ISU infrastructure. In this frame, several research issues attaining to the **architectural design** of the ISU, its **semantic capabilities**, its relation to the **business context**, and finally, its **service provision model** and mechanisms, still remain to be addressed. In fact, the architectural design of the ISU as a “system of systems” can indeed enable interoperability among peer systems, as it does not intervene in vital information or collaboration knowledge exchange, but acts as a mediation mechanism that provides coordination, negotiation and delegation functions. It is also envisaged to be particularly useful and attractive especially to SMEs and start-up companies, focusing on the key services needed by the industry, including also the SME-specific needs and preconditions.

At this point, a gap in literature is identified, regarding the lack of a widely accepted framework of design principles and standards concerning the design, development and actual implementation of an ISU, tackling its research challenges. Until now, there is no concrete framework of guidelines to determine how to design and implement an ISU, what principles should be fulfilled, what steps one should follow, which particular aspects should be tackled and what decisions should be made, in order to select the most appropriate standards and technologies to create a specific ISU that would serve a specific purpose and context. If such a framework existed, the benefit would be two-fold: First, enterprise stakeholders would be practically supported in their effort to design and implement an ISU and secondly concrete foundations would be established for the ISU research field to evolve further.

Therefore, the goal of this paper, via the investigation and distillation of the state-of-the-art in the interoperability domain, either by picking up existing EU directives or by comparing current trends and good practices from other EU research projects, is to finally come up with useful and value-adding guidelines that set the foundations towards the creation of a framework of common standards and principles for supporting the design, development and implementation of an ISU. Special focus is being given to the following **research challenges** (**RC**) set by the EU in the EIRR: **RC1**: the role of semantic and ontologies (T4); **RC2**: the service discovery, brokering, negotiation and mediation (T2.2); **RC3**: role of standards and specifications for interoperability (B3).

2 Methodology

In order to work towards defining a **framework** of basic standards and design principles that govern the design, development and implementation of an ISU the methodology of Fig. 1 has been followed.

The first step includes the Identification of Potential Sources containing raw but still quite useful information, with regard to the ISU. These sources include various suggestions and directions proposed by the European Union (EU), as well as existing ISU approaches and applications developed within relevant research projects, conducted either at EU or national level. The second step includes the investigation of the acknowledged sources, as well as the extraction and collection of all the relevant information especially referring to standards, design principles and research challenges. This approach covers two pillars and serves two specific purposes: on the one hand, it is crucial for ISU designers to follow the general directions of EU, as these are considered to be mandatory. On the other hand, these

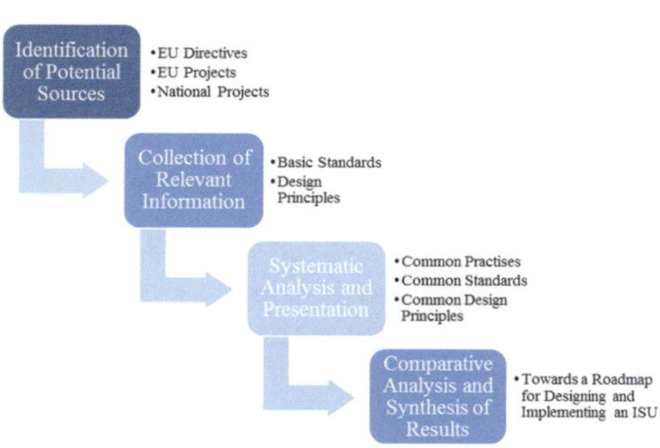

Fig. 1 Towards a framework of ISU standards and design principles

general directions alone constitute a first step towards the standardization of the design principles of an ISU. Being aware of the general techniques and good practices derived from other similar implementations as well as of the general steps and directions followed by the latter, if any, sets the necessary foundations and facilitates the research carried out in the present paper. The third step of the methodology includes the systematic analysis and presentation of the collected information. Notably, in order to have a complete and inclusive overview of the various research projects that have been identified, a suitable project template has been created, covering aspects such as the project's specific area of interest, its objectives, architectural decisions and specifications, its results and generally, any features observed to be common among the projects examined. That way, their systematic comparison was possible, pointing out common standards and design principles, as well as common techniques and good practices applied, and making up the first part of the last step of the methodology, i.e., the Comparative Analysis. The second part has been the Synthesis of Results, where all findings and conclusions were summarized in a framework of guidelines and suggestions, aiming to serve as a supportive roadmap for any interested party wishing to design and implement an ISU.

3 Design Principles and Standards

3.1 Review of EU Directives

In order to provide strategic directions for EI research, EIRR identifies **four Grand Challenges**, one of them pertaining to the definition and promotion of the ISU concept. Notably, in version 4.0 of the Roadmap, interoperability is envisioned as a **utility-like capability** that can be invoked **on-the-fly** by enterprises and support their business activities. Therefore, ISU is envisaged as a **utility-like infrastructure** fulfilling the following initial requirements: It should be simple and easily attainable by enterprises and especially by SMEs, regardless of their capabilities or existing infrastructures. Additionally, it should be all-inclusive for all kinds of enterprises, regardless of their size, business sector or geographical location. Also, the ISU should be available at an affordable initial cost, requiring low maintenance and subsequent investment. Furthermore, it should be of guaranteed quality, as defined by the respective service level agreement, and finally not necessarily controlled, owned or directed by a single private entity.

In this context, a series of more concrete guidelines are further proposed, with regard to the **ISU design principles** (**DP**), the set of basic and value-added interoperability services, the potential business model and ownership status and finally, the regulatory framework for its operation. In this paper, we focus on the design principles and relevant standards. The latter need to encapsulate the above presented rationale, as well as maintain the ISU principal characteristic of being an open infrastructure that supports heterogeneity, flexibility, usability and continuous evolution. The central idea of enabling enterprise and ecosystem interoperability and collaboration must be "designed into" the ISU (**interoperability by design**).

DP1: IT functionalities should be **delivered as** (**web-**)**services** that may reside anywhere on the web and be invoked anytime, regardless of their precise location or means to access them. There should also be two kinds of services: Basic ISU services, allowing end-to-end interoperability, as well as Value Added ISU services, supporting service co-creation and enabling innovation and value creation. Therefore, the ISU should make use of the end-to-end interconnection principle that underpins as well the design of the Internet, and emphasizes functional decentralization, peer-to-peer communication, and intelligent end-points. **DP2**: The ISU architectural design should embrace open standards and specifications, preferably via a **modular-based architecture comparted by a set of modular software components**, **rather than a** hierarchically **layered architecture**. Although this approach comes in contrast with the current logic of the Internet and Web, which require a human end-user in the loop, the key idea is that the ISU must enable and facilitate the exchange of computer-meaningful information and knowledge. **DP3**: The ISU should support **transparency** of modules' interfaces and of the information exchanged among them, enabling thus, its expansion with additional, value-added modules and functionalities, that would be offered as services, smoothly interoperating with the current software solution, without any changes to the core of the system. **DP4**: The ISU should be accompanied by a clearly documented framework of requirements for **message transactions**, so that when new added-value services are added on top of the current ISU-ones, their interoperation and exchange of information will be seamlessly carried out. Within the same framework, the quality of service should be comprehensively defined, documented and guaranteed. **DP5**: Given that the ISU is by nature a system of systems, **scalability** should constitute a native design criterion, ensuring that upon a case of growing input and output end-points of the system across multiple external systems,, the total performance and information propagation will be stable, reliable, determined dynamically and in real-time, both among the existing software components and towards the new added ones.

3.2 Review of Existing ISU Applications

In an effort to validate the abovementioned design principles and approach the aforementioned research challenges of ISUs for Enterprise Applications, a relevant research has been carried out, mainly in the scope of EU (FP7) and national projects. Below, there is a list of the most relevant research projects, focusing especially on the dimensions that are of particular interest to this paper.

- COIN [4]
- iSURF [5]
- SYNERGY [6]
- COMMIUS [7]
- ERMIS [8].

1. **COIN Project** objective was the study, design, development and prototyping of a business-pervasive, open-source service platform (**DP1**, **DP2**), capable of exposing, integrating, composing and mashing up existing and innovative to-be-developed (**DP3**) EI and Enterprise Collaboration (EC) services. This was achieved by applying intelligent maturity models, business rules and self-adaptive decision-support guidelines to guarantee the best combination of the needed services in dependence of the business context, industrial sector and domain, ensuring openness and dynamics of collaboration. The COIN integrated system was a complex cross-enterprise environment, encompassing several different components, platforms and services (**DP5**), constituting an ISU implementation that targeted the settlement of EI/EC issues, encountered by enterprises within the frame of interconnected or collaborative tasks and processes [9]. The COIN ISU was a federation of platforms, services and web interfaces which could be schematically represented as a double cloud (COIN butterfly), made of two distinct clouds, one for service provision (upper cloud) and one for service consumption (lower cloud), supporting the "prosumer" notion; the two clouds tied together through integration software, allowing the communication among the two clouds and the service consumption by other kinds of actors (**DP4**). **COIN highlights**:
   - Service Oriented Architecture (**SOA**) model (SESA reference architecture).
   - **UN/CEFACT** Core Components Technical Specification (**CCTS**) for business document modelling.
   - **XML**-based standards for electronic business messaging.
   - **Semantic Mediation services** for achieving semantic interoperability at different levels of abstraction (Business, knowledge, data, enterprise).
2. **iSURF Project** included the development of a knowledge-oriented, inter-enterprise collaboration environment, in which distributed intelligence of multiple trading partners was exploited in the planning and fulfillment of customer demand in the supply chain. The iSURF ISU has been developed on the basis of the CPFR® (Collaborative Planning, Forecasting, and Replenishment) guidelines [10], aiming at serving as a semantic infrastructure enabling the establishment of interoperability among existing **enterprise legacy applications** (**DP5**) across multiple domains, in order to achieve the required planning data exchange (**DP4**) for deploying a CPFR process within a supply chain consortium [11]. iSURF highlights:

- **Semantic reconciliation** of planning and forecasting business documents exchanged via the development of an ontology, including a common vocabulary for business document exchange.
- iSURF ISU was a fundamental, highly dynamic and flexible building block of the wider iSURF open collaborative supply chain planning environment (**DP2**), accountable for using semantic information and transforming documents from one standard to another.
- Functionalities of the legacy applications are provided **as web services** (**DP1**), facilitating thus the transmission of messages and data between the former and the iSURF components via the Enterprise Service Bus (ESB).
- **UN/CEFACT CCTS**, XML-based **standards** (XML, ebXML, UBL, GS1 eCom) and **BPEL** for specifying actions within business processes with web services.

3. **SYNERGY** This project envisaged the delivery of Collaboration Knowledge services through trusted third parties offering Web-based, pay on demand services, exploitable through ISUs and the develop of dynamic and adaptive Knowledge Management Systems to enable Virtual Organisations (VOs) to collaborate more easily. Its main approach was the semantic, ontology-based modeling of the exchanged information (enabling DP3 and DP4 as described above), implemented via a SoA approach (enabling DP1, 2 and 5, as described above).
4. **COMMIUS** project aimed to provide a utility-like capability for SMEs, capable of guaranteeing a certain set of common rules for doing businesses [12]. The COMMIUS service utility (CSU) was a **reference implementation of the ISU** concept, including services, modules and components to offer functionalities for EI, aspiring to provide SMEs interoperability based on reusing existing applications of electronic communication, such as the e-mail or the internet. From the technical point of view, a COMMIUS service utility provided interoperable interface and concrete functionalities, offered functionalities addressing EI and operated as an independent system. The CSU architecture includes a series of software modules and components providing functionalities for achieving **system, semantic and process** interoperability as well as supporting data management, security and privacy, all of them exposed **as-a-service** (**DP1**), thus being accessible through the Web. Core entities of the COMMIUS architecture could interact with external systems. These included common legacy systems in SMEs as well as specific SME legacy systems and other services supporting the business of SMEs. While the main focus of COMMIUS was to support legacy systems, the CSU also provided means to facilitate the interaction between core COMMIUS elements and other external systems, e.g., Web services-based document repositories. COMMIUS highlights: **UN/CEFACT CCTS and XML**-based standards for business document modeling and electronic business messaging among collaborating partners.

- **BPEL** for business process execution.
- **SMTP** (Simple Mail Transfer Protocol) for sending and receiving mail messages over the internet.
- **SCA** (Service Component Architecture) as a programming model for building SOA applications and ESB (Enterprise Service Bus) for the integration of diverse legacy systems.

5. **ERMIS**, the Greek Government Portal, serves as the one-stop shop of Public Administration web-services, realising the "system-of-systems" principle, by interconnecting on the back-end a large number of services coming from various systems and service providers (national and local administrations, banks, other relevant bodies), offering automated, interoperable and added value digital services, both to citizens and businesses. The ISU infrastructure operating in the back-end, is based on a e-Government Ontology, emphasizing on the formalization and the representation of a number of basic entities, namely: **Services** (digital or conventional); **Documents** (inputs/outputs of a service in electronic or printed format); **Information Systems** (for service provision, including and interconnecting both web and legacy systems); **Public Bodies,** (back-end end-points of services, documents and/or IT systems); **Web Services** (interconnecting IT systems during services' execution); **Legal Framework** (regulating the overall operation); **XML Schemas** defining the structure of the electronically exchanged documents. The aforementioned entities constitute the ERMIS "Interoperability Registry" and are organised in three layers, under a common access control based on the Single Sign On (SSO) principle: **service discovery**, based on web-based and UDDI [13] interfaces; **ontology management**, for process and data modeling capabilities; **information repository**, for interconnected data elements, process models, XML schemas and Web Services descriptions.

4 Analysis—Classification

See Tables 1, 2 and 3.

Table 1 Design principles and research challenges

	DP1: functionalities "as-a-service" (web services)		DP2: modularity	DP3: transparency	DP4: message transactions	DP5: scalability	Semantic mediation (via an ontology)	Service oriented architecture (SOA)	Web service interoperability
	Basic ISU services	Value-added services (service co-creation)							
COIN	√	√	√	√	√	√	√	√	√
iSURF	√	–	√	√	√	√	√	√	√
SYNERGY	√	√	√	√	√	√	√	√	√
COMMIUS	√	–	√	√	√	√	–	√	√
ERMIS	√	√	√	√	√	√	√	√	√

Table 2 Relevant standards acknowledged

DPs and research challenges	Relevant standards
DP1, web services interoperability	SOAP, HTTP, XML, WSDL, XSD, UDDI
Semantic mediation (ontology)	OWL
Business process management/execution (DP3)	Unified modeling language (UML), BPMN, BPEL, BPEL4WS, event-driven process chain (EPC)
Data and business document modeling/transactions (DP4)	UN/CEFACT CCTS, UBL, XML, ebXML, XSD, GS1 eCom, OASIS SET
Modularity scalability (DP2, DP5)	SOA (BPMN, BPEL, BPDM), SCA, ESB

Table 3 Framework proof-of-concept

DPs and research challenges	PLUGIN project
DP1, web services interoperability	SOAP, HTTP, XML, WSDL, XSD, UDDI
Semantic mediation (ontology)	OWL
Business process management/execution (DP3)	UML
Data and business document modeling/transactions (DP4)	XML, XSD
Modularity scalability (DP2, DP5)	SOA (BPMN)

5 Findings: Towards a Framework of Design Principles and Standards

All interested practitioners could benefit from this framework by turning the conceptual ISU model into a boundary object with concrete specifications. Starting from the **context** of an ISU implementation, the **types of services** to be offered, the main **information systems** and components to be involved, one can follow the aforementioned design principles and suitable technologies and standards for the architectural conceptualization and implementation of an ISU.

6 Proof-of-Concept: The PLUG-IN Platform

Based on the aforementioned principles, we instantiated the conceptual architecture of an ISU to implement the PLUG-IN research project [14], which targets the design, development and piloting of a utility software component that will offer the necessary set of services to deliver and sustain interoperability to its enterprises/users across similar as well as differing business domains. PLUG-IN infrastructure supports Service Discovery (e.g., e-Procurement, e-Invoicing), end-to-end interconnection, service orchestration of added value services, semantic mediation, all in a modular and scalable architecture, following all aforementioned DPs and standards (Fig. 2):

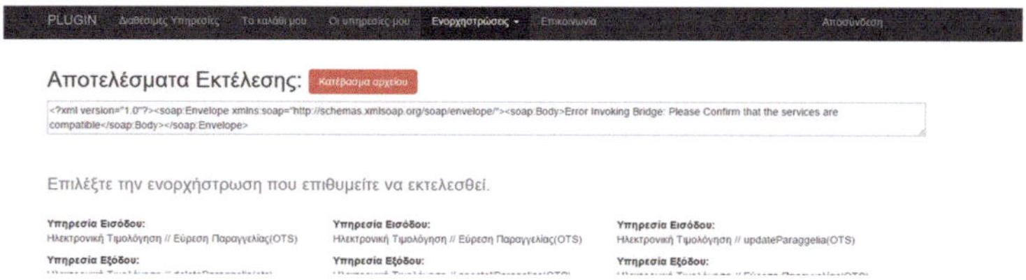

Fig. 2 Example of service orchestration in the PLUG-IN ISU

7 Conclusions—Future Work

The present paper includes a theoretical and empirical approach for the construction of a framework of design principles and standards that govern a conceptual enterprise ISU. The basic conclusion of this study is that the ISU concept is open to several interpretations and may be instantiated through various, even diverse, implementations of different purpose and scale, depending on the particular context and needs that it is required to cover. However, it is concerned as a good starting point for achieving interoperable, "one-stop" and "client-centric" enterprise services, moving from a close-world system design towards an open-world system design, of standards and components, to achieve interoperability by design. Future work could include the instantiation of this framework into a set of reference functional and non-functional specifications and its corresponding proof of concept via a certain implementation, i.e., in our case the PLUG-IN platform. Further work could also include the extension of such design principles, standards and specifications to other domains, like the e-government, for approaching and standardising the Government Service Utility (GSU) via similar frameworks.

Acknowledgements This work has been partially funded by the General Secretariat of Research and Technology (GSRT) in Greece through the Project PLUG-IN: "ICT platform for composing and delivering interoperable electronic transactions for SMEs". The authors wish to acknowledge the GSRT for their support. We also wish to acknowledge our gratitude and appreciation to all the PLUGIN Project partners for their contribution during the development of various ideas and concepts presented in this paper.

References

1. Information Society Technologies. (2006). Enterprise interoperability research roadmap. http://bit.ly/1MPE5AP. Accessed October 25, 2015.
2. European Commission. (2010). A digital agenda for Europe. http://bit.ly/1GEStio. Accessed October 25, 2015.
3. European Commission. (2006). Enterprise interoperability: A concerted research roadmap for shaping business networking in the knowledge-based economy. http://bit.ly/1RyfnbU. Accessed October 25, 2015.
4. COIN Project. http://www.coin-ip.eu/. Accessed October 25, 2015.

5. iSURF Project. http://bit.ly/1KBIjua, http://bit.ly/1Mfmkdy. Accessed October 25, 2015.
6. SYNERGY Project. https://www.synergy-ist.eu, http://cordis.europa.eu/project/rcn/85320_en.html. Accessed October 25, 2015.
7. Commius Consortium. (2009). Commius architecture (final version, deliverable D3.1.2). http://bit.ly/1Ra4LQo. Accessed October, 2015.
8. ERMIS Portal. www.ermis.gov.gr, http://bit.ly/1XuB4O2. Accessed October 25, 2015.
9. Sitek, P., Gusmeroli, S., Conte, M., Jansson, K., & Karvonen, I. (2011). The COIN book-enterprise collaboration and interoperability. Aachen, Germany: Verlag Mainz.
10. European CPFR Insights. (2002). http://bit.ly/1ib699e. Accessed October, 2015.
11. Dogac, A., et al. (2008). iSURF: An interoperability service utility for collaborative supply chain planning across multiple domains. In: *Proceedings of the eChallanges Conference*, Stockholm.
12. Laclavik, M. (2007). COMMIUS: ISU via Email, eChallenges e-2007. http://bit.ly/1GERPSj. Accessed October 25, 2015.
13. Universal Description Discovery & Integration. bit.ly/1Xn8MDU. Accessed October 25, 2015.
14. PLUG-IN Project. (2015). http://plugin-gr.epu.ntua.gr/. Accessed October 25, 2015.

Part VII
Industrial Implementation of Enterprise Interoperability

The Use of Serious Games in Requirements Engineering

Elsa Marcelino-Jesus, Joao Sarraipa, Carlos Agostinho and Ricardo Jardim-Goncalves

Abstract Today we live in a society where time is money. In research and technology development projects, it isn't different, every minute of delay would cause expenses, possible harms or project failures. In this context an efficient requirements engineering would avoid this kind of problems facilitating the accomplishment of the projects accordingly to the plan and technologically reaching the foreseen goals. The objective of this paper is to propose the use of serious games in the support of requirements engineering acquisition and validation. To demonstrate such approach two serious games are presented, which were used in the requirements engineering process of the "OSMOSE" project in two use case scenarios of the automotive and aeronautic industry respectively.

Keywords Requirements engineering · Serious games · Interoperability

1 Introduction

Currently, serious games have been increasingly used by organizations as part of its training or simulation strategies. There are many advantages that can be identified in using Serious Games such as [1, 2]: the basic principles are known from entertainment games; the participants have to actively develop own strategies; mistakes will not have negative consequences in reality and can be used to learn; and knowledge can be practically applied in a realistic setting. Therefore, they work as a "mental contest, played with a computer in accordance with specific rules, that use entertainment to further top government or corporate training, education, health, public policy, and strategic communication objectives" [3].

E. Marcelino-Jesus · J. Sarraipa (✉) · C. Agostinho · R. Jardim-Goncalves
CTS, UNINOVA, DEE, Faculdade de Ciências e Tecnologia,
Universidade Nova de Lisboa, 2829-516 Caparica, Portugal
e-mail: jfss@uninova.pt

© Springer International Publishing Switzerland 2016

K. Mertins et al. (eds.), *Enterprise Interoperability VII*,
Proceedings of the I-ESA Conferences 8, DOI 10.1007/978-3-319-30957-6_21

Despite its popularity, its use is still relatively rare in some sectors, namely in Requirements Engineering (RE). However such condition, the authors applied serious games as a simulation tool to validate and acquire RE in the OSMOSE project.

The OSMOSE project intends to design and develop OSMOsis applications for the Sensing Enterprise (OSMOSE): its main objective is to develop a reference architecture, a middleware and some prototypal applications to interconnect real, digital and virtual worlds in a similar way as a semi-permeable membrane permits the flow of liquid particles through itself. This project defines significant proof-of-concept scenarios and environments in two industrial domains, assessing technology, socio-technical implications, privacy and security issues [4].

Thus, the serious games were used in two real-life industry situations or two cases studies with the objective of validating the requirements identified previously (i.e., during the project development) and if necessary, for the acquisition of new ones. These two mentioned case studies are related to the automotive and aviation industries respectively. The Automotive industry considers the use of OSMOSE to optimise and manage a global manufacturing process of automotive camshafts to produce with less energy, best quality and higher yield. The Aviation industry considers the use of OSMOSE to base on the maintenance operations of a Flight Simulator AW 139 model [4]. This two industries are represented in this project by EPC and AW companies respectively.

In the following section of this paper, it is presented a background observation about RE and serious games. It also intends to clarify how serious games can support RE. In Sect. 3, it is described both developed serious games storylines. Section 4 describes the OSMOSE serious games workshop conducted to play the games to analyse the requirements. Conclusions and prospective work are then presented in Sect. 5.

2 Serious Games and Requirements Engineering

Serious games are characterised by time specificity with fixed starting and finishing points, as well as structured, sequential routines targeted towards achieving specific learning goals to avoid overwhelming the gamer or complicating the gaming process [5]. The creators of this kind of games normally consider three main factors: the starting level of players, the potential evolution and regression trends in the performances of players, and the realistic nature of the game [6]. Consequently, a balance must be established between believability and clarity of the games to prevent uncontrolled adaptations that could rapidly result in the generation of "out of scope" storylines or scenarios [5]. Due to the mentioned advantages and since it was already used in other fields of engineering as in simulation purposes, authors identified its potential for applicability in a RE process as presented later in this chapter.

RE is the science and discipline concerned with analysing and documenting requirements [7], which normally establishes a process to define a set of well-formulated requirements. These must be quantifiable, relevant and clearly detailed. RE acts as the bridge between the real-world needs of users, customers, and other constituencies affected by a system, and the capabilities and opportunities afforded by technologies [8]. Being concerned with the identification and communication of the purpose a specific system, RE helps determining user expectations for a new or modified product/system. It describes the process, in which the needs of one or many stakeholders and their environment are determined to find the solution for a specific problem [9, 10].

In fact, systematic RE and management is a prerequisite for successful projects and products. Despite the fact that more than half of all projects tend to fail or run into difficulties due to inadequate RE [11], its importance tends to be sometimes underestimated; which can lead to errors or omissions in requirements specification. Hence, OSMOSE derived an pervasive methodology that applies proven practices, methods and tools for helping, engineers and others, driving the RE process.

Figure 1 illustrates the common phases of a RE process namely the requirements elicitation, analysis, specification and validation, which are the basis of the OSMOSE requirements methodology defined by Marcelino-Jesus in [7], accomplished with a serious games phase.

Requirements elicitation is the act to determine or obtain the relevant requirements for the development of a solution, which should bring the greatest possible benefit concerning the goals and motivations of the involved stakeholders, assuming they are correctly understood [10].

Requirements analysis in a broad sense is related to the first step in the system design process, where requirements should be clarified, categorized and documented to generate the corresponding specifications [12]. However, inside a RE overall process, requirement analysis is also about reviewing, and to analyse them in detail, negotiating with stakeholders on which requirements are to be considered [13].

Requirements specification describes the requirements in an appropriate and unambiguous form [14]. This phase prepare requirements into a readable and

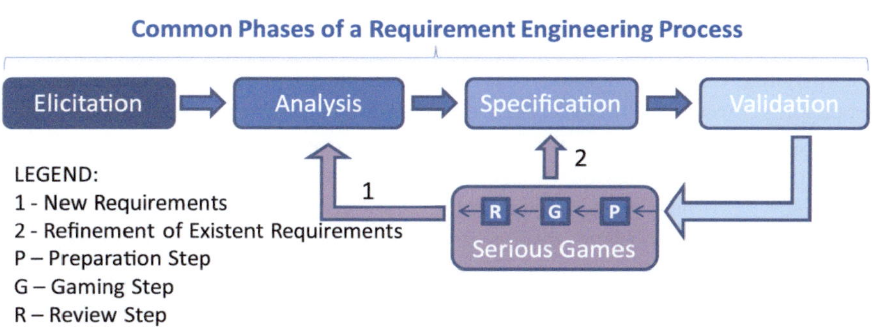

Fig. 1 Requirement engineering (based in [7])

understandable form to anyone that didn't participate in the elicitation and/or analysis processes.

Requirement validation is to review or validate requirements for clarity, consistency and completeness [13]. Requirements validation phase is used to identify the faults in the determined requirements, as the specified requirements have to accurately express the stakeholder's needs [14].

It is in this validation phase that an extra validation of requirements is conducted in a kind of a feedback cycle, which encloses the serious games (as illustrated in Fig. 1). Such validation cycle can be repeated till all the requirements are agreed and considered as enough for the development plans. New requirements identification could result from each serious game phase. If so, these new requirements have to follow to the analysis phase of the RE methodology (option 1 in Fig. 1). If just a redefinition of existent requirements are proposed, this means that such requirements just have to proceed again to the specification phase of the RE methodology (option 2 in Fig. 1).

The process of Serious Games within RE can be composed of three steps as defined in [10] and illustrated in Fig. 1: (1) Preparation step; (2) Gaming step and (3) Review step. The preparation step deals with the identification of the main focus of the project developments, which usually is discussed in meetings where "to be" scenarios are defined. It should rule the main scenario(s) to be followed in the serious game. The preparation step represent the moment where the games are developed accordingly to the identified scenarios. Finally, the gaming step comprises the gaming practice that could happen in a particular event as an workshop. As a result of such gaming practices, desired changes in particular parameters or variables followed in the game can be discussed among the end-users or participants, which could end up in the next step as new derived requirements. The review step represents the gaming performance or results analysis. It could end in the update of existent requirements or in the identification of new ones. Particular documentation to accomplish the formalisation of these requirements are expected, which includes some aspects of requirements analysis and specification.

3 Storylines of the OSMOSE Serious Games

In OSMOSE we have applied serious games in RE, reducing the tediousness associated normally to the task. At the first moment, it seems unrealistic to expect that a game could transform a RE process into a fun activity, but it can definitely help with motivation making it less cumbersome and perhaps time consuming (if we exclude the time for the game development). By applying the project use cases and scenarios to the game design in such way to facilitate testing scenarios ambiance and use cases functionalities it becomes both possible and useful. Consequently by playing the game will be able to test the requirements and verify if new requirements should be taken in consideration.

The serious games preparation step dealt with the preliminary identification of needs identified during early project meetings with the end users for the definition of as-is and to-be business use cases and related scenario descriptions. Afterwards, based on such information were defined the games' scenarios and storylines for its further development. Two games, one per industry (i.e., EPC and AW) were developed. The gaming step comprised the act of playing the developed games, which in the OSMOSE was through a specific serious games workshop. In this workshop, the stakeholders played the game and checked the game ambiance in comparison to the to-be scenarios to which the OSMOSE platform would be able to accomplish. This helped to have some discussions to validate the requirements defined before and elicit new hidden ones. The review step represented the analysis of the reports and discussions made during the workshop to validate the existent requirements and elicit new ones if appropriated.

In the following two sub-chapters both EPC and AW serious games, storylines, objectives and other main characteristics are presented. These two games were developed in the Unity 5.0 platform.

3.1 "EPC" Camshaft Game

The objective of the "EPC" camshaft game is to build a camshaft able to work till the year 2050 with the lowest possible level of costs (Fig. 2).

In the game, to accomplish the above objective, the player has to choose the best data integration approach and the best evaluation manufacturing process. Both choices represent specific manufacturing costs, which directly influences the life

Fig. 2 EPC serious game scene

duration of the camshaft being produced and its costs. Such values accordingly to each choice made has a specific value or cost, as represented in the game diagram illustrated in Fig. 3. For example, if the player chooses "Digital Traceability" as the data integration approach, it represents 10 years of extension in the camshaft life and an additional cost of 1000€.

If not chosen the "Human Checking" as the evaluation manufacturing process, the camshaft is sent for exploitation. Then, at its exploitation place, the player can define the camshaft revision scheduling program, by choosing when, and how many

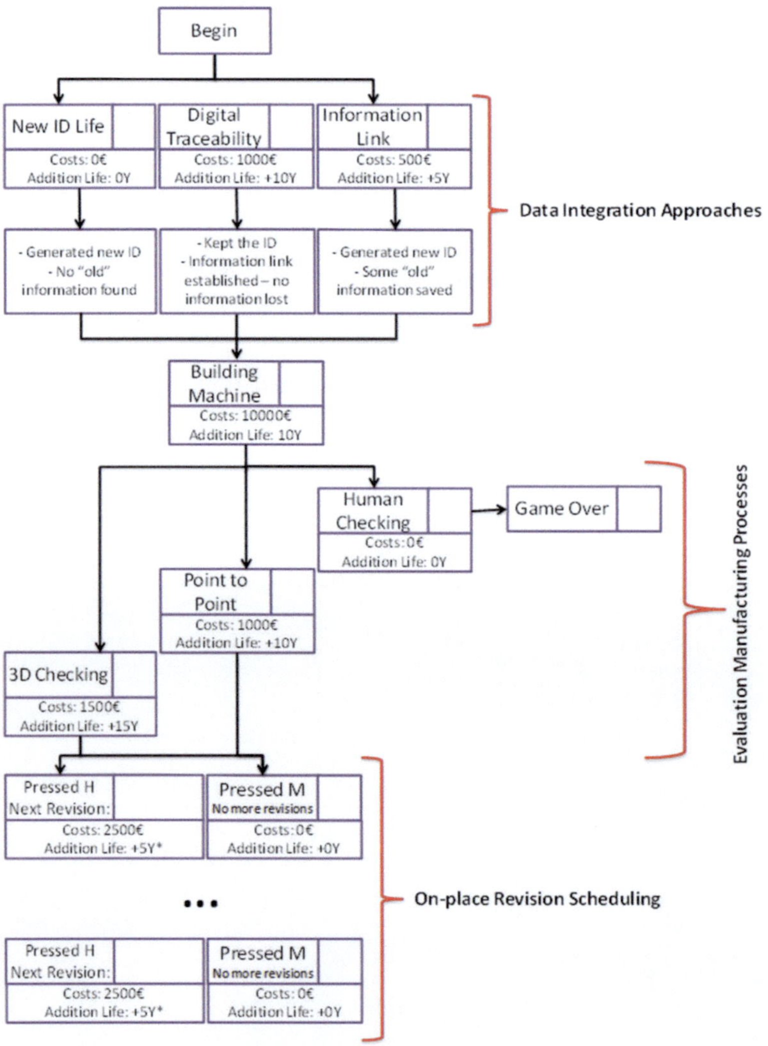

Fig. 3 Flow chart of the main player decisions in the EPC game

times the camshaft will have a monitoring or revision after its installation, or none at all. If any, he or she defines the date (year) of the next revision. Each monitoring or revision procedure schedule represents 2500€ of additional costs, but at the same time gives 5 years of extension counting from the inspection date. The player can choose to have how many revisions he/she think is appropriate, that's why the diagram in Fig. 3 have "…" in the "on-place revision scheduling" part. The main catch is that if the player chooses the best approaches the camshaft does not need any revisions. If it had cheaper procedures in the manufacturing procedure it could still reach the year 2050, but in that case it has to have one or more revisions, which obviously will increase its price/costs. All the choices, and its related figures will be reported at the end of the game.

Additionally, this game introduces a random variable, which represents the cases of customers that in real life, do not use the camshaft properly. This means that even if the player chosen the best manufacturing procedures and approaches the camshaft could still not reach the expected year (e.g., 2050). But if the player chooses the "Digital Traceability" and later the "3D Checking" options at the manufacturing phase of the game it represents that EPC system will have all the details about the camshaft. This means that EPC has enough information to complain about the bad behaviour of the customer, saying that if the camshaft didn't worked till the expected year it was because of their bad use or so.

The mentioned EPC game can be played through a regular web browser, which URL is: http://gris-dev.uninova.pt/osmose/EPC.html.

3.2 "AW" Flight Training Game

In the "AW" flight training game, the player experiments the role of a pilot (trainee) and a manager. As a trainee, the player has to identify snags during the flight simulation, which afterwards has to report them. As a manager, the player has to validate pilot snags from a list. As in the real life, if the pilot does not report snags or if the manager does not validate correctly their existence, additional expenses will happen to the AW. Thus the main objective of the game is to test the real scenarios that happen daily at AW and try to avoid extra expen (i.e., bad identification or validation of snags).

The story of the game occurs in the galaxy where a regular helicopter is represented as a "galaxycopter" Fig. 4. The first main scene happens in a simulation room, this means inside a "galaxycopter", where the main goal, as mentioned before is to identify snags that could occur during a hypothetical rescue mission to one planet (i.e., Earth and Mars).

Four types of snags can be randomly generated in the game, whose costs for fixing them are presented in Fig. 5 diagram. If the player do not notice the first snag, which is the only snag that always happen, and that is related to the coordinates system (the "galaxycopter" is sent to a wrong place/planet), the game will

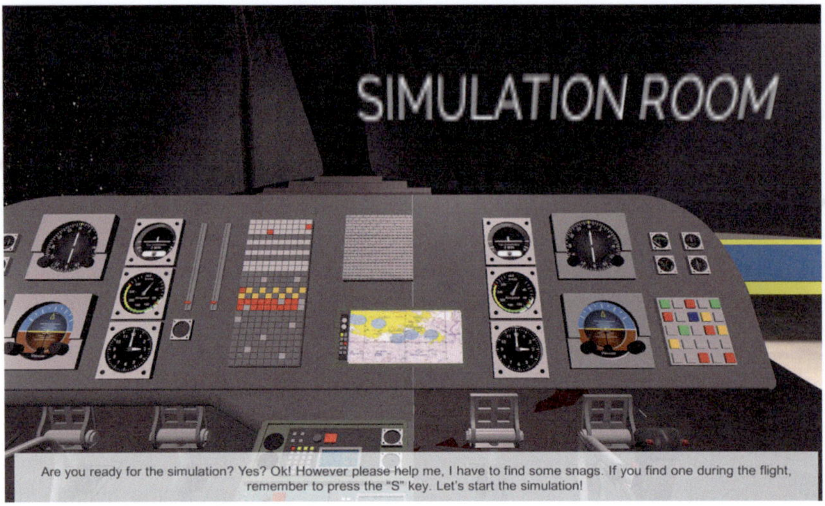

Fig. 4 AW serious game scene

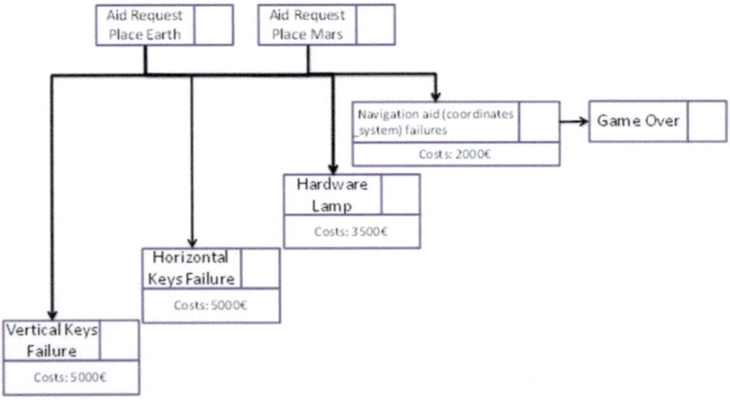

Fig. 5 Flow chart of the possible snags able of happening during the AW game

over. After have completed the rescue simulation, the pilot (player) is sent to the "TOTTEM Room" to perform the snags report.

After this, the player is sent to the "Manager Room", where he performs the role of the AW manager. This means that he has to validate from a list of snags, the ones that really happened during the game, illustrating this way a normal snags validation procedure.

Afterwards is generated and shown a report of the costs related to the actions performed by the player, which also represents the efficiency level of the player in the role of a pilot (trainee) and as AW Manager. This report follows some rules and costs. There are five different situations, which represent costs. These situations are related to:

1. SNAGS and costs associated for issue fixing and training re-scheduling.
2. Snag issues solved that were not necessary (invalid snags).
3. Snags do not solved that caused disruptions of functioning (not identified snags).
4. Snags solved, but that the pilot didn't identified (request additional people effort discovering them).
5. Snags that were identified by the pilot but that didn't happen and also were not validated by managers (snags not identified at all, thus conducting to training sessions disruption).

Additionally, as in the EPC, this game can be played at the following link: http://gris-dev.uninova.pt/osmose/AW.html.

4 OSMOSE Serious Games Workshop

Accordingly to Westra in [5], the serious games are characterized in two types: (1) the game playing, and (2) the role-playing. The first replicates the experiences of users in environments that mimic real life, while the second focuses on experiences of interactions among users based on defined roles as surrogates non-stakeholder experiencing the role of a stakeholder, or a stakeholder playing the role of a user.

The OSMOSE serious games workshop took place during the periodic general meeting of the project in June 2015 at Lisbon. During the workshop, it was performed both mentioned gaming and role-playing approaches. Hence participants also assumed the roles of other stakeholders (role-playing). By such collaboration, the OSMOSE partners analysed the scenarios from a different perspective, which helped to fundament different views of them.

Thus, each stakeholder played at least 3 times each game consuming about 10 min per game. They were divided in groups of two, accompanied by a tutor, which registered the choices (steps taken), doubts and results achieved during the game. For each game played a different form was filled-up with such information, which in later analysis helped to compile the results, and elicitate new requirements.

As a result of such analysis, a list of the possible new requirements or user stories identified were presented in the room, and that are presented in Tables 1 and 2.

A further analysis of the new created requirements at Lisbon workshop it was also conducted. Its overall conclusion was that through the serious games it was possible to approve the 24 AW existent requirements and extract 2 new ones (ids: 1487, 1488). Later, both were confirmed and improved, making a total of 26 user stories

Table 1 AW raw requirements

AW requirement	New user story
ID—1487	As a "AW maintenance manager" I want "to keep track about hardware fault history" So that "the system can alert for a possible future HW fault before it happens and I could be prepared on time"
ID—1488	As a "AW manager" I want "a program/device to save automatically a snag by clicking/pressing in a specific key/button" So that "later training pilots don't need to insert all the information about this snag in the Totem software, (it would facilitate snags report—semi-automatic report)"

Table 2 EPC raw requirements

EPC requirement	New user story
ID—1483	As a "EPC manager" I want "the OSMOSE platform could receive information about the camshaft installation from the customer" So that "the quality control of the camshaft and consequently its maintenance could be better scheduled, and I could verify if there is any bad installation of the camshaft by the customer"
ID—1485	As a "EPC manager" I want "the OSMOSE platform to be able to generate reports or specific evaluation data at the end of each camshaft production" So that the quality control could be improved and it becomes possible to identify each camshaft warrant/life duration"
ID—1491	As a "EPC manager" I want "the OSMOSE system facilitate the costs calculation of the various manufacturing processes" So that "could better choose what process to follow (e.g., 3D checking versus human checking)"
ID—1492	As a "EPC manager" I want "the OSMOSE system integrate scheduling features" So that "the maintenance planning could be better performed and dependent to each particular camshaft (building) characteristics"

for the AW industry scenario. In the same process, 20 EPC existent requirements or user stories were validated and 4 new ones came out of the workshop (ids: 1483, 1485, 1491, 1492). However, only the first two from these four were validated resulting in a total of 22 requirements for the EPC industry scenario.

The requirements mentioned above can be consulted at the wiki webpage: http://gris-dev.uninova.pt/osmose/index.php/Main_Page.

5 Conclusions and Future Work

The use of serious games in RE has proven to be a very participative form of involving different types of actors (managers, users, developers, etc.) in the requirements validation. In the OSMOSE project was presented two serious games, one for each scenario, and the results achieved with them, enabled not only to elect new "hidden" requirements but also to confirm the existing ones.

Additionally, the serious games also helped to illustrate the OSMOSE generic business applications for the liquid-sensing enterprise concept, which became fundamental for dissemination purposes.

Consequently, authors considered, despite of the big effort (in resources) for developing serious games that they are essential in research projects developing. Such affirmation resulted by the fact that serious games can be used to help in validation of the RE process, its related technical and business scenarios, and at the same time used for dissemination purposes with a big impact.

As future work, authors intend to use the serious games approach in others research projects with the same objective but also as an additional training object to facilitate knowledge transfer to the projects' stakeholders.

Acknowledgments The research leading to this result has received funding from the European Union 7th Framework Programme (FP7-ICT) under grant agreement of the OSMOSE—OSMOsis applications for the Sensing Enterprise, project ID nr. 610905, (http://www.osmose-project.eu/).

References

1. Thoben, K.-D., Schwesig, M. (2007). Simulation gaming to train engineers to efficiently collaborate in manufacturing networks. In M. Taisch, K.-D. Thoben & M. Montorio (Eds.), *Advanced manufacturing—An ICT and systems perspective*. Taylor and Francis.
2. Jarke, M. (2007). Serious games: The impact of pervasive gaming on business. In *Proceedings of the 8th Internationale Tagung Wirtschaftsinformatik—Band 1, Universitätsverlag Karlsruhe* (pp. 15–16), Karlsruhe, Germany.
3. Wikipedia. (2015). *Serious games*. Retrieved from the web at July 2015: https://en.wikipedia.org/wiki/Serious_game
4. OSMOSE project on a page. Retrieved from the web at September 2015: http://www.osmose-project.eu/
5. Westra, J., Hasselt, H.V., Dignum, V., Dignum, F. (2008, December). On-line adapting games using agent organizations. In *IEEE Symposium on Computational Intelligence and Games*, (pp. 15–18, art. no. 5035646, pp. 243–250), Perth, Australia.
6. Andrade, G., Ramalho, G., Gomes, A. S., Corruble, V. (2006). Dynamic game balancing: An evaluation of user satisfaction. In J. E. Laird & J. Schaeffer (Eds.), *AIIDE* (pp. 3–8). The AAAI Press.
7. Marcelino-Jesus, E., Sarraipa, J., & Agostinho, C. (2014). A requirements engineering methodology for technological innovation assessment. *Advances in Transdisciplinary Engineering*, 577–586.
8. Kaur, R., & Singh, T. (2010, October). Analysis and need of requirements engineering. *International Journal of Computer Applications (0975–8887)*, *7*(14).

9. Nuseibeh, B., Easterbrook S. (2000). Requirements engineering: A roadmap, international conference in software engineering, In *Proceedings of the Conference on the Future of Software Engineering* (pp. 35–40). Limerick, Ireland: ACM Publishers.
10. Deliverable 51.2. (2012). Methodology for requirements engineering and evaluation in manufacturing service ecosystems. MSEE—Manufacturing Services Ecosystem. Delivered at 31/03/2012.
11. Vector Consulting Services. (2014). Retrieved January 2014: https://vector.com/vc_training_requirements_engineering_en.html
12. Tran, E. (1999). Requirements and specifications. Available at March 2014: https://www.ece.cmu.edu/~koopman/des_s99/requirements_specs/
13. Bavani, R. (2010). Requirements Engineering = Elicitation + Analysis and Negotiation + Specification or Documentation + Validation! Software Engineering. Available at March 2014: http://se-thoughtograph.blogspot.pt/2010/01/requirement-engineering-key-aspect.html
14. MSEE Project Consortium, Deliverable 51.1. (2011). Applicability of existing user requirements approaches in MSEE. MSEE—Manufacturing Services Ecosystem.

Knowledge-Based System to Enhance Coordination of Hospital Practitioners: A Case Study

D. Guillon, E. Vareilles, G. Marques, P. Gaborit and M. Lauras

Abstract The work reported in this paper aims to design a knowledge-based system in order to improve the first contact between an expectant mother and the hospital practitioners in charge of pregnancy following-up (midwives, gynecologists). Our work focuses on the appointment making process which is most of the time carried out by auxiliary nurses of the gynecology and obstetric department. Generally, these auxiliary nurses have only a few minutes to collect information from expectant mothers, to identify the risk level of the pregnancy, and to make an appointment with the relevant and available practitioner. Two main difficulties have been identified in this process: the call pick-up rate and errors in making appointment. These two difficulties have become for us two principal areas for improving the service quality given to patients. We have extracted and formalized some knowledge as a constraint satisfaction problem. This knowledge-based system is currently tested in the gynecology and obstetric department of a French hospital.

Keywords Practitioners coordination · Knowledge-based system · Constraint programming · Obstetric application

1 Introduction

For an expectant mother, the relationship with the hospital where she wants to give birth is something critical. The initial step of this relationship starts with the first call which aims to plan the timeline of pregnancy monitoring. The pregnancy monitoring mainly depends on the risk of the pregnancy (which can be low-risk or high-risk), on the medical history of the expectant mother (first pregnancy, miscarriages, etc.), and on some specific cases (anonymous childbirth, patient under

D. Guillon · G. Marques
Toulouse University Hospital Centre, Toulouse, France

D. Guillon · E. Vareilles (✉) · P. Gaborit · M. Lauras
University of Toulouse, Mines Albi, France
e-mail: elise.vareilles.emac@gmail.com; elise.vareilles@mines-albi.fr

© Springer International Publishing Switzerland 2016
K. Mertins et al. (eds.), *Enterprise Interoperability VII*,
Proceedings of the I-ESA Conferences 8, DOI 10.1007/978-3-319-30957-6_22

18). Nevertheless, in the French hospital which is our case study, there is no completely dedicated people in charge of answering the phone in gynecology and obstetric department. In addition to their other tasks, these agents have to pick up the phone and be able, in a few minutes, to collect relevant information from the expectant mothers, to identify the proper pregnancy following-up (series of appointments) and make an appointment with the correct and available practitioner. Therefore, the call pick-up ratio in such a department is very low, frequently longer than needed (around 5 min to have the time to plan several meetings) and some errors in making appointment frequently occur (appointment with an irrelevant or unavailable practitioner). To improve the making appointment process, we have extracted and formalized as a constraint satisfaction problem some knowledge acquired by the agents dedicated to the phone appointment scheduling. Our areas for improvement are: enhance the pick-up ratio by mastering the call time, and in the same time, avoiding errors in making appointment and improving the relevance of the taken appointment. The rest of the paper is organized as follow. First, the knowledge of gynecologic and obstetric appointment setup is described. In a second part, constraint satisfaction problem is introduced and the constraint-based model is presented. Then some conclusions and perspectives are drawn.

2 Analysis of the Appointment-Making Process

Although at first glance an appointment-making seems easy and simple, it is a complex process which considers many factors. After having observed and analyzed the appointment-making process in a French hospital department, we have identified its various components. Then we can define an appointment as the result of the combination of a practitioner, a type of consultation, a time slot, a place and eventually some prerequisite exams, as illustrated in Fig. 1. An appointment is

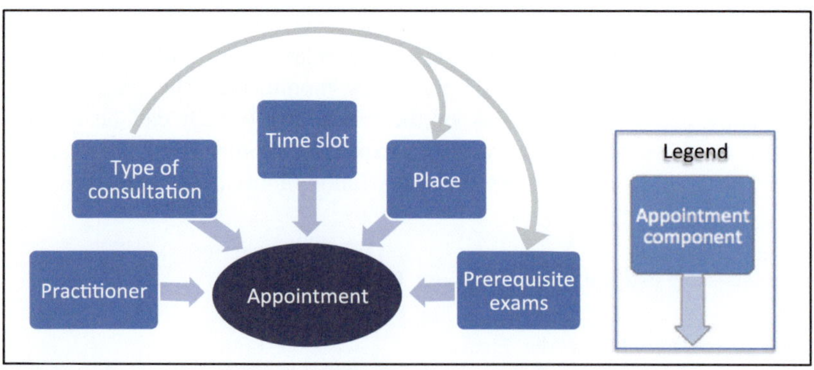

Fig. 1 Appointment components

properly made when each item is relevant and properly defined. This definition can be applied to any medical appointment.

The place and the prerequisite exams are linked to the type of consultation. Then, when the type of consultation is identified, the place and the prerequisite exams are automatically deduced. For instance, a 6th month antenatal consultation (type of consultation) will take place in a consulting room (place) after an ultrasound scan (prerequisite exam), while a preparation for birth (another type of consultation) is a common class which takes place in a dedicated room (another place without prerequisite exam).

The time slot depends on the type of consultation and on the practitioner. A software already in place allows the agents to identify an available time slot for each pair (type of consultation, practitioner). But the chosen time slot also depends on the patient availabilities. As this is most of the time the case, we will suppose here the "earlier due to patient availabilities" rule is applied.

Then, we are looking to identify only two components which are the proper type of consultation and the relevant practitioner (since the others thereunder). The phone interaction between the agent and the expectant mother will allow it, as shown is Fig. 2.

In order to make the relevant appointment, the agent has to ask each expectant mother a series of questions in order to establish her profile. The set of questions is always the same but the order may vary depending on the agent and some are needed only in particular situations. Each of them has her/his own way to ask them. These questions concern mainly the expectant mother and her health condition:

- Some of the questions aim to estimate the due date and schedule all the meetings:
  - What is the date of the last period?
- Some are indented to estimate the pregnancy risk:
  - Have you suffered miscarriages?
  - Are you currently taking medication or treatment?
  - Is it a twin pregnancy?
  - Is it a first pregnancy?
- In case of a non-first pregnancy, some additional questions about the previous deliveries have to be asked:
  - What type of delivery: vaginal, instrumental extraction or caesarean section?
  - Was your previous pregnancy healthy or not?

With all the answers, the agent is able to determine the needs of the expectant mother, the due date, the type of consultation and the skills of the practitioner (midwives or gynecologists). There are several types of consultations, among others:

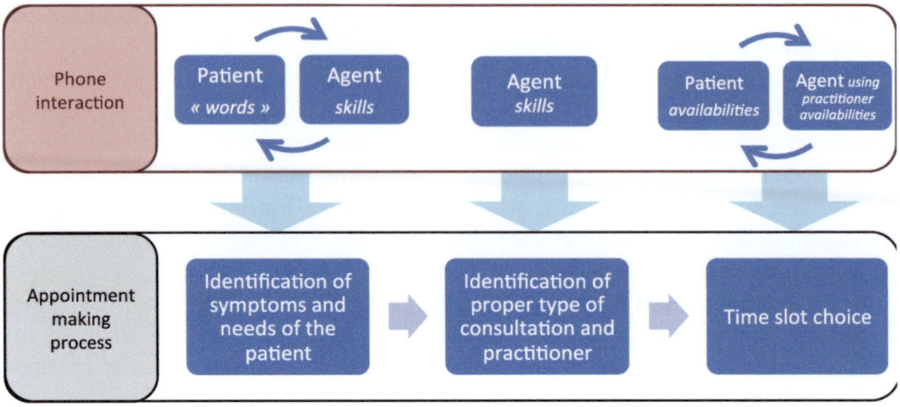

Fig. 2 Appointment making process

- those related to healthy pregnancies,
- those related to pathological pregnancies,
- those related to birth preparation courses, …

Finally, the last factor to be taken into account is the practitioner availability, in order to make the appointments.

3 Constraints Satisfaction Problem

The appointment setup can be considered as a routine process as the series of questions are always the same (with some optional). The corresponding knowledge can be therefore extracted and formalized. The care path of the expectant mother is personalized (or configured) regarding her own story. Many authors such as [1–3] have shown that configuration could be efficiently modeled and aided when considered as a Constraints Satisfaction Problem (CSP). A CSP is a triplet {X, D, C} where X is a set of variables; D is a set of finite domains (one for each variable) and C a set of constraints linking the variables [4]. The variables can be either discrete or continuous. The constraints can either be of compatibility, when defining the possible or forbidden combinations of values for a set of variables (lists of compatible values, mathematical expressions), or of activity, when allowing the activation of a subset of variables and constraints [5].

The appointment setup is interactive, meaning that the expectant mother and the agent dialogue in order to determine the relevant type of consultation and therefore to fix the appointments. We have seen that the set of relevant questions can be modified depending on the answer given by the future mother: "Is it a first pregnancy?" implies that some specific questions about the previous pregnancies and deliveries have to be asked. Not all the questions are relevant at the beginning [6]. To be compliant with the natural interactivity, we have chosen to use the filtering techniques of constraints programming, such as arc-consistency [7]. Constraints are not "oriented", meaning that a value given to any variable of the problem can be propagated, through the constraints, on the remaining variables.

Translating the questionnaire as a CSP means that each of the question is associated with a discrete variable. With these questions, the system is able to deduce a unique value (considered as a result) to the variables representing the patient needs or symptoms, the type of consultation and the practitioner, who is going to look after the expectant mother during her pregnancy (midwive or gynecologist). In our model, some variables are always present in the questionnaire:

- date of the last period (to deduce the delivery date and then ensure that a place will be available. Indeed, each month, the hospital can take charge a limited number of deliveries): yes, no
- miscarriages: yes, no
- first time mother: yes, no
- medication or treatment: yes, no
- expectant mother needs: healthy pregnancy, 1st appointment for pathological pregnancy, 2nd appointment pathological pregnancy, etc.
- consultation type: 1st time gynecologist, 2nd time gynecologist, midwive, etc.
- practitioner: Dr. Baby, Dr. Child, Ms. Wise, Ms. Owl, etc.

Others variables are only taken into account depending on some previous answers, such as "Was your previous pregnancy healthy or not?". In that case, the corresponding variable Previous pregnancy is inactive and will only be activated if needed:

- Inactive previous pregnancy: healthy or not
- First time in this hospital: yes, no

The logical link between existence of questions into the questionnaire is formalized with an activity constraint. In that case, if the variable first pregnancy is valuated to no, the variable previous pregnancy is being activated and the agent, answering the phone, can give it a value.

The links between the variables of the questionnaire and the expectant mother needs are formalized as compatibility tables. For instance, an healthy first pregnancy is only compatible with the need "healthy pregnancy". If it is not the first pregnancy, the expectant mother needs depend on the fact that the mother has already been followed in this hospital or not (First time in this hospital valuated to yes). This knowledge is formalized as presented in Table 1.

Table 1 Questionnaire and Expectant mother needs compatibility constraint

Healthy pregnancy	First appointment for the following-up of this pregnancy	Expectant mother needs
Yes	–	Healthy pregnancy
No	Yes	1st RDV pathological pregnancy
No	No	2nd RDV pathological pregnancy

Table 2 Expectant mother needs and Type of consultation compatibility constraint

Expectant mother needs	Type of consultation
Healthy pregnancy	Midwives
1st appointment pathological pregnancy	1st time gynecologist
2nd appointment pathological pregnancy	2nd time gynecologist
....	....

To each expectant mother need, is associated one type of consultation. The type of consultation is formalized as a discrete variable:

- Type of consultation: 1st time gynecologist, 2nd time gynecologist, midwives

This knowledge, linking the expectant mother needs and the type of consultation, is formalized as presented in Table 2.

Each practitioner has her/his own skills and is able to carry out some specific type of consultation. This knowledge is formalized as presented in Table 3.

In our case study, there are:

- 11 variables which belong to the questionnaire. 6 of them are mandatory, 5 of them are optional and depend on the answers given to the first six ones;
- 6 values to characterize the expectant mother needs;
- 8 values to characterize the type of consultation;
- 30 different practitioners with specific skills;
- 16 activity constraints to modify the questionnaire by adding relevant questions;
- 6 compatibility tables linking questionnaire to the expectant mother needs;
- 2 compatibility tables linking the expectant mother needs to the consultation type, and the consultation type to the practitioners list.

This formalized model results from a six-month knowledge extraction and validation process in a French hospital department.

The interactive aiding appointment setup consists in giving a value to a variable v or limiting the domain of a variable v belonging either to the questionnaire or the practitioners (in order to take into account their unavailability). The modification of the domain of v is propagated through the constraints network to the other variables by retrieving all the values from their domain which are now inconsistent with the reduced domain of v. This mechanism, repeated several times, progressively

Table 3 Type of consultation and practitioner compatibility constraint

Type of consultation	Practitioner
Midwives	Ms. Wise
Midwives	Ms. Owl
1st time gynecologist	Dr. Baby
2nd time gynecologist	Dr. Baby
2nd time gynecologist	Dr. Child
...	...

restricts the solution space in order to reach a consultation type and the short list of available and relevant practitioners.

4 Conclusion and Future Work

The work presented in this paper concerns the enhancement of the phone appointment setting thanks to a knowledge-based system. By formalizing the knowledge acquired by the agents dedicated to the phone appointment setting and modeling the appointment setting by a constraint satisfaction problem, we aim to develop a tool improving the relevance of the taken appointment and enhancing the coordination between all the practitioners. Also, by capitalizing the knowledge of the current agents, we will facilitate the formation of newcomers. And last but not least, the tool will allow mastering the time of conversation.

Directly link to the software able to identify available time slot for each pair (type of consultation, practitioner), the tool will be very helpful to configure every expectant mother's care path. The interoperability between these two tools has to be studied and developed with regards to SOA paradigm.

This analysis can be applied to any medical appointment making. In the targeted French hospital, the method used to build this tool is currently extended to another scope, the gynecology appointments, and will be extended to all the appointments taken in the hospital. If this pilot works, the agents will be formed to apply this method to other departments.

Acknowledgments The authors wish to acknowledge all the doctors and midwives of the CHU de Toulouse for their contribution during the extraction and knowledge validation phases. All the knowledge collected and formalized is their.

References

1. Tsang, E. (1993). *Foundations of constraints satisfaction*. London: Academic Press.
2. Sabin, D., & Freuder, F. (1996). *Configuration as composite constraint satisfaction* (pp. 28–36). Technical report FS-96-03, Workshop on configuration. Boston: AAAI Press.
3. Djeffal, L., Goncalves, G., & Babes, M. (2006). Constraint satisfaction problem based modelling and value ordering based heuristic for nurse scheduling. *Journal of Applied Computer Sciences, JACS, 14*(2), 7–28.
4. Mittal, S., & Falkenheiner, B. (1990). Dynamic constraint satisfaction problems. In *Proceedings of the 8th National Conference on Artificial Intelligence* (pp. 25–32). Boston: AAAI Press.
5. Montanari, H. (1974). Networks of constraints: Fundamental properties and application to picture processing. *Information Sciences, 7*, 95–132.
6. Ventilii, M. (2011). Dynamic constraint satisfaction approach to hospital scheduling optimization, Università degli Studi di Padova, Master Degree in Computer Science Engineering.
7. Mackworth, A. K. (1977). Consistency in networks of relations. *Artificial Intelligence, 8*, 99–118.

Towards a Flexible Gamification Model for an Interoperable E-learning Business Process Simulation Platform

Sarah Zribi, Tom Jorquera and Jean-Pierre Lorré

Abstract E-learning, technology-assisted education has become a recognized and widespread educational tool in today's society. This transition allowed to the domain to import ideas from other domains of the digital world. Among those, the concept of gamification has quickly become one of the central tools to for keeping learners engaged. This paper proposes a gamification approach for collaborative simulation-based learning. It introduces a simple and flexible model for introducing gamification mechanisms for collaborative simulation-based learning. The work is applied to a business process simulator developed in the context of the Learn PAd project aiming at providing a learning platform for workers in public administrations.

Keywords Business process · Business process simulation · E-learning · Gamification mechanisms · Interoperability · Scoring mechanisms · Virtual certificate reward

1 Introduction

In modern society Public Administrations (PAs) are undergoing a transformation of their perceived role from controllers to proactive service providers, and are under pressure to constantly improve their service quality while coping with quickly changing context (changes in law and regulations, societal globalization, fast technology evolution) and decreasing budgets. The Learn PAd project[1] aims at building a holistic e-learning platform for PAs that enables process-driven learning and fosters cooperation and knowledge sharing.

[1]Learn PAd Project. www.learnpad.eu.

S. Zribi (✉) · T. Jorquera · J.-P. Lorré
Linagora, 75 Route de Revel, 31400 Toulouse, France
e-mail: szribi@linagora.com

© Springer International Publishing Switzerland 2016

K. Mertins et al. (eds.), *Enterprise Interoperability VII*,
Proceedings of the I-ESA Conferences 8, DOI 10.1007/978-3-319-30957-6_23

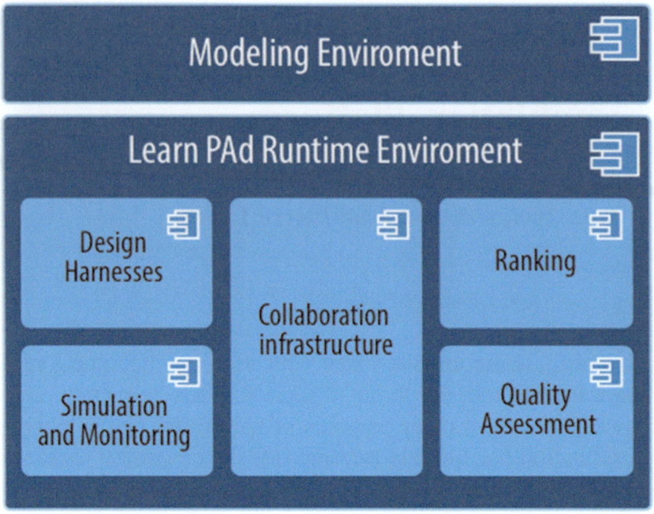

Fig. 1 Learn PAd architecture

Learn PAd technical innovation is based on four pillars: (i) a new concept of model-based e-learning (both process and knowledge), (ii) an open and collaborative e-learning content management, (iii) an automatic, learner-specific and collaborative content quality assessment, and finally, (iv) an automatic model-driven simulation-based learning and testing. In Fig. 1, we depict the overall architecture of the project supporting this approach.

This paper focuses precisely on the simulation and monitoring component and proposes a simulation platform providing a flexible and interoperable simulation environment with support for collaboration. The simulation engine supports both the software emulation of the involved parties, and the provisioning of dedicated means for gathering learners willing to train on the business process by interacting with other learners. The infrastructure also includes monitoring feature providing feedbacks for the evaluation of learners, business processes, and associated learning contents. Model-based learning simulation can be considered as a collaborative game where a team of any number of learners works together in order to achieve a common objective. The main objective is to provide facilities useful to learners for improving their knowledge about a given business process in an enjoyable and entertaining context. The remainder of this paper is organized as follows. In Sect. 2, we sketch an overview of the proposed e-learning simulation architecture. Section 3 defines briefly the gamification and presents two gamification mechanisms that can be considered within the simulator to provide engaging with a learner. Thereafter, Sect. 4 illustrates an implementation of practical use case study. Finally, the last section concludes and gives insights of future works.

2 Overall E-learning Simulation Architecture

The e-learning simulation platform allows multi-sessions business processes simulation and can engage civil servants in both individual and collaborative simulation allowing them to train on the business process tasks by interacting with other learners. Moreover, the simulation environment includes an event-based monitoring framework aiming at providing feedbacks for the learner evaluation and checks if the KPI are fulfilled by the learner while executing the simulation of the BP. Figure 2 presents the high level architecture includes Business Process simulation engine and the monitoring infrastructure. The simulation framework includes 7 main components:

- **SimulationGUI** is in charge of the interactions between civil servants and simulator's components.
- **PersistenceLayer** stores the status of the simulation at each step in order to give the civil servant the ability to stop it and restart when needed.
- **RobotFramework** allows simulating the behavior of civil servants by mean of robots.
- **SimulationEngine** is the core component. It enacts BP and links activities with corresponding civil servants or robots. In the context of the Learn PAd project, we consider the open-source activiti [1] as the BPMN engine.

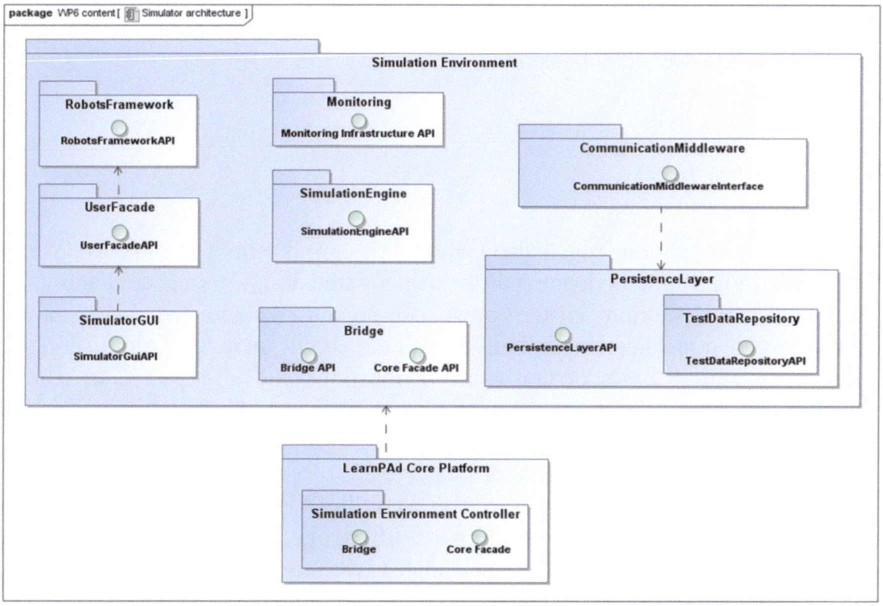

Fig. 2 E-learning simulation architecture

- **Monitoring** collects the events occurred during the simulation and infers rules related to the BP execution. Additionally, the monitor component also interacts with the LearnPAd Core Platform in order to provide off-line civil servant assessment scores.
- **CommunicationMiddleware** provides event-based communication facilities between the simulation components according to the publish/subscribe paradigm.
- **UserFacade** is in charge of encapsulating civil servants or robots in order to make the learner interaction transparent to the other components of the architecture.

Each simulation component is exposed as a service and provides an API as unique point of access. The simulation platform interacts with the rest of Learn PAd components by means of the Learn PAd Core Platform and specifically through the Bridge component (including the Bridge and the Core Facade interfaces).

3 Gamification Challenges for an Interoperable E-learning Simulation Platform

The current section describes gamification mechanics that complements the logical architecture of the e-learning simulator in order to provide an entertaining and engaging with a learner in a simulated work context. To this end, we start by defining gamification in the first sub-section before presenting two gamification techniques in the second sub-section.

3.1 Gamification

Since 2010, gamification (called also Gameful design) is a hot topic that everybody talks about. In [2], authors define it as the use of game design elements in non-game contexts. This definition is the most comprehensive and the best known. Gamification is a fast growing initiative, with the aim of increasing motivation and participating among users of non-game applications and is expected to revolutionize all aspects of life in the not too distant future [3, 4]. Its adoption in both industry and research has been widespread over a range of different domains (medical, rehabilitation, education, etc.).

Gamification in learning and teaching according to Kapp in [5] is the use of game-based mechanics and game thinking to engage, motivate action, promote learning and solve problems. In fact, they relate to the promotion of learning using the inherent problem-solving skills and engagement that interactive games facilitate [2].

3.2 Gamification Mechanics for the E-learning BP Simulation Platform

Thereafter, we present two important gamification elements that we can find for learning purposes and ensure more engagement with learners: *Game Score* and *Virtual Certificate Reward.*

Game Score. It is important for learners to see success displayed incrementally. To this end, we integrate in the platform a scoring mechanism in order to generate ranking. Two types of score are identified: Session Score and Global Score.

- Session Score is calculated for each learner participating in a given simulation session. This score is calculated using a weighted sum of scores attributed to the Civil Servant for each task of the Business Process realized during the simulation. Considering "n" the number of tasks executed by the civil servant during the learning session and "P" the weight of the task, the session score is computed as follow:

$$session\_score = \sum_{i=1}^{n} task\_score_i \mathrm{P_i} \quad (1)$$

Each task of the Business Process is associated with a weight specified as a metadata. These metadata are attributed in the BP definition and defined by the modeler. The calculation of the score's task is based on several criteria, namely number of attempts, Success/Fail and finally KPI (e.g., response time). Formula 2 allows calculating this score.

$$task\_score = success * \left(\frac{1}{nb\_attempts} + \sum_{i=1}^{k} \left(\frac{expected\_KPI\_value}{observed\_KPI\_value} \right) \right) \quad (2)$$

where "k" is the number of KPI considered in the evaluation of the civil servants performances.

- Global Score represents the overall score obtained by the learner in all the executed business processes. It is defined as the sum of the best scores obtained for each simulated Business Process (formula 3).

$$globale\_score = \sum_{i=1}^{j} \max(session\_score_i) \quad (3)$$

Virtual Certificate Reward. While some commercial games may use real gifts (e.g., money, gift cards, etc.), many games motivate players with virtual rewards as certificates. In the E-learning BP simulation platform, learners who satisfy some conditions can be automatically awarded by the platform with a specific certificate. Being awarded a certificate is important to the learner since it gives to him/her additional rights. For example, a learner with a good game score in some Business Processes simulations and with considerable contributions to the simulation platform is awarded the "expert" certificate. This allows him/her to check answers from other learners for validation during a learning simulation. Moreover, the platform constantly monitors the activity of learners. When one of them satisfy the required obtaining conditions, the platform assigns him/her the corresponding certificate automatically and informs him/her about the new deserved rights.

4 Applications

The municipality of Senigallia is a port town on the Adriatic coast. It is located in the Marche Region and in the Ancona Province. It directly offers to its citizens the SUAP ("*Sportello Unico Attivitá Produttive*"). This service refers to the activities that the Italian administration has to put in place in order to permit to entrepreneurs to set up a new company. The SUAP implements the front end of a complex BP involving several public administrations. In particular, the BP permits to satisfy all the necessary obligations related to the establishment of a new business.

Focusing on competences of SUAP office workers, in order to execute the activities of SUAP case some skills and knowledge of referring laws are a need. It is true in particular in tasks in which the human performer needs a strong knowledge of the law in order to check the completeness of the request. Depending on a case, also knowledge of third parties PAs and their aims. In fact, they are always involved in the BP, and sending a request to a wrong organization implies waste of time.

The BP of "*Titolo Unico*" scenario is modeled using BPMN 2.0 language. Four participants are involved:

- *Entrepreneur* is the citizen that has to certify the start of an activity.
- *Municipality of Senigallia*, different unit of the municipality are involved in the BP, they are represented as BPMN 2.0 Lane.
  - SUAP Office;
  - Other offices, they are other offices of the municipality that are delegated to check some part of the documentation sent by the Citizen Applicant. They are not generalizable since they change case by case (the complete set of offices than can be involved.
- *Third Party PA*, it is an external PA that is involved in the BP. It is delegated to do some checks. More than one external PA can be involved in the "SCIA Commerciale" then, we type this participant using Multi-Instances BPMN 2.0 marker.

- *Service Conference*, it represents the group of people that are involved in a meeting to discuss about the case. They all decide if the request of the citizen is feasible.

An excerpt of this Business Process is shown on Fig. 3. This excerpt contains the part of process involving the Municipality.

Such process can be loaded in the simulator to be executed for training purpose. After a process has been instantiated, the simulation is launched and the users can access it using a web interface as shown in Fig. 4.

At connection, the interface shows the user the tasks to be completed (if there is any at this point of the process). For each task, the interface will display its name, a description, some information about time on task and number of attempts, as well as a form to be filled to complete the task. On the side, the interface show the other participants to the simulation, as well as the user ongoing session score, updated after each task completion. Regarding the list of participants, note that the number

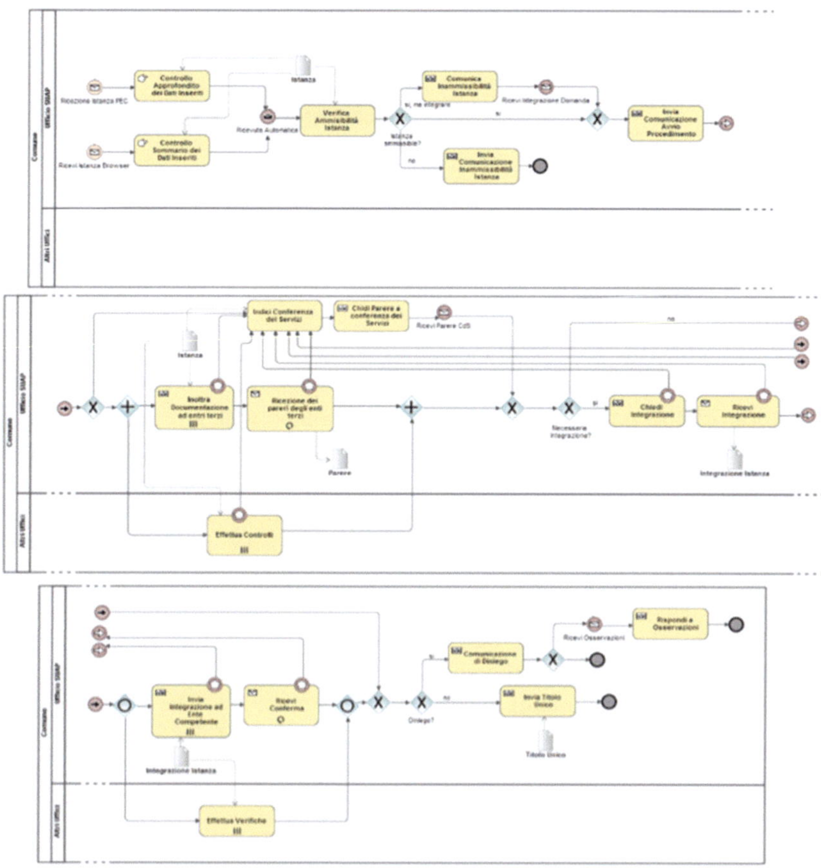

Fig. 3 SUAP process excerpt

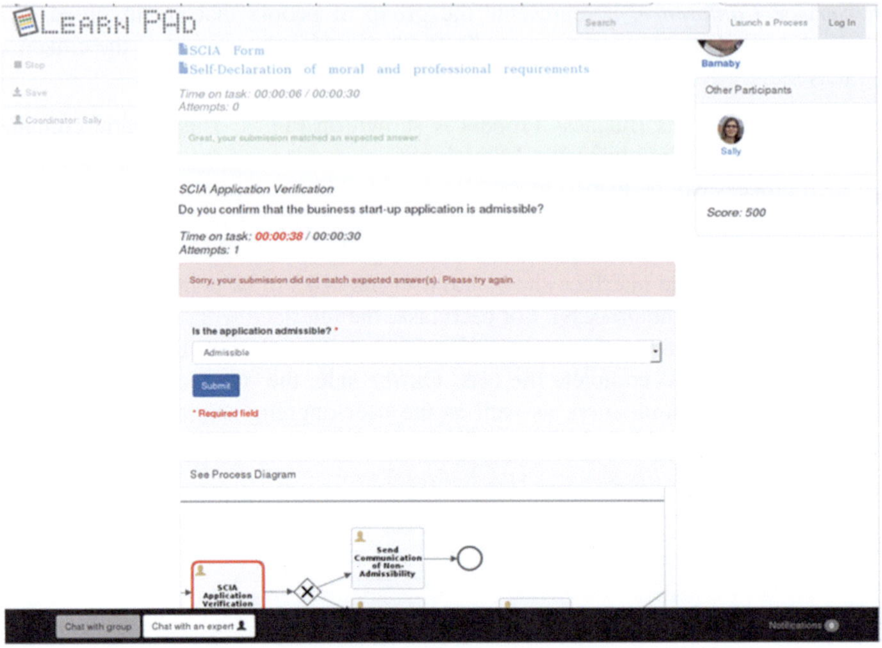

Fig. 4 Screen capture of the simulator web interface

of participant may differ from the number of roles in the process being simulated. For training purposes, some participants may assume several roles during the simulation.

After submitting the task, if all the entries are correct, the simulator will indicate that the task has been validated, and will display new tasks corresponding to the continuation of the process. Otherwise the simulator will indicate that the submission is incorrect and the user will have to try again, and the number of attempts will be incremented.

The simulator validates the tasks by comparing the outputs submitted by the learner to expected output values contained in a knowledge base. If they all match, the task is validated. Otherwise, the submission is rejected and the user is instructed to make a new attempt.

In some cases, the knowledge base may not contain matching entries for a given set of inputs. In this situation, the simulator notifies an assigned expert about the new, undiscovered configuration, and pause the simulation. The expert must analyze the situation and either validate the proposed output of the user, or inform the knowledge base of what would be a valid set of output for those specific inputs.

5 Conclusion

This paper presents an e-learning simulation platform providing a flexible and interoperable simulation environment with support for collaboration which originality relies on its ability to offer to learners to simulate Business Processes in an enjoyable and entertaining platform through the use of gamification. First, it provides the logical architecture of the simulator. Then, it describes two gamification mechanics that complement the architecture of the simulator. The framework is currently under evaluation inside the Learn PAd project. However, the preliminary results provided in [6, 7] evidenced positive feedbacks, especially concerning the possibility of executing collaborative simulation and providing learners assessment. Moreover, in order to provide a common and integrated infrastructure, some components of the platform have to be finished. In addition, the game score calculation may be refined, for example by considering the complexity of the chosen Business Process path. Besides, other gamification techniques can be explored in future applications.

Acknowledgments This work has been partially supported by the EC framework by the Model-Based Social Learning for Public Administrations project (EU FP7-ICT-2013-11/619583). The work reflects only the author's views. The community is not liable for any use that may be made of the information contained therein.

References

1. Activiti BPM Platform. Available at: http://www.activiti.org/
2. Deterding, S., Khaled, R., Nacke, L., & Dixon, D. (2011). *Gamification: Toward a definition*. Presented at the Computer Human Interaction, Vancouver, British Columbia, Canada: ACM.
3. Chatfield, T. (2010). *Why games are the 21st century's most serious business*. London: The Royal Society for the Encouragement of Arts.
4. Schell, J. (2011). *The pleasure revolution: Why games will lead the way*. Google Tech Talk.
5. Kapp, K. M. (2012). *The gamification of learning and instruction: Game-based methods and strategies for training and education*. New York: Wiley.
6. Calabro, A., Lonetti, F., & Marchetti, E. (2015). KPI evaluation of the business process execution through event monitoring activity. In *2015 International Conference on Enterprise Systems (ES)* (pp. 169–176). IEEE.
7. Calabro, A., Lonetti, F., & Marchetti, E. (2015). Monitoring of business process execution based on performance indicators. In *2015 41st Euromicro Conference on Software Engineering and Advanced Applications* (pp. 255–258). IEEE.

Part VIII
Collaborative Supply Networks Interoperability

Towards a Methodology to Support the Strategies Alignment Process in Collaborative Networks

B. Andres and R. Poler

Abstract Over the last years, collaborative networks (CN) have been widely studied due to the benefits obtained by the enterprises that take part in current global and dynamic markets. Enhancing enterprises flexibility and agility is one of the challenges associated to overcome continuous and unforeseen changes. The enterprises flexibility is enhanced by rapidly sharing information and knowledge, and the quickly adaptation of the objectives and strategies, when sudden changes occur, in order to keep working in the same direction without losing the common vision. This paper focuses on addressing the strategic alignment process, having a global view of all the strategies formulated in the CN, with the main aim of identifying the ones that have higher levels of alignment. In the light of this, a methodology to deal with the strategies alignment process is proposed, as a complementary mechanism to the model and tool previously propose, in the CN context.

Keywords Agility · Flexibility · Interoperability · Strategies alignment · Negotiation

1 Introduction

Nowadays, collaborative networks (CN) of enterprises are embedded in dynamic, continuously changing and unpredictable environments. The efficient collaboration boosts enterprises flexibility and agility, features that allow them to easily adapt to new situations. In a CN, each enterprise defines its own objectives and strategies,

B. Andres (✉) · R. Poler
Research Centre on Production Management and Engineering (CIGIP),
Escuela Politécnica Superior de Alcoy, Universitat Politècnica de València (UPV),
Centre D'Innovació I Investigació, Calle Alarcón, 03801 Alcoy, Spain
e-mail: bandres@cigip.upv.es

R. Poler
e-mail: rpoler@cigip.upv.es

© Springer International Publishing Switzerland 2016

K. Mertins et al. (eds.), *Enterprise Interoperability VII*,
Proceedings of the I-ESA Conferences 8, DOI 10.1007/978-3-319-30957-6_24

and when unexpected events occur, collaborative enterprises have to react by adapting the initial objectives and, therefore, adjusting the strategies activated, with the main aim of keep working in the same direction. In order to increase the enterprises' effectiveness in response to the effects of potential disruptive events, the collaborative establishment of the strategies alignment process is considered as relevant in the literature [1]. Within this process, as it is performed in a collaborative way, the exchange of information and knowledge and the adoption of interoperable systems within the enterprises are key issues.

Having a set of strategies formulated by each of the enterprises of the CN, the strategies alignment is defined as the decision-making process in which the decision-makers select those strategies that, if activated, positively influence the majority of objectives defined by each of the enterprises of the CN. So that, the selected strategies are aligned when each activated strategy has positive influences on the majority of objectives defined not only in the same enterprise in which the strategy is formulated, but also in the majority of the objectives defined by the rest of the network partners.

Enterprises taking part in a CN are characterised by being autonomous and heterogeneous [2], therefore achieving an agile alignment of strategies is a difficult task, due to this heterogeneity is translated to the formulation of strategies, each one dealing with different types of objectives. Accordingly, collaborative enterprises could reach a situation in which, the strategies selected might be misaligned, and conflicts among the objectives defined could arise. Collaboration infrastructures are required to support the strategies alignment process, which should include interoperability issues, information and knowledge sharing, safe communications, negotiation and mediation mechanisms, intra-inter organizational collaboration tools and services, and technological approaches [3].

This paper proposes a methodology, as an additional mechanism to the model, the method and the tool proposed in [1, 4], to support enterprises to collaboratively carry out the process of strategies alignment.

In the light of this, the paper is describes, in Sect. 2, the problem of strategies alignment, in the CN context. After that, in Sect. 3, an additional mechanism to the model and tool previously proposed, consisting of a methodology is described, with the main aim of supporting the strategies alignment process within a CN. Finally, conclusions and research lines are described in Sect. 4.

2 Problem Description

Enterprises willing to collaborate must overcome a set of barriers not only associated with the establishment of identified collaborative processes [5] (e.g., products design, demand forecasting, operations planning, replenishment, uncertainty management, share costs and profits, scheduling, information exchange, interoperability, etc.), but also when defining compatible goals, activating complementary strategies [6] or aligning their core values [7, 8]. Focusing on the strategies alignment process, the mere consideration of all the enterprises' objectives when deciding which strategies are the best ones to carry out will allow achieving higher levels of adaptability, agility, and competitiveness [9], strengths that are specially valued in current turbulent contexts and dynamic markets.

The autonomy and heterogeneity that characterises the enterprises, belonging to a CN, implies that each one defines its own objectives, so that each one formulates its own strategies, which are also characterised by being heterogeneous. Contradictions and misalignments among the strategies formulated could appear, triggered when a strategy activated in one enterprise negatively influences the majority of the objectives defined by other network partners. This situation may lead to the dissolution of the CN. In previous works, the strategies alignment process has been addressed, proposing a mathematical model [4, 6], in which the collaborative process of strategies alignment is represented. In the strategies alignment model a set of parameters characterising the strategies (strategy *s* formulated by enterprise *i*: str_{is}) and the objectives (KPI *k* formulated by enterprise *i*: kpi_{ik}), as well as the inter and intra-enterprise influences established between the strategies and the objectives, are defined ($val\_str_{is}\_kpi_{ik}$ and $val\_str_{is}\_kpi_{ik}$). The model output consists of the decision variables: number of units of strategies (u_str_{is}) and time of activation of strategies (ti_str_{is}). A tool, consisting of a simulation software based on system dynamics method [1] is used to solve the proposed model. An optimisation experiment build in the simulation software generate the alternative of solutions, as regards the values of the decision variables, that allow to maximise the network performance (Δkpi_{net}).

As a next point, this paper focuses on the development of a methodology, used as a complementary mechanism of the model and the tool previously proposed to support the strategies alignment process. The objective is to guide enterprises in (i) the process of the network characterisation, (ii) the definition of the collaboration level established, (iii) the process of data collection required to solve the strategies alignment model, (iv) the identifications of the interoperability levels that characterise the data exchange, and (v) the assessment, analysis and negotiation of the alternative of solutions obtained as a result of the implementation the existing strategies alignment model [1, 4], allowing collaborative enterprises to select and agree an alternative of solution that best fits to all partners.

3 Strategies Alignment Methodology (SA-Met)

A methodology to support decision makers in the process of selecting aligned strategies, among the enterprises belonging to a CN, is proposed, so that an aligned solution that best fits the all the network enterprises is identified.

This section describes in detail the 12 phases that make up the proposed methodology. The Strategies Alignment Methodology (SA-Met) is developed from a perspective, flexible enough, so that it can be adapted to any CN independently of the sector or the enterprises participating in the strategies alignment process. Through the implementation of the SA-Met, potential users will find support to implement the model, method and tools developed for the strategies alignment process [1, 4, 6]. The main aim is to provide support to the decision making as regards the strategies to be activated, which will be characterised by having higher levels of alignment. The designed SA-Met promotes the exchange of information, in an interoperable way, and supports the negotiation and collaboration process to deal with the strategies alignment, within a CN. Accordingly, in the SA-Met different levels of collaboration are considered. The enterprises will opt for a one or another collaborative scenario depending on the information exchanged. Moreover, each collaborative scenario will follow a particular the negotiation process to select and agree an alternative of solution resulting from the implementation of the strategies alignment model. The twelve phases forming the proposed methodology are briefly described next:

Phase 1. Definition of the Collaborative Network. The network of enterprises object to study is defined. The enterprises willing to collaborate, when carrying out the strategies alignment process, are identified. The enterprises have to decide the context in which the strategies alignment process is applied. Identifying: (i) the type of relationship established among the partners belonging to the defined network which can be non-collaborative scenario (NCS) or collaborative scenario (CS); (ii) the type of Decision-Making, which can be centralised or decentralised [10]; and (iii) the Topology of Network that the enterprises, which can be Hierarchical Network (HN) or Non-Hierarchical Network (NHN) [5].

Phase 2. Roles Identification. (i) Network Manager: this role is held by a consultant or expert in the CNs discipline. The network manager must be an expert in the strategies alignment process; and (ii) Enterprise Manager: this role is held by the person or people linked to the decision making of the activation of strategies in en each enterprise

Phase 3. Performance indicators definition (KPIs). Each enterprise defines its own objectives and the data as the KPIs used to measure the level of achievement of these objectives is gathered:

- Objective x defined by each enterprise i: o_{ix}.
- Define and enumerate the performance indicators k that measure the attainment of the previously defined objectives: kpi_{ik}.
- Estimate the parameter Δkpi_{ik}^{max} as the maximum increase of kpi_{ik} estimated by the enterprise i, for the performance indicator k. The performance indicators can be defined in any unit because once defined, there will be homogenised considering the parameter Δkpi_{ik}^{max}.
- Determine the parameter $Threshold\_kpi_{ik}$ as the value from which the associated kpi_{ik} is influenced when a strategy is activated. Below $Threshold\_kpi_{ik}$ the influence of str_{is} is not observed, from $Threshold\_kpi_{ik}$, the influence exerted by the strategy is considered.

- Determine the parameter Δkpi_{ik}_min as the minimum level of increase for the kpi_{ik} that the enterprise is willing to accept, once the $Threshold\_kpi_{ik}$ parameter is computed.
- Define the parameter w_{ik} as the weight (relevance) that kpi_{ik} has for enterprise *i*.

Phase 4. Strategies formulation. In this phase the strategies formulated in each enterprise are identified. The strategies are formulated, in each enterprise, with the main aim of reaching the objectives defined. The set of formulated strategies are potential to be activated in the future. The model and the tool proposed (system dynamic-based simulation software), will support the decision of which strategies to activate, amongst all the formulated, to obtain higher levels of alignment.

The data, associated to the strategies, required to feed the SAM are listed below:

- Strategy *s* formulated in each enterprise *i*: str_{is}. A short description will be required.
- The strategies str_{is} can be decomposed in units of strategy u_str_{is} to be activated. Definition of the parameter u_str_{is}, if applicable, as the units of strategies to be activated. A short description will be required.
- Define the parameter c_str_{is} as the cost, in monetary units, of activating one unit of strategy u_str_{is}.
- Define the parameter $d_1\_str_{is}$ as the delay (in time units) of influence of the strategy u_str_{is}. The time period between the initial time of activation of u_str_{is} (ti_str_{is}) and the time when the kpi_{ik} is started to be influenced by the activated str_{is} ($d_2\_str_{is}$) [t.u.].
- Define the parameter $d_2\_str_{is}$ as the time period between the u_str_{is} starts to influence the kpi_{ik} until the maximum level of influence in is achieved ($inf\_str_{is}\_kpi_{ik}$), [t.u.].
- Total duration of u_str_{is} ($d_4\_str_{is}$) [t.u.].
- Define the parameter b_i, in monetary units, as the budget that each enterprise owns to invest on the activation of the strategies.

Phase 5. Identification of the Collaboration Level. Depending on the information exchanged, the collaboration levels are identified as:

- Level 1 of Collaboration: Enterprises only exchange information as regards the KPIs defined and enumerated kpi_{ik}. The parameter w_{ik} is also exchanged. Nevertheless at this stage of collaboration the parameters $Threshold\_kpi_{ik}$, Δkpi_{ik}^{max} and Δkpi_{ik}_min are supposed by each the enterprise implementing the strategies alignment model, considering them as 0. Each enterprise estimates the values of influence by only considering the information of the KPIs. Enterprise i estimates ($val\_str_{is}\_kpi_{ik}$ and $val\_str_{is}\_kpi_{jk}$) and Enterprise j estimates ($val\_str_{js}\_kpi_{ik}$ and $val\_str_{js}\_kpi_{jk}$). The values of influence estimated, by each enterprise, are not exchanged.
- Level 2 of Collaboration: The enterprises exchange information as regards (i) the KPIs and the parameters that characterise them, and (ii) the number of strategies (only the ID of the strategies, not the definition) and the parameters

that characterise them. The value estimated for the budget is also exchanged. Each enterprise estimates the values of influence that its own strategies have on its own KPIs: *Enterprise i* estimates $val\_str_{is}\_kpi_{ik}$ and Enterprise j estimates $val\_str_{js}\_kpi_{jk}$. Both companies separately estimate the cross-impact of the strategies and KPIs. Enterprise i estimates $val\_str_{is}\_kpi_{jk}$ and Enterprise j estimates $val\_str_{js}\_kpi_{xk}$. In this case, the network manager (if required), according to the expertise and the knowledge acquired, can assess the enterprises on estimating the values of influence All the values as regards the values of influence estimated by each enterprise are exchanged; moreover the parameters defining the KPIs and the strategies are also exchanged.
- Level 3 of Collaboration: The enterprises exchange information as regards the (i) KPIs defined and the parameters that characterise them, and (ii) the definition of the strategies formulated and the parameters that characterise them. The value estimated for the budget is also exchanged. On the one hand, each enterprise estimates the intra-enterprise values of influences of its own strategies and KPIs ($val\_str_{is}\_kpi_{ik}$ and $val\_str_{js}\_kpi_{jk}$). On the other hand, the network enterprises jointly estimate the cross-impact (inter-enterprise values of influence) of the strategies and KPIs. *Enterprise i* and *Enterprise j* agree the values of influences of $val\_str_{is}\_kpi_{jk}$ and $val\_str_{js}\_kpi_{ik}$.

Following the framework for enterprise interoperability proposed in the ISO standard 11354 [11], interoperability is defined across four levels, each with different concerns (i) business, (ii) Processes, (iii) Services and Applications and (iv) Data. For all the collaboration levels defined, data interoperability is considered. Interoperability at data level is concerned with finding and sharing information coming from heterogeneous data bases, and which can moreover reside on different machines with different operating systems and data base management systems (DBMS). It is concerned with the ability to exchange both non-electronic data (documents) and machine transportable data (data files, data stored in a database) and use the data/information exchanged. Data interoperation may occur when two partners simply exchange two data files [12].

Phase 6. Estimation of the Values of Influence. When a formulated strategy str_{is} is activated, the objectives defined receive positive or negative influences, increasing or decreasing the values associated to the KPIs. This phase provides support to the enterprises in the estimation of the values that define the influences received in the KPIs when one unit of strategy (u_str_{is}) is activated. In the light of this, the strategies alignment model proposes the parameter $val\_str_{is}\_kpi_{ik}$ to determine the influence that the activation of a specific strategy str_{is} has on a defined kpi_{ik}. The values related with the parameter $val\_str_{is}\_kpi_{ik}$ are estimated differently, depending on the collaboration level adopted, identified and in phase 5.

Phase 7. Information Gathering. In order to gather all the information retrieved in the Phases 3, 4 and 6 a template has been designed, named SAM Complete Template. This template will be distributed among the enterprises participating in the strategies alignment process in order to gather all the information required to feed the SAM. The SAM Complete Template will gather information as regards:

(i) Strategy ID, (ii) Units of Strategy Definition (*u_str*$_{is}$), (iii) Units of Strategy Cost [m.u] (*c_str*$_{is}$), (iv) Strategy delay [t.u] (*d1_str*$_{is}$), (v) Time that the strategy takes to generate the maximum increase in the KPI [t.u] (*d2_str*$_{is}$), (vi) Total Strategy Length [t.u] (*d4_str*$_{is}$), (vii) KPI ID, (viii) KPI Definition (kpi_{ik}), (ix) KPI weight (w_{ik}), (x) KPI Threshold value (*Threshold_kpi*$_{ik}$), (xi) KPI_min (kpi_{ik}*_min*), and (xii) KPI_max (kpi_{ik}*_max*).

Phase 8. Introduction of the data in a data base. A database that stores all the information required to feed the strategies alignment model are gathered in a *Microsoft Access Database* specifically designed.

Phase 9. Strategies Alignment Model automatic creation. The Strategies Alignment GENerator (SAGEN) is an application that automatically generates the strategies alignment model in SD simulation software. The database built in Microsoft Access Database contains the necessary information to automatically build the strategies alignment model in SAGEN application. SAGEN automatically generates an XML file containing all the data as regards the parameters and the structure required in SD to build the strategies alignment model in the simulation software.

Phase 10. Generate solutions. The optimisation experiment integrated in the proposed tool (simulation software) generate the solutions, as regards the decision variables *u_str*$_{is}$ (units of strategies to activate) and *ti_str*$_{is}$ (the range of time in which to initiate the activation of a strategy) that optimise the network performance, kpi_{net}.

Phase 11. The process of negotiating the solutions. The process of negotiating the solutions will be initiated provided that the enterprises, participating in the strategies alignment process, establish collaboration (collaborative scenario, CS) and disagree with the optimal solution given by the simulation software, in which the strategies alignment model is solved. Normally, when the enterprises decide to apply a "Centralised" *decision* for the application of the strategies alignment model, the solution adopted by all the enterprises will be the optimal solution given by the simulation software. When the Type of decision carried out in the application strategies alignment model is "Decentralised", a negotiation process will be commonly required. In the decentralised scenario, there will be as many models as network enterprises. Each network partner implements within a CS the strategies alignment model, SAM^{CS_i}. Each enterprise will select the alternative of solution, that bets fits to its requirements, from all the set of solutions provided by the optimisation experiment computed in the simulation software, in which the strategies alignment model is solved. The network partners' exchange the alternative of solutions generated through the resolution of the strategies alignment model. The enterprises develop iterative processes for exchanging information, which facilitate the negotiation and the achievement of agreements to support the process of identifying what are the strategies to activate and in which time frame. In each iterative process, each network enterprise analyses the proposals made by the other network partners. The partners negotiate the alternative of solution that generates a performance as close to the optimum for each partner. Minimum and maximum boundaries of performance have to be defined by each enterprise (acceptance criteria), and the alternative of solution will be among these margins defined. The negotiation process will stop when the enterprises (i) achieve some goal or (ii) when arrive to the minimum number of proposals and counterproposals, previously agreed.

Next, an example of a negotiation process carried out by the enterprises that decide to establish the Level 1 of Collaboration (NP_CS1) is described (see Fig. 1). The NP_CS1 will allow enterprises to negotiate the solutions (values of the decision variables $u\_str_{is}^{CS1}$, $ti\_str_{is}^{CS1}$) obtained in the application of the strategies alignment model. As observed in Fig. 1, the negotiation process starts with the calculation of the strategies alignment model in the NCS (SAM^{NCS_i}) (Step 1), in which each enterprise of the network individually applies the strategies alignment model by only taking into account the information about their own objectives and strategies, and obtains: (i) the values of the KPI at enterprise level: *Enterprise i* kpi_i^{NCS} *and Enterprise j* kpi_j^{NCS}; and (ii) the solution as regards the strategies to be activated and the time of activation: *Enterprise i* $u\_str_{is}^{NCS}$, $ti\_str_{is}^{NCS}$ and *Enterprise j*: $u\_str_{js}^{NCS}$, $ti\_str_{js}^{NCS}$. These values will be compared with the values obtained in the calculation of the SAM in the considering the CS1 (SAM^{CS1_i}), in which each enterprise implements its own SAM^{CS1_i} using the information estimated (Step 4). The solution of the SAM^{CS1_i} corresponding to the alternative *n* is obtained, including: (i) the values of the KPIs at enterprise level: *Enterprise i*: $kpi_i^{CS1\_n}$ and *Enterprise j*: $kpi_j^{CS1\_n}$; and (ii) he solution as regards the strategies to be activated and the time of activation: *Enterprise i*: $u\_str_{is}^{CS1\_n}$, $ti\_str_{is}^{CS1\_n}$, and *Enterprise j*: $u\_str_{js}^{CS1\_n}$, $ti\_str_{js}^{CS1\_n}$. The resolution of the SAM^{CS1_i} provides a set of solutions. One or more solutions will be optimal, while others are close to the optimal (considering them suboptimal solutions). Each solution provided by SAM^{CS1_i} will be considered as an alternative. There will be as much alternatives as solutions "*n*" is the maximum number of alternatives of solution. The comparison between the alternative selected in the CS1 and the solution provided in the NCS will allow defining the acceptance criteria (e.g., $kpi_{net}^{CS1\_n} \geq kpi_{net}^{NCS}$) (Step 5). The parameter α/β will respectively acquire the value 1 if the criteria choice defined by *Enterprise i/j* is fulfilled and 0 if the criteria choice defined by *Enterprise i/j not* is fulfilled. A collaborative solution will be reached after a set of iterations is carried out (Step 7, 8 and 9 of Fig. 1), considering that both enterprises reach the acceptance criteria previously defined; or when the stopping rule of the iterative process, previously defined (Step 6) is met.

Phase 12. Solutions Assessment. The main aim of this phase is to identify potential appearing conflicts when activating certain strategies. Possible misalignments and negative-influences appearing in the alternative of solution selected are to be identified and analysed. Focusing on the enterprises that generate misalignments in the activation of their strategies, which generate negative influences in the defined KPIs, reducing the performance levels. A sensitivity analysis is also performed, in which the information to be analysed is altered to identify what is the effect, if exist, observed in the optimised solution obtained in the strategies alignment model. The purpose of sensitivity analysis is to determine the effects

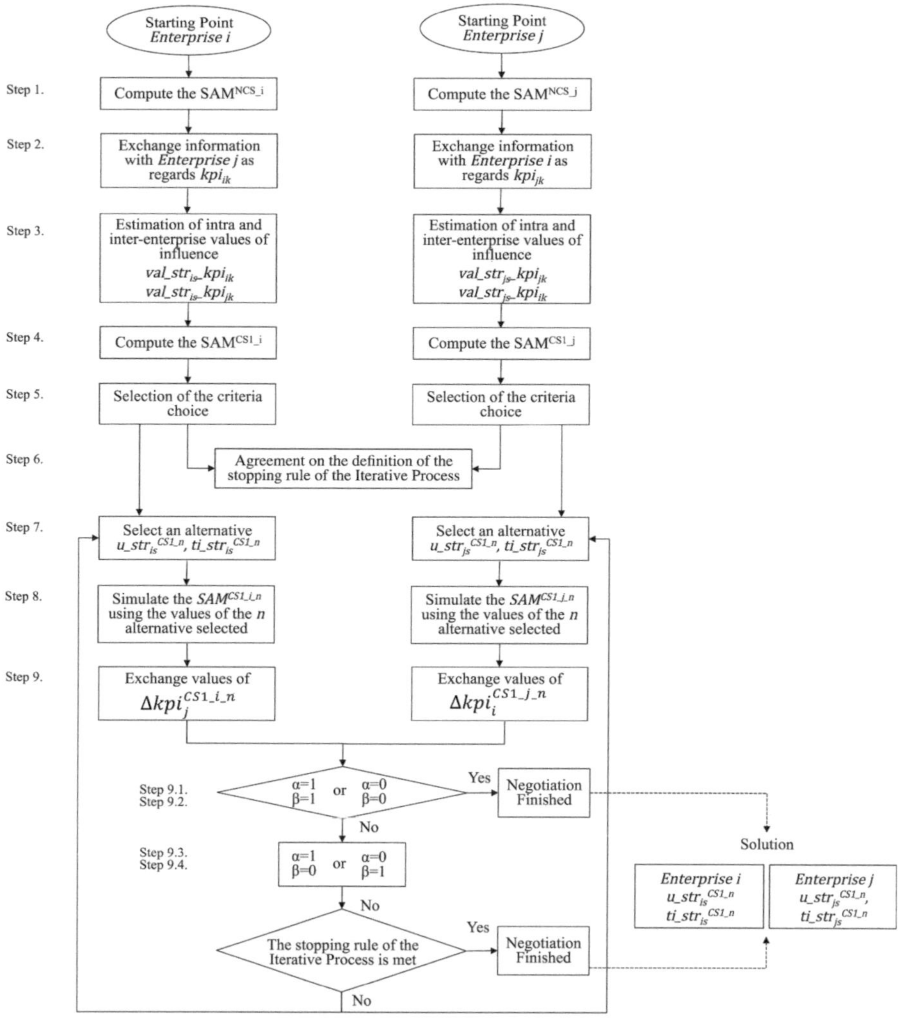

Fig. 1 Scheme of the negotiation process for the level 1 of collaboration

undergone in the dependent variables when the values assigned to the independent variables change. The value associated with the parameter $val\_str_{is}\_kpi_{ik}$ could be often estimated, by the enterprises, in an imprecise way. Since generally decision makers (enterprise managers) might not be able to correctly determine the exact value of the parameter $val\_str_{is}\_kpi_{ik}$, it is important to analyse how the value of the parameter $val\_str_{is}\_kpi_{ik}$ influences on the optimised solution, resulting from the implementation of the strategies alignment model in the simulation software. That is, to examine into which extent a small change in the value defined for $val\_str_{is}\_kpi_{ik}$ provokes a change in the final solution obtained in the optimisation experiment.

4 Conclusions

The participation in CN increases the enterprises agility in current dynamic and changing environments. Nevertheless, potential benefits of collaboration are limited if enterprises do not align their strategies. As a result, a new research challenge has appeared to address the decision-making process of selecting aligned strategies, from a collaborative perspective. In terms of practical implementations, the authors have already a formulated mathematical model to represent the strategies alignment process [4, 6] and have also proposed a tool based on a simulation software using the system dynamics approach [1]. Nevertheless, as a complementary mechanism, a methodology was considered necessary to support the enterprises on implementing the proposed model and tool. Future research lines lead to implement in a real network the contributions as regards the model, methodology and tool, allowing identifying the drawbacks in terms of interoperability when enterprises exchange the information required to solve the strategies alignment process.

Acknowledgments This work was supported in part by *Programa Val i + d para investigadores en formación* (ACIF/2012/006).

References

1. Andres, B., & Poler, R. (2015). Collaborative strategies alignment to enhance the collaborative network agility and resilience. *IFIP Advances in Information and Communication Technology*.
2. Camarinha-Matos, L. M., & Afsarmanesh, H. (2005). Collaborative networks: A new scientific discipline, *Journal of Intelligent Manufacturing*, 439–452.
3. Camarinha-Matos, L. M. (2014). Collaborative networks: A mechanism for enterprise agility and resilience. In *Enterprise Interoperability VI. Interoperability for Agility, Resilience and Plasticity of Collaborations* (pp. 3–11).
4. Andres, B., & Poler, R. (2015). Modelling the Strategies Alignment process in the Collaborative Network context. In *ICIEOM-CIO-IIIE International Conference. Engineering Systems and Networks: The Way Ahead for Industrial Engineering and Operations Management*.
5. Andres, B., & Poler, R. (2015). Models, guidelines and tools for the integration of collaborative processes in non-hierarchical manufacturing networks: a review. *International Journal of Computer Integrated Manufacturing*.
6. Andres, B., & Poler, R. (2014). Computing the strategies alignment in collaborative networks. In K. Mertins, F. Bénaben, R. Poler, & J.-P. Bourrières (Eds.), *Enterprise interoperability VI* (pp. 29–40). Cham: Springer International Publishing.
7. Bititci, U., Turnera, T., Mackaya, D., Kearneyc, D., Parunga, J., & Waltersb, D. (2007). Managing synergy in collaborative enterprises. *Production Planning and Control: The Management of Operations, 18*(6), 454–465.
8. Macedo, P., Abreu, A., & Camarinha-Matos, L. M. (2010). A method to analyse the alignment of core values in collaborative networked organisations. *Production Planning & Control, 21*(2), 145–159.
9. Poler, R., Carneiro, L. M., Jasinski, T., Zolghadri, M., & Pedrazzoli, P. (2012). *Intelligent non-hierarchical manufacturing networks. Networks and telecomunications series.* New York: iSTE WILEY.

10. Alemany, M. M. E., Alarcón, F., Lario, F. C., & Boj, J. J. (2011). An application to support the temporal and spatial distributed decision-making process in supply chain collaborative planning. *Computers in Industry, 62*(5), 519–540.
11. International Standard Organization, "ISO 11354-1-2011(en)". (2011). *ISO OBP. NoE: IST-508 011 network of excellence, interoperability research for networked enterprises applications and software (2003).* http://www.interop-noe.org
12. Athena. (2007). The Athena interoperability framework. In *3rd International Conference on Interoperability for Enterprise Software and Applications (I-ESA'07).*

Meta-Modeling of Collaborative Supply Chain

Zheng Jiang, Jacques Lamothe and Frédérick Bénaben

Abstract Due to the restricted resources, European SMEs hardly have access to advanced management systems and collaborative tools currently. The European project C2NET aims at building a novel cloud architecture to provide SMEs affordable tools to help them to overcome the current economic crisis and to enhance their competitiveness in the global economy. This paper mainly focuses on capitalizing on the previous works done by CGI-IO team and proposing meta-model that is derived for supply chain collaboration modeling and is the basis of a reference ontology for the C2NET project. The meta-model is mainly divided into three sub-systems: the context system, the objective system and the partner system.

Keywords Collaboration ontology · Supply chain management · Meta-model

1 Introduction

European SMEs (Small and Medium Enterprises) hardly have access to advanced management systems currently and to collaborative tools due to their restricted resources. C2NET project is going to create cloud-enabled tools, which SMEs could afford, in order to help them to overcome the current economic crisis and to enhance their competitiveness in the global economy [1]. For supporting the SMEs supply network optimization of manufacturing and logistic assets based on collaborative demand, production and delivery plans, C2NET Project will provide a scalable

Z. Jiang (✉) · J. Lamothe · F. Bénaben
Mines Albi, University of Toulouse, Campus Jarlard,
Route de Teillet, 81000 Albi, France
e-mail: zheng.jiang@mines-albi.fr

J. Lamothe
e-mail: jaques.lamothe@mines-albi.fr

F. Bénaben
e-mail: frederick.benaben@mines-albi.fr

© Springer International Publishing Switzerland 2016
K. Mertins et al. (eds.), *Enterprise Interoperability VII*,
Proceedings of the I-ESA Conferences 8, DOI 10.1007/978-3-319-30957-6_25

real-time architecture, platform and software to allow the supply network partners to gather data of the supply network and optimize the manufacturing/logistics assets, etc.

Based on the requirements captured and prioritized from pilot industries, it will be performed a study leading to the definition of the Reference Model that fulfills all the requirements, and covers all the relevant actors. The obtained Reference Model will be the general framework where all C2NET solutions must fit and interact. One of the main goals of this task is to define a meta-model (describing collaborative concepts), relevant and compliant with the uses that will be made of this Reference Model in the whole project. Here in this paper, we are going to propose such a dedicated meta-model.

2 Bibliography

The main point of identifying key concepts and relationships is to look at existing ontologies and the dedicated meta-models. If one fits some of the requirements, it can be directly included. On the other hand, the comparison of existing ontologies enables to identify key concepts by two ways:

- Concepts that are present in several ontologies must be key.
- Original concepts that only appear in some ontology must be discussed in order to identify if it is important for C2NET.

During the project, a series of standards and ontologies have been reviewed. It can be noticed that some are oriented on the collaborative process level, others on the supply chain level:

- Collaborative process level: BOWL [2], Collaboration ontology [3], MIT process Handbook [4] and KAEC [5].
- Supply chain level: Enterprise Ontology [6], TOVE [7], MSE [8],
- IDEON [9], VE [10], SCOR [11], Eaglet [12], Supply chain ontology [13], Transport Ontology [14], Grubic et al. ontology [15].

2.1 *The Collaboration Meta-Ontology*

The proposed principle is to start from a Generic Collaboration Meta-Ontology (Generic CMO), with the idea to develop the CMO in layers (Fig. 1).

The core is centred on a metamodel related to the general concepts of collaboration in social systems. This "kernel" is common to all collaborative situations whatever the application domain is. The idea is to limit this kernel to a restricted series of concepts associated to the management of a collaborative process with rules that implement basic agility requirements.

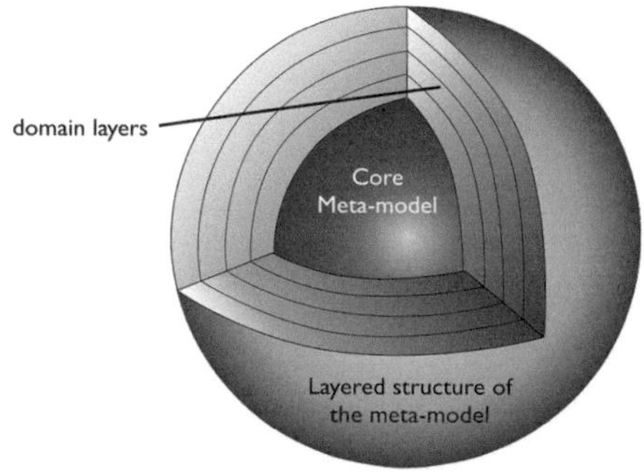

Fig. 1 Structuration of the collaboration meta-ontology

The domain layer is associated to supply chain collaborative planning. It allows defining all the concepts associated to supply chain collaboration, inheriting these concepts from the previous layers and being populated from various complementary standards and already existing ontologies. Here also, several sub-layers might be distinguished. This layer should also fix the semantical issue associated to the data. Indeed, it should define ontology of data that can be used as input or output in the various activities of supply chain planning.

An application layer can be associated to each pilot. It introduces the specific concepts associated to a given network and associates them with the domain layers. In particular, it enables an alignment of vocabulary between partners (in their planning processes and in their information systems) and the reference in the C2NET platform.

2.2 Analysis of Existing Literature

The core meta-model answers "scope and purpose" that we defined in as at a collaborative process level, the reference model should support a federated approach of interoperability of partners' processes, include a reasoning process for supporting the design of a collaboration and model collaboration context to help in detecting changes.

When mapping the fundamental concepts of the previous ontologies an alignment of concepts can be identified as shown in Table 1.

From the analysis of these general collaboration ontologies, it results that they can have two main perspectives. Some are oriented in the capitalization of taxonomy of process models and their decomposition: BOWL and MIT Process Handbook. Others are oriented on the design of a collaboration process. They therefore include a reasoning process for supporting the design. Here the Collaboration Ontology is oriented on the collaboration of people: it was designed for supporting the virtual organization in the network for developing. Considered resources are data, document or group of people. While the KAEC model was

Table 1 Alignment of concepts between the collaboration ontologies

Concepts	BOWL	Collaboration ontology	KAEC	MIT process handbook
Context	Supervised criteria	Collaboration objective	Environment/environment Component/characteristics	
Goal	Business goal	Colaboration criteria	Objective, performance objective	Goal
Task	Task	Activity	Capacity/activity	Process
Process	Plan/method	Process/procedure	Pattern/process	Process/bundles
Actor	Actor	Participant/group/collaboration clients	Collaborative network/partner/mediator	
Flow	Product/thing/document/flow pattern/control flow	Components/data	Flow	Resource
Resources	Hierarchical task network ontology	Groups	Resource	Resource

designed for supporting the design of collaboration of organizations. In particular, it identifies a particular partner, the mediator, whom role is to orchestrate the collaboration of the other selected partners. Moreover KAEC develops a view on the context of the collaboration, which is helpful for detecting changes in this context and thus adapting the collaboration goals.

Figure 2 recalls the KAEC Meta-model [5].

The Collaborative Situation meta-model has been organized in five complementary packages:

- Context package including components and characteristics of the considered environment, and also opportunities or threaten specific to these environment characteristics.
- Objective package contains the common facts and objectives, which has to be managed by the collaborative network.
- Partner package externalizes the resources and know-how of the partners. It includes capabilities, patterns, instructions, resources (people, material, information, etc.), flows (links among capabilities) and mediator, which orchestrate the various business processes.
- Behavior package characterizes the specific operations. These operations are deployed to specialize the collaboration. This package includes the business activities or processes, and further more the associated events.
- Performance package evaluates the comprehensive performance of the collaboration network by comparing dedicated KPIs and the expected performance objective to the measures on the field.

The different systems are showed in following proposal.

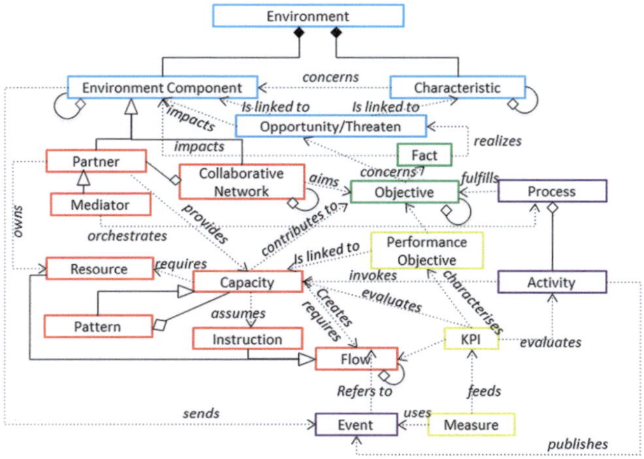

Fig. 2 The KAEC meta-model for the core

3 Proposal

In a literature review, Grubic and Fan [13] reviewed ontologies dedicated to supply chain management. We selected the ontologies to be analysed in the cited in this paper and in new ontologies that cite that paper.

Daniele and Fereira Pires (2013) identify 5 basic concepts that has to be integrated in the logistic ontology [16]:

- Activity represents the relevant actions in order to achieve logistics and provides value for customers.
- Actor denotes individuals or companies that could be a provider or demander of activities and operate those activities on related resources.
- Physical Resource denotes the objects that are used in the activities.
- Location denotes the geographical area used to define the place relevant for the activities.
- Time denotes the start/end time or the time interval associated to activities.

3.1 Mapping of Concepts

In the core collaborative meta-model, there are five systems, which are: context system, objective system, partner system, behavior system and the performance system. We therefore map the concepts of the existing "supply chain" ontologies and standards with the core (Table 2).

It appears that most of the ontologies have introduced detailed concepts for the partner view. The environment is mostly limited to the location of things and the modelling of the market, customer and demand. The concept of opportunity and Threaten are poorly developed, even if there exists a vast literature on Supply chain Risk Management and resiliency. Finally the objective view introduces the concepts of goal and strategy in some models. But this strategy is never associated to an analysis or a vision of the environment.

In these environment, the SCOR has a specific place:

- First it has developed a principle of configuration of the model. At level 1 generic concepts are introduced. They make it possible to model the perimeter of the supply chain. Then, this supply chain is configured at level 2 depending on some strategies. Level 3 details the concepts of level 2.
- SCOR has developped a rich partner view but it also has formalized many KPI and metrics for the Performance view.
- SCOR is known and used by practionners and legacy systems developpers.

In C2NET, the pilots used the SCOR principles to describe their actual processes.

Table 2 Alignment of concepts between the supply chain ontologies and the core concepts

System	Core concepts	Enterprise ontology	TOVE	MSE	IDEON	VE	SCOR	Eaglet	Supply chain ontology	Transport ontology	Grubic et al.
Context	Environment component	Market, customer, competitors	Order ontology		Environment, competitors	Customer	World map	Location	Supply chain structure	City, connection point/link,	Location, customer
	Characteristic	Needs, image			Observation					Geographic element, infrastructure	
	Opportunity/ threaten	Potential sale				Enquiry			Purpose		
Objective	Objective	Purpose, objective, vision, mission, goal	Organization goal	Project, strategy	Goal, strategy, objective	Customer order, order/product	Level I process, production strategy, performances attribute		Objective, strategic purpose		Strategy, request
	Fact										
Partner	Partner	Partner	Organization ontology	Enterprise		Unit, supplier		Agent	Party	Operator	Entity
	Collaborative network	Partnership, corporation		Extended enterprise	Partner organization	Virtual enterprise			Supply chain		
	Resource	Person, machine, resource, skill, asset	Resource ontology, transportation ontology, inventory ontology	Resource, facility, factory, shop, cell, station	Resource, person, skeleton plan, information product, document, physical material, human resource, role	Resource capacity	Skill	Equipment	Resource	Transport line/network, vehicle type	Asset, resource, facility

(continued)

Table 2 (continued)

System	Core concepts	Enterprise ontology	TOVE	MSE	IDEON	VE	SCOR	Eaglet	Supply chain ontology	Transport ontology	Grubic et al.
	Capacity	Activity/process specifications	Manufacturing activity ontology		Process	Planning, factory, factory service	Lv2&3 process		Activity	Transport mode	Function
	Flow	Plan		Flow		Plan, quotation, product, order	Input and outputs	Thing	Transfer object	Vehicle journey	Document, product, service, program, requirements
	Instruction		Product requirements constraints				Good practices			Journey pattern	Policy, rule, practice, guideline
Performance	Perf objective		Cost and quality		Assessment		KPI		Performance	Price	
	KP1						Metrics		Performance metric		Metric
	Measure										
Behaviour	Event	Event	Time ontology		Event, condition			Event			
	Activity	Activity, pre-post conditions	Activity ontology		Activity, decision activity, plan		Lvl3 process		Activity	Journey	Function
	Process	Process		Process	Process		Lvl2 process				Process

For the 3 levels of processes, between level 1 processes and level 2 processes there is a selection principle, which is the production strategy. It includes "make-to-stock", "make-to-order" and "engineer-to-order". This production strategy concept is of importance in the supply chain domain, as it can influence the objective of the supply chain, the way of managing resources and capacities. And also the input and output concept is important to mention since it describes the relationship between level 3 processes and the flows. During the mapping of SCOR concepts, the concept of "level 1 process" is considered as an objective. The point is that it describes the flow that the supply chain has to manage. For the performance attributes, there are 3 levels of metrics that could measure the performance of each attributes, the performance of each process or activity, and then the performance of the whole supply chain. For the people, the most worthy to talk about is the human skill, which could describe the required capacity of the people in the collaboration.

We therefore selected to use the SCOR model as a backbone of the meta-model and then introduce other concepts when particular deficiencies are detected.

3.2 *Description of the Supply Chain Layer*

Three types of links are used in the following models:

- Inheritance: ——▷
- Composition: ◆——>
- Aggregation: ◇——>
- Association: *does* ··········>.

If no name appears, the name is just "concerns".

The global view of the layer is shown in Fig. 3.

All the relations do not link only each view to the core. Some of the relations therefore link concepts of the supply chain level:

A level 1 process is associated to Goods and Areas because it consists in doing something to some goods and somewhere.

A level 2 process is the result of the configuration of a level 1 process because of a strategy. Level 2 and level 3 Metrics evaluate level 2 and level 3 processes.

Transportation fleet is the resource used by an actor to move things along a transportation link using a transportation mode. Products are a kind of Goods.

A customer is a kind of actor that personalize a market.

An actor can have a plan that is the fore-cast of the trend of a market.

3.2.1 The Context System

As defined in various supply chain meta-models, the environment is modeled with the environment components of Market and Area. Various markets can take place in an Area. And conversely a Market is associated to one Area and various Goods.

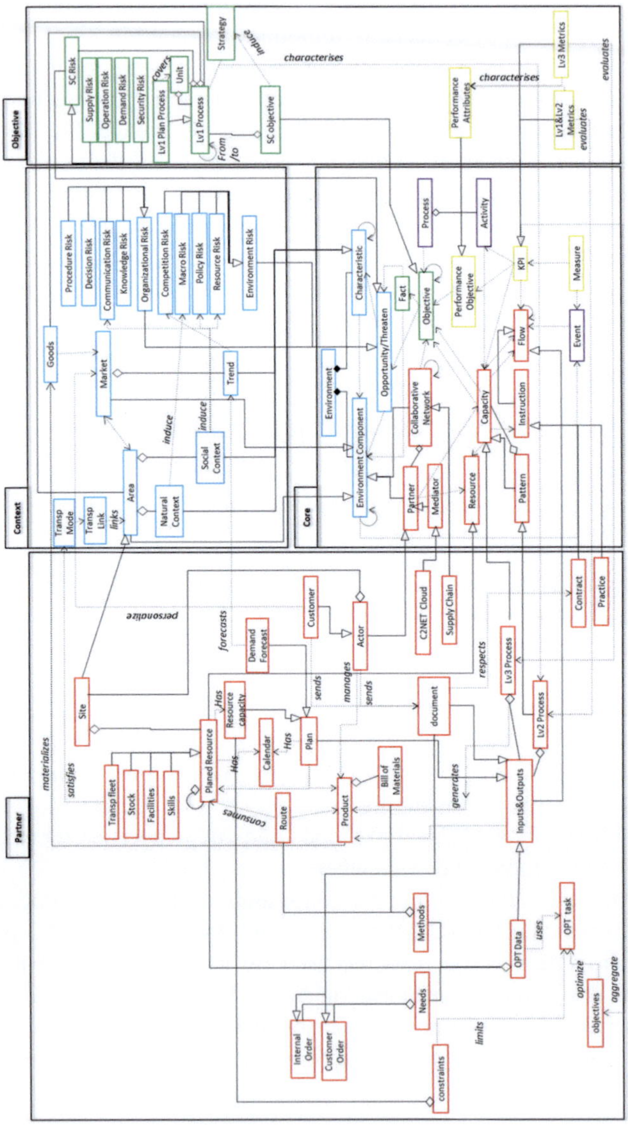

Fig. 3 The global view of the layer

Moreover the transportation ontology suggests to add several concepts in order to model transportation issues. Here we simply retain the transportation links between two areas and transportation modes accessible for each link. If complex transportation issues have to be introduced all the transportation ontology concepts should be linked to the C2NET meta-model. In order to add context information to the description of the environment, various concepts have been added as characteristics of the environment components. This concerns first natural context and social context that can be associated to an Area. The concept of Trend is used to characterise a Market. If big uncertainties exist a market may have several potential Trends. These characteristics are linked to various risks. Supply chain risk management is one of the major evolutions of supply chain in the recent years. This translates the desire of managers to control the sources of uncertainties in the supply chain. For this ontology, we retained a categorization of different supply chain risks introduced by Manuj and Mentzer [17] that distinguishes Environment risks, organizational risks, and Network risks as shown in Table 3. We refer to Manuj and Mentzer for the definition of the various Risks. It has been considered that Organizational Risks and Environment Risks are influenced by Area and Market characteristics. While Network Risks are associated to supply chain operations that are described in the objective view.

3.2.2 The Objective System

This view has to formalize the objectives when developing cooperation within a supply chain. The main objective of a supply chain is the customer satisfaction. Nevertheless, customer satisfaction is too general to be an objective since every supply chain has the same objective. In this view we have retained the configuration principle of the SCOR Model. At the level, the idea is to define the supply chain as a series of interconnected level 1 processes: the objective is that these processes have to jointly satisfy customers. At least the objective is to deliver something to someplace.

Table 3 The categorization of risks in supply chain context (Manuj and Mentzer 2008)

Category	Type	Example
Organizational risks	Procedure risks	Lack of formal procedure or quality control system
	Decision risks	Bureaucracy
	Communication risks	Cultural differences
	Knowledge risks	Lack of training
Network (SC) risks	Supply risks	Disruption of supply
	Operation risks	Breakdown of operations
	Demand risks	Variation in demand
	Security risks	Theft, infrastructure breakdown
Environment risks	Marco risks	Economic shifts
	Policy risks	Laws
	Competition risks	Uncertainty about competitor's action
	Resource risks	Lack of HR or technology

One particular "Level 1 Process" type if of particular interest for C2NET, the "Level 1 Plan Process". A "Level 1 Plan Process" has a decisional scope that can concern several "Level 1 Process" of types Source, Make, Deliver or Return. In order to model this scope the concept of Unit is introduced. There exists another relation between two level 1 processes: from/to enables to model the chaining of processes. Finally as suggested in the Manuj and Mentzer (2008), network risks are directly associated to the existence of "Level 1 processes".

The SCOR Model has also introduced the concept of Performance Attribute and Metrics.

As seen in various ontologies, the strategy must be considered because it orientates the choice of objectives for the supply chain. These strategies impact the processes. For example, in the SCOR Model strategies of "make to order/engineer to order/Make to stock" are used to configure a Level 1 process and obtain a Level 2 process. In the meta-model this is modelled with a characterisation relation between an "Level 1 process", a "strategy" and a "Level 2 process".

3.2.3 The Partner System

This view is the most detailed in the supply chain ontologies. The supply chain is the collaborative network of actors that use the platform in order to perform their collaborative planning process. The "C2NET Cloud" orchestrates this collaborative process and thus is the mediator between the actors. An actor is a partner that has a collection of sites and manages some products. A product is generated throughout a Level 2 or Level 3 Process. It is also the materialisation for a partner of a Good that is identified in the context view. Different types of attributes are associated to a product: its management unit (kg, number, m^2, etc.), lot size, box size, pallet size. The decomposition of a product, into components and raw materials is expressed with its Bill of Materials that lists the item and quantity of each required per unit of the product. Each actor' site contains "planned resources" of various kinds (transportation fleet, stocks, facilities and Human Skills) that are applied to products under the respect of route.

At the activity level, the vocabulary of the SCOR (SCOR11) is retained for Level 2 processes, Level 3 processes and Practices.

A Level 2 or Level 3 process requires or generates Flows of Input and outputs that in general concern some product. Two kinds of Input/output are distinguished: a document and a plan.

4 Conclusion and Perspectives

In this paper, we introduced the mapping between the core concepts and the concepts from existing ontologies and the proposal of the collaborative supply chain layer is introduced which is a layer around the core collaboration meta-model.

Also we adapt this to a real pilot in C2NET, which is the PFDC (Pierre Fabre Dermo Cosmetics) use case. Dealing with the constrained place, it is not presented in the paper. It shows that we can model a collaborative supply chain.

This meta-model is a first version that will be developed and amended in the future. This will first happen because of precisions in the industrial requirements, and secondly because of the deeper formalization of the C2NET cloud architecture and the modules specifications.

Acknowledgments This work has been partly funded by the European Commission through the Project C2NET: Cloud Collaborative Manufacturing Networks (Grant Agreement No. 636909). The authors wish to acknowledge the Commission for their support.

References

1. C2NET Proposal, SEP 210155140 SignedEC.
2. Ko, R. K. L., Lee, E. W., & Lee, S. G. (2012). Business-OWL (BOWL); A hierarchical task network ontology for dynamic business process decomposition and formulation. *IEEE Transactions on Services Computing, 5*(2), 246–259.
3. Knoll, S. W., Plumbaum, T., Hoffmann, J. L., & De Luca, E. W. (2010). Collaboration ontology: Applying collaboration knowledge to a generic group support system. In *Group Decision and Negotiation Meeting* (*GDN*) (pp. 12–26).
4. Malone, T. W., Crowston, K., & Herman, G. A. (2003). *Organizing business knowledge: The MIT process handbook* (642 p). Cambridge: MIT Press.
5. Lauras, M., Benaben, F., Truptil, S., Lamothe, J., Mace-Ramete, G., & Montarnal, A. (2014). A meta-ontology for knowledge acquisition and exploitation of collaborative social systems. In: *Proceedings of 2014 International Conference on Behavior, Economic and Social Computing* (*BESC*) (pp. 1–7).
6. Uschold, M., King, M., Moralee, S., & Zorgios, Y. (1998). The enterprise ontology. *The Knowledge Engineering Review, 13*(01), 31–89.
7. Fox, M. S., Barbuceanu, M., & Gruninger, M. (1996). An organisation ontology for enterprise modeling: Preliminary concepts for linking structure and behaviour. *Computers in Industry, 29*, 123–134.
8. Lin, H. K., & Harding, J. A. (2007). A manufacturing system engineering ontology model on the semantic web for inter-enterprise collaboration. *Computers in Industry, 58*, 428–437.
9. Madni, A. M., Lin, W., & Madni, C. C. (2001). IDEONTM: An extensible ontology for designing, integrating, and managing collaborative distributed enterprises. *Systems Engineering, 4*, 35–48.
10. Soares, A. L., Azevedo, A. L., & de Sousa, J. P. (2000). Distributed planning and control systems for the virtual enterprise: Organizational requirements and development life cycle. *Journal of Intelligent Manufacturing, 11*(3), 253–270.
11. Huang, S. H., Sheoran, S. K., & Keskar, H. (2005). Computer-assisted supply chain configuration based on supply chain operations reference (SCOR) model. *Computer and Industrial Engineering, 48*, 377–394.
12. Geerts, G. L., & O'Leary, D. E. (2014). A supply chain of things: The EAGLET ontology for highly visible supply chains. *Decision Support Systems, 63*, 3–22.
13. Grubic, T., & Fan, I.-S. (2010). Supply chain ontology: Review, analysis and synthesis. *Computers in Industry, 61*(8), 776–786.

14. de Oliveira, K. M., Bacha, F., Mnasser, H., & Abed, M. (2013). Transportation ontology definition and application for the content personalization of user interfaces. *Expert Systems with Applications, 40*(8), 45–59.
15. Grubic, T., Bastl, M., Fan, I.-S., Harrison, A., & Templar, S. (2010). Towards the integrative supply chain model. *International Journal of Logistics Research and Applications, 13*(1), 59–73.
16. Daniele, L., & Ferreira Pires, L. (2013). An ontological approach to logistics. Enterprise Interoperability, Research and Applications in the Service-oriented Ecosystem (pp. 199–213). ISTE Ltd, John Wiley & Sons, Inc.
17. Manuj, I., & Mentzer, J. T. (2008). Global supply chain risk management strategies. *International Journal of Physical Distribution & Logistics Management*, *38*(3), 192–223.

Seamless Interrelation Between Business Strategies and Tactical Planning

Frank-Walter Jaekel, Simon Schlosser, Boris Otto, Dobrila Petrovic and Ali Niknejad

Abstract There is an increasing pressure on manufacturing enterprises to innovate in their business model due to the accelerated change of market factors (e.g., digitalization, globalization, increasing customer individual demands, etc.). An important problem is the efficient, comprehensive and rapid transformation of a strategic idea to the tactical planning and its realization. Considering the variety of internal and external factors that have an impact on the global production network. In order to support decision makers with these complex tasks, the paper proposes an approach that facilitates more informed and faster decisions on new business opportunities. The approach incorporates an information structure allowing tracking and evaluating decisions from the first idea to its realization.

Keywords Business model · Strategic decisions · Tactical planning · Global production network · Decision monitoring

1 Introduction and Motivation

The approach presented in this paper has been elaborated in the scope of a European project under the vision 'to start with a new business idea in the morning and have all required data and networks understood by the afternoon'. Of course, the vision is quite ambitious and, in fact, the target is to considerably reduce the time it takes for a business idea to be implemented. The challenge is twofold: (1) to carry out a clear

F.-W. Jaekel (✉)
Corporate Management, Fraunhofer IPK, Berlin, Germany
e-mail: frank-walter.jaekel@ipk.fraunhofer.de

S. Schlosser
University of St. Gallen, St. Gallen, Switzerland

B. Otto
Technische Universität Dortmund, LogistikCampus, Dortmund, Germany

D. Petrovic · A. Niknejad
Faculty of Engineering and Computing, Coventry University, Coventry CV1 5FB, UK

© Springer International Publishing Switzerland 2016

K. Mertins et al. (eds.), *Enterprise Interoperability VII*,
Proceedings of the I-ESA Conferences 8, DOI 10.1007/978-3-319-30957-6_26

transformation from a strategic idea to the tactical planning, (2) to obtain feedback via business indicators such as potential market share, costs and time.

- Observations made in several previous projects and discussions with industrial partners suggest that gaps between the strategy and real-life implementation can easily appear. Examples of such challenges are as follows:
- The performance indicators (KPIs) to measure strategic objectives are not interconnected with the indicators defined in the tactical planning. Therefore their contributions to the targets are not transparent.
- The owner of the strategy is not clear, and, the strategy has not been communicated clearly. Therefore, each organization unit created their own strategy.
- The strategy became obsolete. The strategy is known but the organization or the environment, such as the market, changed faster than the strategy.
- The hierarchical structuring of the strategy was unclear, or, the development of the strategy was not structured at all. The company strategy was not well broken down to the organization units and staff.

As a result, it is difficult to understand the impact of changes in the strategy on the tactical planning or vice versa. This is represented in Fig. 1 with the broken chains between the strategy and the tactical planning. Moreover, a similar issue exists between the tactical planning and the Global Production Network (GPN) because of rapid changes in demands during the detailed planning of the global production network.

In general, the strategy needs to be set with a clear understanding of how to follow it and how to realise it. This can be seen as a prerequisite to understand a new business idea in the context of the business strategy.

This is especially important if the business includes several partners and the intention is to use or adapt a GPN for its realisation. Therefore, existing methods in the strategic and tactical area has been analysed to create a suitable package of methods and applications to support a coherent decision making process.

It is also necessary to have a common knowledge base across the different views between strategy and tactical planning to be able to interconnect the different methods and applications.

The paper describes an approach to interlink the strategic decision making with the tactical planning derived from existing methods. The target is to overcome the gaps between the different decision making processes to establish or update a global production network.

Fig. 1 Typical situation in industry between strategy and its realisation

2 Related Approaches and Foundations

This section provides an overview of existing relevant approaches. Particularly, the focus is on the relationships that have been identified during literature analysis of business model, strategy and tactics.

2.1 *Strategic Decisions and Business Models*

Many different definitions of the term and the core concepts of business models are stated in the literature [1–5]. A broad overview and analysis of 33 business model definitions is provided in [6]. The great variety of scientific and practitioner's contributions to these research themes underlines the importance of business modelling.

In particular, a business model can be understood as the reflection of an enterprise's realised strategy and demonstrates how an enterprise operates and creates value in its value network [7, 8]. In order to consolidate the variety of definitions, there has been some efforts in creating a unified business model vocabulary [1].

In this paper we adopt the perspective on business models described in [7]: A business model consists of different choices on how the enterprise must operate in order to create value and be competitive, and the consequences of these choices. Particularly, choices can be categorised regarding policies, assets and governance. While policy choices as well as asset choices are directly concerned with tangible resources such as key partners, manufacturing plants or distribution centres in a GPN, governance choices deal with the contractual arrangements (e.g., owning a production machine or leasing it from a third party). Each choice has some consequences: for example, the pricing policy might have an impact on the customer segment, the location of manufacturing facilities, the selection of key partners such as suppliers and on target markets.

A study of literature, and, also the discussion with our industrial partners has led us to use the business model CANVAS (BMC) [9–11] approach as a basis for the business model definition. It is a well-known and widely used approach to create business models.

Strategy comprises of high-order choices that create an initial system which aligns the value creation, product/service, sales/marketing and revenue logic [7, 12]. Each strategic decision has consequences that can put some constraints on other strategic and/or tactical choices. The set of strategic choices taken and the resulting consequences constitute the business model to be employed. Following this conceptual view, tactics are understood as the residual choices of the enterprise which are constrained by the boundaries that are spanned by strategic decisions [7].

2.2 Tactical Model and Enterprise Model

The tactical model describes the actual steps needed to achieve the strategic decisions. In the scope of the FLEXINET project, the tactical model covers the realisation planning of the strategy. This requires the planning of required tasks related to the responsibilities and monitoring of those tasks. In the case of a GPN it covers the processes, organisation structure and infrastructure of the network with the different actors. But it also requires aspects of the internal structures of these actors. Therefore, model fragments are developed to represent individual actors as building blocks in the GPN [13]. The model fragments are targeting [13]:

- To reduce the effort needed to model the process.
- Frontloading (experiences from the past), using company specific fragments of GPN partners which are already known in more detail.
- To ensure activities and their relationships are coherent.
- To support the incremental growth of the data related to a new business, especially using new data to immediately revise the previous analysis on the basis of the predefined indicators and their relations to the data sources.

The model fragments are also related to enterprise objectives and performance indicators as well as risk factors. These building blocks provide a detailed understanding of the GPN actors' inputs, outputs and inner functions. They include information about risk factors and strategic scenarios. The coherence of all the elements are ensured by utilising an underlying ontology for GPNs developed in FLEXINET [14, 15]. The inner structure of the building blocks is built using enterprise model elements.

Enterprise modelling related to the GPN ontology approach is used to integrate different enterprise views and to accelerate the process of creating or adapting GPNs to new business ideas. From a conceptual viewpoint, the elements of the strategic management are mapped to classes, objects and properties of the enterprise model.

The GPN design is affected by different choices of business model, for example choices about business partners, activities, resources, channels, or customers. These business elements will directly appear in other views of the enterprise model such as the process view or the organisation view. From here, the details of the GPN can be defined, and, in the next step the flows between the GPN partners can be analysed.

The roots of Enterprise modelling are in the systems theory. Enterprise modelling is the process of building models of whole or part of an enterprise with process models, data models, resource models, control models as well as other enterprise related models. In ISO 19440:2007 TC 184/SC 5 Enterprise integration—Constructs for enterprise modelling as well as in POP*, a result of the ATHENA-IP project [16], an enterprise model consists of different dimensions such as product, decision, organisation and process dimension. Further dimension can be added on demand, such as extensions related to sustainability [17]. The ISO 19439 and the ISO 19440:2007 provide definitions which are used within this paper.

Among the different enterprise modelling methods that conform to the ISO19440, the Integrated Enterprise modelling (IEM) [18, 19] has been selected. This selection was mainly due to the availability of the software support within the project partners; one FLEXINET industrial partner already has previous experiences in using the method. Both, the method and the software package, enable the used constructs to be flexibly extended. As the method is object-oriented, extension of used constructs are done by adding new classes and attributes. This was a necessary feature that enables further views, such as BMC, to be included. It is assumed that enterprise modelling can be used to glue the different methods and related models together.

2.3 Objectives and Indicators

The definitions of enterprise model's objectives, indicators and drivers used in the paper are derived from a method called ECOGRAI [20]. It allows the progress of performance indicators to be evaluated, from a starting value to a target value. The derived description of an indicator covers the following aspects [13]:

- Indicator Name is a unique identifier of the indicator.
- Purpose represents a description of the indicator.
- Format stands for the possible values such as integer, text, real, enumeration.
- AS IS value is the current value of the indicator.
- Information needed to evaluate the indicator e.g., the parameters.
- Calculation Processing represents the evaluation method for the indicator.
- Required evolution (Target) represents the value to be achieved.
- The owner (Who measures) points to the responsible organization unit.
- Period is the time span required to evaluate the indicator.
- Actions to react depending on the value of the indicator such as corrective actions.

In the early stages of new businesses and the development of processes from scratch, most of the values related to indicators are just approximations. However, these must be updated whenever more accurate data is available.

3 The Approach to Strategic/Tactical Modelling

The work intends to provide a guideline approach to operationalise a strategy by the selection of methods and tools. This selection can be seen as an initial set which can be extended with further methods and tools on demand.

The selection follows the decision making path from the strategy to the design of the GPN (Fig. 5). It provides the relevant methods and models for every aspect, including "strategy development", "business modelling", "tactical planning" and the final decision of an adequate "global production network". The interconnection of the methods relies on a common enterprise modelling approach which provides different views on the same structure. However, this requires the support of the FLEXINET ontology, which provides the foundation for the interrelation of different concepts used in the methods.

The methods currently considered are BMC for business modelling, IEM for enterprise modelling and an approach derived from ECOGRAI to model objectives, drivers and indicators. Further approaches are also considered but not finally integrated in the paper which are balance score cards (BSC) and business rules to support strategic decisions.

A major but quite often underestimated concept is the "Objective". Therefore, the definition and documentation of the objectives together with indicators and drivers are specifically considered. The objectives with sub-objectives and indicators guide the further process of model building.

The strategy provides a vision statement which defines perspective and scope for objectives and business models. This is therefore the initialisation of the approach between strategic view and the tactical view (see Fig. 2). Indicators are not described within this figure, but, they are related with the objectives. Indicators are described, as detailed in ECOGRAI, by providing start, target and current value but also with weights, responsibilities, purpose and evaluation horizons. The indicators are interlinked with the business model and the description of objectives (see Fig. 3).

Fig. 2 Model related to objectives and drivers

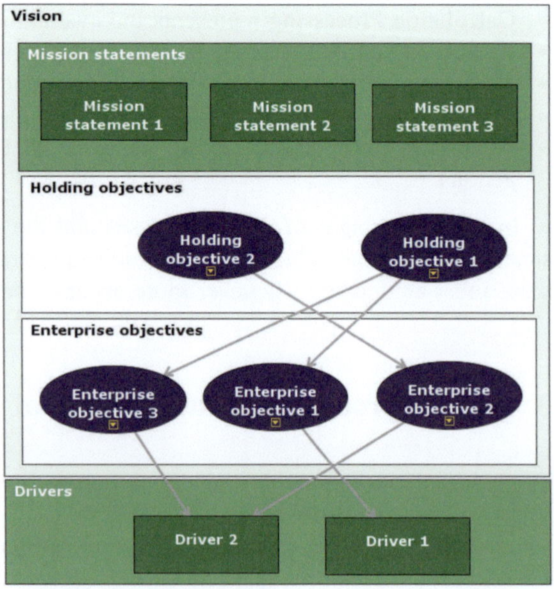

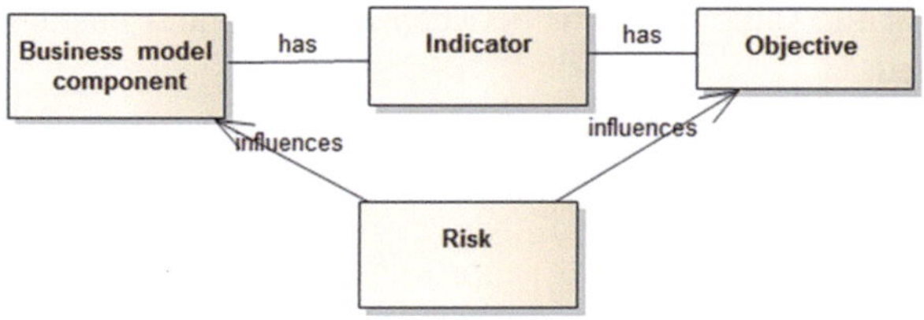

Fig. 3 Principle relation between objective and business model component

Component	Scenario 1	Scenario 2
Key activities	▪ source ▪ make ▪ deliver ▪ plan	▪ Monitoring of licence usage ▪ Development of updates
Customer relationship	▪ Call center ▪ Web	▪ Licence contracts
Channels	▪ Market place	▪ Business partners
Customer segments	▪ Internet community	▪ Licence broker
Cost structure	▪ Cost of service provision	▪ Cost of licence provision
Revenue streams	▪ Selling of services	▪ Selling of licences
Value proposition	▪ Product/Service idea	▪ Product/Service idea
Key resources	▪ Software infrastructure	▪ Licence agreement
Key partnership	▪ Subsidiaries ▪ Main suppliers	▪ Franchisee ▪ Licensee

Fig. 4 Morphologic box for business model scenarios

On the basis of a general description of business models [21], and, also based on the experiences of industrial partners, business model CANVAS [9] has been selected to group the business model into different business model components. A morphologic box approach is used to allow the selection of different business model alternatives. The alternatives are represented in terms of different scenarios (see Fig. 4). Scenario 1 illustrates a simple example of potential business model elements for each of the business model components.

The different BMC components are interconnected with the objectives but also with other elements of the enterprise model. It is assumed that an objective is described and measured by key performance indicators, which are in turn related to business model components. This describes how a business model component contributes to the objective (see Fig. 3). An objective has indicators and the business model component has indicators. A business model component contributes to an objective if the same indicator between objective and business model component is affected.

In addition, risk aspects are also related to the objectives as well as to the business model components. This can be later used for the analyses of the business model.

Within the enterprise modelling approach, the objective is modelled as a "control" object, or, in IEM terms, represented as an "Order" object. The "key activities" are modelled as processes. "Key resources" are represented as a set of resources which are interrelated with the processes. "Key partners" are objects of the type organisation. In this way, each of the business model elements can be mapped into IEM.

The target of this approach is to provide a model which can indicate the interrelationships between the enterprise objectives, the indicators, the related business model and the process structure to realise the objectives.

The general approach starts from a strategic viewpoint with high level objectives and follows a sequence via business model, planning of the realisation of the strategy to a related GPN (see Fig. 5). It requires feedback loops in each of the states and extensions of the different models.

The presumption is that a business idea exists and an initial agreement to follow this idea has been agreed on. Now, the business model is considered by taking into account the previously defined strategic targets of the organization. These results in a second stage of defining detailed objectives related to the new business idea and considering the business model. The results are a set of potential business ideas and objectives. These are the basis for evaluation of the business models and the selection of the most adequate models. This is the foundation to derive a model of the GPN (see Fig. 6). The figure includes 3 software application developed in the FLEXINET project. ODIM stands for "Objective, Driver, Indicator Model" and MBV stands for "Morphologic Box View", used for business models, while, EM fragments provides a building block library for GPN process models. In fact the

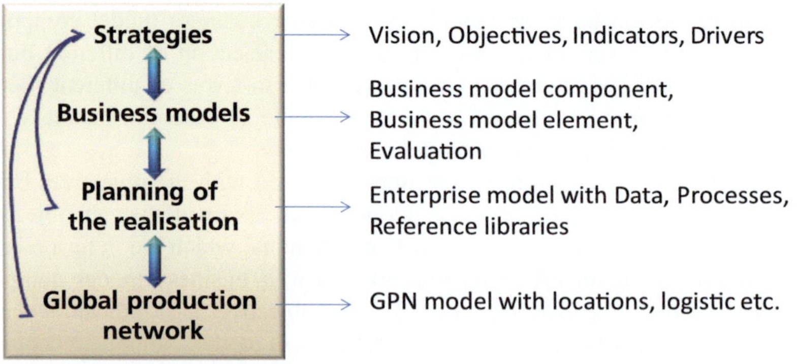

Fig. 5 Approach from strategy to GPN

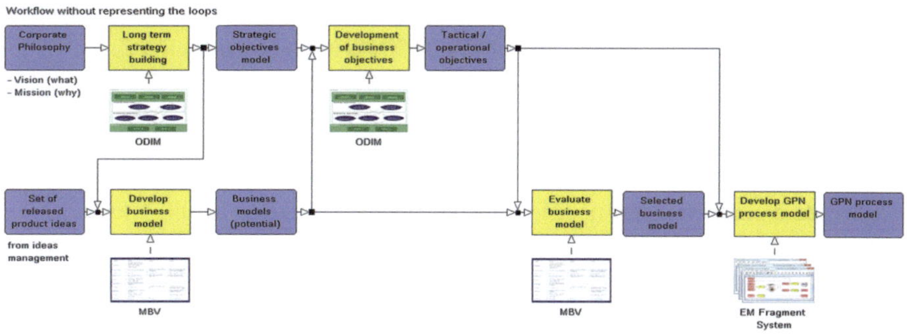

Fig. 6 Process model of the use of the approach and related tools [13]

figure illustrates the workflow in a straight forward way. The feedback loops are not shown but they exist for each of the steps.

The development of business objectives can influence the long term strategy. Also, the business model evaluation can influence the tactical objectives. This is important to ensure the coherence between the different steps from strategic planning to the tactical planning. In Fig. 5 it is also called planing of the realisation.

The approach is currently in the test phase and first impressions are that it provides a better understanding of the different aspects that are needed to be taken into account during the implementation of a new business idea. However it also illustrates that the strict distinction between the development of objectives and the business model is difficult and it is more adequate to work on both models simultaneously. This is now being further evaluated in the development of new business models by the industrial partners.

4 Conclusion

The paper provides an insight into a part of research and development work in a European project with a specific focus on the seamless connection of strategic and tactical decisions to realize new business ideas within a GPN. In general, the topic has been proved interesting in discussions with several industrial companies, particularly with regard to acceleration and monitoring of their decision processes in early phases of the product development process. However more detailed tests are required in the future, especially considering the evaluation of their business models.

The project is ongoing and experiments with the use cases of industrial users are currently in progress. These works extend the existing end user scenarios of the project. The target is the implementation of the approach into the end user companies. However, during the application of the approach, some adaptations are expected, especially in relation to the end user views. As an example, better visibility of the relationships between objectives and business models has been already requested by the industrial partners. This also shows the importance of the software applications.

The software application prototypes are now in the test phase. A specific focus is to prove the methods and the whole approach within industrial case studies. This will also result in the necessary fine-tuning of the modelling approach. Especially, the evaluation functionality needs to be extended by covering further methods used in the project such as balance score cards and business rules.

An expected outcome of the tests is also to get a better idea of the degree of acceleration and correctness delivered by the approach.

Acknowledgments This work has been partly funded by the European Commission through the Project FLEXINET: Intelligent Systems Configuration Services for Flexible Dynamic Global Production Networks (Grant Agreement No. 608627). The authors wish to acknowledge the Commission for their support. We also wish to acknowledge our gratitude and appreciation to all the FLEXINET project partners for their contribution during the development of various ideas and concepts presented in this paper.

References

1. Mettler, T. (2014). Towards a unified business model vocabulary: A proposition of key constructs. *Journal of Theoretical and Applied Electronic Commerce Research, 9*(1), 5–6.
2. Teece, D. J. (2010). Business models, business strategy and innovation. *Long Range Planning, 43*(2–3), 172–194. doi:10.1016/j.lrp.2009.07.003
3. Timmers, P. (1999). Business models for electronic markets. *Electronic Markets, 8*, 3–8.
4. Alt, R., & Zimmermann, H. -D. (2001). Business models. *Electronic Markets, 10*, 3–9.
5. Hedman, J., & Kalling, T. (2003). The business model concept: Theoretical underpinnings and em-pirical illustrations. *European Journal of Information Systems, 12*, 49–59.
6. Scheer, C., Deelmann, T., & Loos, P. (2003). *Geschäftsmodelle und internetbasierte Geschäftsmodelle-Begriffsbestimmung und Teilnehmermodell*. Mainz: Johannes Gutenberg Universität.
7. Casadesus-Masanell, R., & Ricart, J. E. (2010). From strategy to business models and onto tactics. *Long Range Planning, 43*(2–3), 195–215.
8. Österle, H., & Blessing, D. (2000). Business engineering model. In H. Österle & R. Winter (Eds.), *Business engineering: Auf dem Weg zum Unternehmen des Informationszeitalters* (pp. 61–81). Berlin: Springer.
9. Hauksson, H., & Johannesson, P. (2014). Metamodeling for business model design: Facilitating development and communication of business model canvas (bmc) models with an omg standards-based metamodel. In *the 8th International Workshop on Value Modeling and Business Ontology (VMBO 2014)*. Berlin, Germany.
10. Bouwman, H., De Reuver, M., Solaimani, S., Daas, D., Iske, P., & Walenkamp, B. (2012). Business models tooling and a research agenda (pp. 1–28).
11. Osterwalder, A., & Pigneur, Y. (2010). *Business model generation: A handbook for visionaries, game changers, and challengers*. Hoboken: Wiley.
12. Matzler, K. (2013). Business model innovation: Coffee triumphs for Nespresso. *Journal of Business Strategy, 34*(2), 30–37.
13. FLEXINET (2015) Deliverable D4.2: Impact analysis and assessment of new business models. Public deliverable on November 21, 2015.

14. Urwin, E. N., Palmer, C., Cutting-Decelle, A. –F., Sánchez Cid, F., Pinazo-Sánchez, J. M., Pajkovska-Goceva, S., et al. (2014). Reference ontologies for global production networks. In *Proceedings of the International Conference on Knowledge Management and Information Sharing, KMIS 2014* (pp. 133–139).
15. Young, R. I. M., Popplewell, K., Jaekel, F. W., Otto, B., & Bhullar, G. (2015). Intelligent systems configuration services for flexible dynamic global production networks (FLEXINET). www.flexinet-fof.eu. Accessed June 30, 2015.
16. ATHENA-IP (2014). Deliverable DA1.3.1: Report on methodology description and guidelines definition. http://interop-vlab.eu/ei_public_deliverables/athena-deliverables/list-of-public-deliverables-submitted-during-the-athena-project. Accessed December 14, 2014.
17. Oertwig, N., Wintrich, N., & Jochem, R. (2015). Model-based evaluation environment for sustainability. In *Global Conference on Sustainable Manufacturing (GCSM) 2014, Procedia CIRP 26* (S. 641–645). Johor Bahru, Malaysia.
18. MO^2GO (2015). www.moogo.de. Accessed June 30, 2015.
19. Mertins, K., & Jaekel, F. -W. (2006). MO^2GO: User oriented enterprise models for organizational and IT solutions. In P. Bernus, K. Mertins & G. Schmidt (Eds.), *Handbook on architectures of information systems* (2nd ed., S. 649–663). New York: Springer.
20. Doumeingts, G., Clave, F., & Ducq. Y. (1995). ECOGRAI—a method to design and to implement performance measurement systems for industrial organizations—concepts and application to the maintenance function. In A. Rolstadås (Ed.), *Benchmarking—theory and practice*. New York: Springer.
21. Osterwalder, A., Pigneur, Y., & Tucci, C. L. (2005). Clarifying business models: Origins, present, and future of the concept. *Communications of the Association for Information Systems*, *16*(1).

A Competition Model for the Offer: An Experiment of Management Simulation

C.M. Dias Junior, D. Oliveira, A.M. Ramos and R.T.O. Lacerda

Abstract It is observed that contemporary learning tools have their logic of development grounded in learning and not merely in the traditional teaching practice. In other words, learning happens by the experimental action of the individual considering the conceptual basis that this individual holds. Taking into account the high complexity of the problems, especially those related to the field of management, this study proposes a model of competition that contextualizes the practice of business decision-making. Therefore, this study presents a conceptual modeling simulation with the construction of an algorithm that enables this modeling by describing its application, taking as reference an exploratory methodology with a qualitative and quantitative approach. The results revealed to be satisfactory, inasmuch as they prove the interdependencies between the competitive variables chosen to determine the performance of a list of companies.

Keywords Simulation · Decisions · Management · Model

1 Introduction

The context of the development of competencies inherent of the organizational management activity has been, for a long time, instituted on theoretical orientation arising from the business practice established as a reference standard. The Organizational Theory comprises a framework of successful experiences of reports that, supposedly, seeks to provide undergraduate and graduate students, specifically to sciences such as Business, Accounting, Economy, and Production Engineering, conceptual elements that might format professional practice on management.

C.M. Dias Junior (✉) · D. Oliveira · A.M. Ramos · R.T.O. Lacerda
Administration Science Department, Federal University of Santa Catarina,
Florianópolis 88040-900, Brazil
e-mail: claudelino.junior@ufsc.br

© Springer International Publishing Switzerland 2016

K. Mertins et al. (eds.), *Enterprise Interoperability VII*,
Proceedings of the I-ESA Conferences 8, DOI 10.1007/978-3-319-30957-6_27

Nevertheless, the desired practice of the skills developed theoretically in these sciences is limited to stimulate insights on students under a limited interface between the knowledge produced in universities and the real needs of the business practice. One of the recurring attempts to solve this variety of problems is the redesign of the contents of the curricula of these undergraduate and/or graduate courses. However, the obstacles arising from this redesigning are substantial, such as adjusting to the different expectations of the market that is going to absorb the professional and the legal requirements of professional regulatory institutions.

It is observed that the competency-based formation allows for an emphasis in the context of work, acknowledging the combination of theory and practice. Therefore, it is possible to be simulated in computational environments. Thus, it is worth mentioning that the use of electronic learning environments, grounded by methodological artifacts, can stimulate the development of competencies that are required for managers.

2 The Decisions Simulator

There have been several attempts to approach the development of individual competences that can ensure better performance to private organizations, especially to those that enhance the conception and the implementation of successful business strategies, which are only achieved by improving the quality of management decisions.

To this end, the environment of the decisions of the organizational managers holds a significant percentage of accountability of them with the maintenance of the economic value on the time of the resources of material nature and intangible under their management.

In order to remodel the applicability of the theoretical basis suggested as “best practices” in the management area, simulation exercises have been used with very positive results, while preparing “experientially”, from the use of IT resources (Information Technology), the future decision-maker.

Perrenoud [1] has a position in favor of the connection of the development of competence and the acquisition of contextualized knowledge; “the more complex, abstract, mediated by technology, supported by systemic models of reality the actions are considered, the more in-depth, advanced, organized and reliable knowledge they require” (our translation).

The individual qualification of the exercise of a competence developed by the simulation is not a new practice. Their uses are common and known in the fields of medicine, aviation, and so on, where it implies to state that it constitutes the best possible testing environment prior to the effective professional practice.

Medeiros and Schimiguel [2] state that in addition to the benefits that the playful elements promote to the learning process, to the decision making, on the cognitive development, and on the stimulation of creativity, the use of games in the teaching-learning process may stimulate a higher interest by the students, taking them from the passivity inherent to traditional models of teaching, creating an active process of knowledge construction, and contributing to decrease school evasion.

Given the increasingly growing of the required level of qualification of business decisions, much due to the growing complexity of work in very specific business contexts where the process of internationalization of the economy positions itself as an authoritative factor for companies.

In the meantime, it is relevant and timely the possibility that the simulation provides to experience the quality of the decisions of future and current managers with the concurrent applicability of conceptual basis more adjusted for achieving results.

The practice of management builds its reference model, since the 50 s, from a logic of allotment of objectives of interacting systems and with related purposes.

Therefore, the Systemic Theory has been the basis for building models of intentional simplification of the reality, considering that its assumptions sustain the sense of interdependence of the parts that comprise a living system. This same understanding covers the socially organized systems, i.e., the companies.

The social organizations that are seen as companies can be compared to living organisms that need a continuous stream of inputs in order to ensure their survival, from the generation of a surplus of energy, which consequence on those represents the profit with the process of commercializing the object of their production, i.e., a good or a service.

Thus, production of goods and services in a business context these days makes use of massive utilization of IT to create value in elements with tangible representation (physical). In this way, good or more technologically advanced service that meets or creates a new nature of different needs is more likely to survive in competitive markets.

3 The Simulation Reference Model

The reference model in the area of strategy is represented mainly by Porter's contribution [3] where in his classic "Competitive Strategy" institutes a list of external forces that affect the performance of the organization, originally called the 5 (five) Porter forces, as follows: suppliers, competitors, customers, substitute products, and new products. Later, in a redesign of his work, in "Competitive Advantage" (2004, our translation), he acknowledges that there is a sixth force, which is the "Government" [4].

The transposition of some of these forces in a dynamic competitive model enables the creation of a simulation environment very close to the action of competitors and of clients' choices for a good in a real market, while being able to establish relationships of dependency between some variables, in this case price, quality, and demand.

Consequently, in order to build a simulated competition environment, that represents the dynamics of an oligopolistic market and within a random number generator and still considers the exercise of decision making, it is necessary to be able to represent the approximate behavior of this market.

4 Methodology

The conceptual model for the simulated competitive environment is based on the construction of a mathematical algorithm that represents the process of manufacturing and marketing of a good, taking into account the business activities of a number of agents (competitors- players) in an oligopolistic market, each responsible for a list of intervened decisions, such as price choices, the volume of production, the quality to be offered, and the gross revenue.

This research has an exploratory nature and the approach is qualitative and quantitative. The dynamics of the relationship between the variables of the competition model makes use of management of basis and it is built on equations that represent the competition between competitors, considering the disturbance and the interference of the decisions from one of the agents over others.

4.1 Algorithmic Construction of the Competition Model

The reference of the conceptual transposition implementation is Porter's [3]. It is transcribed by means of algorithm that considers that the demand reacts to the quality and to the price differently, i.e., the demand is taken as high when the quality of the good produced is higher than the quality of the good produced by the competitors considering the same price. On the other hand, the variable price for the products with the same quality tends to be the product differentiation factor, meaning that the cheapest product will be the most competitive.

It is possible to verify whether the sensitivity of the demand is given by the relationship between the variables "price" and "quality". For example, a sensitive demand to price suggests increased susceptibility to variations observed on price, i.e., it depends on the pricing practices of all competitors. In the same way, a demand sensible to quality is subject to higher investments in R&D by the same set of concurrent activities.

5 The Case Study (The Simulation)

The case study departs from a hypothetical situation of competition in a given market, represented by an oligopoly and with the performance of 5 (five) manufacturing companies in the same product with identical technical specifications and quality levels for this given moment in time T0 (Quarter 0). Moreover, these five companies still share 1/5 (20 %) of market share, when new decisions related to price and investments in R&D start to be taken in T1 (Quarter 1).

While the competition between the 5 (five) competitors during the estimated time of the passing of 1 (one) year is evaluated, representing a logic of interaction between their decisions and the return of their investments in R&D, each competitor aims to reach the rating of the product and the increasing of the volume of the quarterly sales in order to enable the monetary returns in time.

The competition model for the companies considers that the demand is sensitive to quality, establishing preliminarily equal prices to the same product offered by competitors. The formulation of this scenario intends to determine the behavior of the demand for the same product after 4 (four) quarters of decisions (1 year management) regarding the definition of new prices and possible investments in R&D, that potentially alter the perception of the quality of the products of each one of the competitors and enable the increasing of the prices initially practiced. Thus, the investments in R&D traditionally have its effects perceived in the long term, due to the variation of the qualitative perception of the product from each of the competitors will be given by a scale ranging from level 2 (commodity) to level 4 (potential product) over the simulated quarters, corresponding to qualitative level rise of the proportion of 20, 30, and 50 % of investment in R&D every quarter, according to Sauaia [5].

Therefore, in order to develop the possibility of response of the demand according to the offered quality, it is observed that the construct or model seeks to determine the sensitivity of the demand with regard to the price and quality according to the following:

T0—the five companies practice the same prices without investments in R&D, each one has a proportionally equal market share, corresponding to 50,000 units sold, with gross revenue of $ 10 million;
T1—two of the five companies begin to differentiate themselves by the quality of the product from the investments in R&D, respectively corresponding to an amount of $ 300,000 and $ 250,000 for companies 1 and 4 keeping the prices used at T0;
T2—the same Companies (1 and 4) keep their investments in R&D at the rate of $ 250,000 for the Company 1 and $ 300,000 for the Company 4, with the same prices as those practiced in the previous quarter. Furthermore, considering the action of competition, the Company 3 and the Company 5 invest respectively $ 550,000 and $ 600,000 in R&D;
T3—the Company 1 increases its price by 10 % and the Company 4 by 15 %. At the same time, Companies 3 and 5 keep their prices. On the other hand, the Company 2 reduces its price in 7.0 % in order to recover loss of market share;
T4—Companies 1, 3, 4 and 5 decrease their investments in R&D, i.e., all choose to invest $ 100,000, and increase their prices by 5 % having as reference the prices practiced in T3, and Company 2 chooses to invest a percentage 5 % of the actual revenue in T3.

6 Analysis of the Results

While each set of quarterly decisions were taken by companies 1, 2, 3, 4 and 5, the results demonstrate, preliminarily, that in the course of the four quarters, the interdependence of variables contained in the model (price, quality, market share and R&D) and that they condition the business performance of each of the competitors.

When verifying the increase of the percentage in the market share for companies 1 and 4 in T1, respectively 21.03 and 20.73 % (see Fig. 1), it is suggested that such increases are due to the impact of their investments in R&D. Their policies of differentiation of their products have guaranteed them perceived quality levels of 2.10 and 2.08, respectively, and, consequently, higher monetary returns in terms of gross revenue (see Fig. 2). Even though they have opted to absorb the costs of quality of delivery to the market not passing on to prices that remain at $ 200 per unit, at least during the first two quarters of decisions.

It is observed in Fig. 3 that while equivalent amounts of investment in R&D for companies 1 and 4 at the end of T3, their Gross Revenue approximate ($ 12,018,380 and $ 11,758,980) and their respective market shares (21.85 and 20.45 %) are different, highlighting the mutual interference of the variables price and perceived quality for that moment, considering that the unit price of the Company 1 was $ 220, and of the Company 4 was $ 230, with quality levels of 2.76 and 2.73, respectively. It shows that the market chooses the best benefit ratio found in the product from Company 1.

Moreover, it is possible to observe that the Company did not opt for a constant qualification of its product, in this case Company 2, which noticed its gross revenue

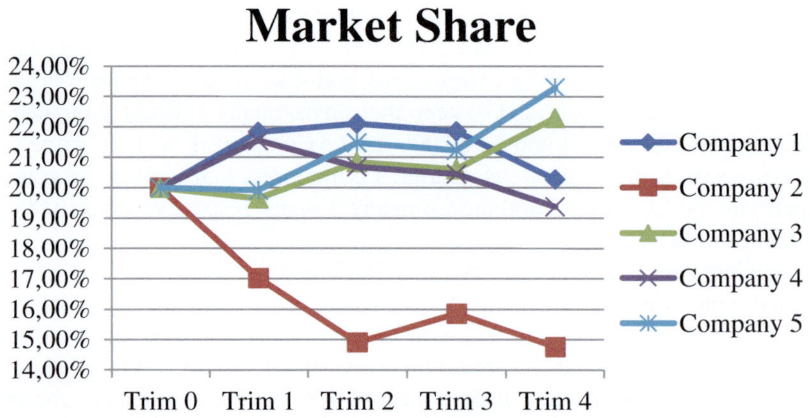

Fig. 1 Quarterly market share

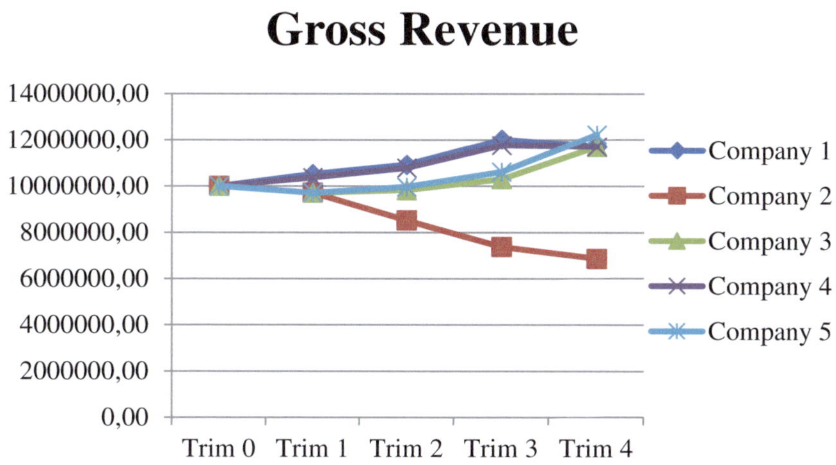

Fig. 2 Gross revenue quarterly sales

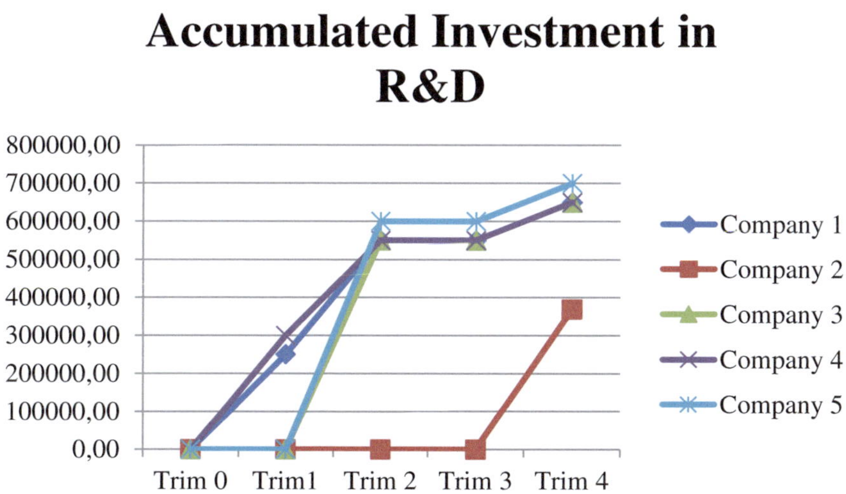

Fig. 3 Investment amount in R&D

and its market share decrease over the four quarters of decisions. Thus, the Company 2 initially had a gross revenue of $ 10 million in T1, rising to approximately $ 6,864,393 in T4 and an initial market share of 20 % decreased to 14.76 % in T4, even practicing the lowest average price among all other competitors, being $ 193 (see Fig. 4).

Finally, considering the better performance of supply between competitors, the Company 5 appears with the highest level of product quality in T4 (3.13),

Fig. 4 Prices quarterly committed

representing an amount of investment of approximately $ 700,000 (see Fig. 3), a better gross revenue of approximately $ 12,278,269.60 (see Fig. 2), a market share of 23.29 % (see Fig. 1), and one of the most competitive prices for the quarter ($ 210).

7 Conclusions and Considerations

The algorithm model used has presented itself completely assertive in all the proposed simulated quarters, as well as the analysis of the results demonstrates the effective interdependence of decision variables taken as reference, in this case, price and amount of investments in R&D, showing how business activity can create a differentiated strategy in its offer of a product.

It is not taken into consideration in this article the changing balance of the assets of each of the competing companies and, therefore, the change of their economic value. However, it is reiterated that what is sought with the model at this point is to enable the trial of decision making with few decision variables in the formulation of a commercial offer.

It is relevant to mention that the model does not include limits for investments in R&D, i.e., the extent to which the consumer market will satisfactorily answer to these investments. However, it takes the level of quality as a differentiating factor for gaining market share, indicating a positive relationship between the existence of competitors offering the same product and consumption.

Additionally, when the price increases or decreases, the consumer looks for the best relationship of benefits, as in the real business competition environment, requiring a greater demand for the products from the companies presenting the best offers.

References

1. Perrenoud, P. (1999). *Construir as Competências desde a Escola*. Porto Alegre: Artes Médicas Sul.
2. Medeiros, M. O., & Schimiguel, J. (2012). Uma Abordagem Para Avaliação De Jogos Educativos: Ênfase No Ensino Fundamental. *Revista Novas Tecnologias na Educação, 10*, 1–10.
3. Porter, M. (2000). *Estratégia competitiva: Técnicas Para Análise de Indústrias e da Concorrência* (2ª ed.). Rio de Janeiro: Editora Campus.
4. Porter, M. (2004). *Vantagem Competitiva: Criando e Sustentando Um Desempenho Superior* (27ª ed.). Rio de Janeiro: Editora Campus.
5. Sauaia, A. C. A. (2010). *Laboratório de Gestão: Simulador Organizacional, Jogo de Empresas e Pesquisa Aplicada*. São Paulo: Editora Malone.

Author Index

© Springer International Publishing Switzerland 2016

K. Mertins et al. (eds.), *Enterprise Interoperability VII*,
Proceedings of the I-ESA Conferences 8, DOI 10.1007/978-3-319-30957-6

MIX
Papier aus verantwortungsvollen Quellen
Paper from responsible sources
FSC® C105338

If you have any concerns about our products,
you can contact us on
ProductSafety@springernature.com

In case Publisher is established outside the EU,
the EU authorized representative is:
Springer Nature Customer Service Center GmbH
Europaplatz 3, 69115 Heidelberg, Germany

Printed by Libri Plureos GmbH
in Hamburg, Germany